Regulations Regarding Living Organ Donation in Europe

Leonie Lopp

Regulations Regarding Living Organ Donation in Europe

Possibilities of Harmonisation

Springer

Leonie Lopp
Faculty of Law
University of Münster
Münster
Germany

D6

Erster Berichterstatter: Prof. Dr. Thomas Gutmann
Zweiter Berichterstatter: Prof. Dr. Andreas von Arnauld
Dekan: Prof. Dr. Thomas Hoeren
Tag der mündlichen Prüfung: 13.11.2012

Zugl.: Münster (Westf.), Univ., Diss. der Rechtswissenschaftlichen Fakultät, 2012

ISBN 978-3-642-33798-7 ISBN 978-3-642-33799-4 (eBook)
DOI 10.1007/978-3-642-33799-4
Springer Heidelberg New York Dordrecht London

Library of Congress Control Number: 2013935606

Printed on acid-free paper

Springer is part of Springer Science+Business Media (www.springer.com)

Foreword

This dissertation was accepted by the University of Münster in the summer of 2012. I hope my detailed work on the topic of living organ donation helps make society more aware of the problem of organ shortage and the possibility of living organ donation. I also hope the possibilities presented here to improve the situation serve as an encouragement.

This dissertation has been written under the supervision of Prof. Dr. Thomas Gutmann. I want to thank Prof. Dr. Thomas Gutmann for his ongoing help and valuable advice throughout the entire duration of this dissertation. His support made a significant contribution to the development and completion of this dissertation. Dr. Bijan Fateh-Moghadam was also a great help; he answered several questions, especially with regard to the situation of living organ donation in the United Kingdom. In addition, I want to thank Prof. Dr. Andreas von Arnauld for his written report.

The research leading to these results was funded by the European Commission Seventh Framework Programme (FP7/2010–2012) under the Grant Agreement 242177 Living Organ Donation in Europe (EULOD project). I am grateful to all members of the EULOD project for being open to questions and for helping me to understand the different national legal systems. A special thanks goes to Prof. Dr. Willem Weimar and Ms. Frederike Ambagtsheer for organising the project in such a great way. Furthermore, I also want to thank the members of the ELPAT Working Group on Legal Boundaries, with whom I discussed the several significant points of this work during two EU-funded ELPAT working group meetings in Sofia (2010) and Berlin (2011). ELPAT is the European platform on ethical, legal and psycho-social aspects of organ transplantation. It is an official committee of the European Society for Organ Transplantation (ESOT). The legal experts from this committee, who are cited several times and mentioned in the references, provided me with valuable information about LOD. I want to thank them for sharing their knowledge with me.

Finally, I would like to thank Nicolas Rücker, Oliver Lopp and Rebecca Walsh for reading parts of my dissertation several times and making helpful suggestions. I would also like to thank my parents for their ongoing support and encouragement.

Düsseldorf, June 2012 Leonie Lopp

Contents

Table of Abbreviations

Action plan	Action plan on organ donation and transplantation
Additional Protocol on Transplantation of Organs	Additional Protocol to the Convention on Human Rights and Biomedicine concerning Transplantation of Organs and Tissues of Human origin
Art.	Article
BGHSt	Decisions of the German Federal Supreme Court for criminal cases [Entscheidungen des Bundesgerichtshof in Strafsachen]
BVerfGE	Decisions of the German Federal Constitutional Court [Entscheidungen des Bundesverfassungsgerichts]
cf.	Confer, Latin "compare"
CFI	Court of First Instance
CFREU	Charter of Fundamental Rights of the European Union
CFR-LOD	Common Frame of Reference for European Laws on living organ donation (CFR-LOD)
Commission staff working document, Accompanying document to the Communication	Commission staff working document, Accompanying document to the Communication from the Commission to the European Parliament and the Council, Organ Donation and Transplantation: Policy Actions at EU Level, Impact Assessment

Communication from the Commission	Communication from the Commission to the European Parliament and the Council – Organ Donation and Transplantation: Policy Actions at EU Level
Convention on Human Rights and Biomedicine	Convention for the Protection of Human Rights and Dignity of the Human Being with regard to the Application of Biology and Medicine
Directive 2004/23/EC	Directive 2004/23/EC of the European Parliament and of the Council on Setting Standards of Quality and Safety for the Donation, Procurement, Testing, Processing, Preservation, Storage and Distribution of Human Tissues and Cells
Directive 2010/53/EU	Directive 2010/53/EU of the European Parliament and of the Council of 7 July 2010 on Standards of Quality and Safety of Human Organs Intended for Transplantation
Directive 2010/53/EU	Directive 2010/53/EU of the European Parliament and of the Council of 7 July 2010 on Standards of Quality and Safety of Human Organs Intended for Transplantation
e.g.	Exempli gratia, Latin "for example"
EAC	Treaty establishing the European Atomic Energy Community
ead.	Eadem, Latin "the same (man)"
EC	Treaty establishing the European Community
ECJ	European Court of Justice
f.	Following (pages)
i.e.	Id est, Latin "that is"
ibid	Ibidem, Latin "the same place"
LKD	Living kidney donation
LOD	Living organ donation
No.	Number
p.	Page
pmp	Per million people
resolution on organ donation and transplantation	Resolution on organ donation and transplantation: Policy actions at EU Level (2007/2210 (INI))

resolution on the Action plan	Resolution of 19 May 2010 on the Commission Communication: Action plan on Organ Donation and Transplantation (2009–2015): Strengthened Cooperation between Member States (2009/2104 (INI))
Sec.	Section
TEU	Treaty on European Union
TFEU	Treaty on the Functioning of the European Union
Vol.	Volume
WHO	World Health Organization

Introduction

Transplantation medicine is well developed and the best possibility to cure organ failure. However, too few donor organs are available[1] to entirely utilise the medical progress made in the field of organ transplantation.[2] This causes death every single day. In 2007, about ten people died in Europe every day due to the unavailability of donor organs[3] and about 40,000 people waited for a donor organ.[4] In 2010, even more people waited for a kidney or liver transplant, while less than 20,000 kidney transplantations and about 6,500 liver transplantations were performed during the same year.[5] In Germany, right now about 12,000 people are listed as waiting for an organ transplantation.[6]

The rate of organ donation can be increased by making an effort to ensure that more organs are donated post-mortem or by increasing the rates of living organ donation (LOD).[7] Even though opinion surveys show that the majority of the population is willing to donate organs after death, the actual donation rates are lower.[8] Since it cannot be assumed that the waiting time for donor organs could be

---

[1] Abbub-Filho et al., in Gutmann et al. (ed.) (2004), p. 149; Broelsch, in Broelsch (ed.) (2006), p. 31; Coppen (2010), p. 9; Esser, in Höfling (ed.) (2003), p. 201; First, Vol. 29 Transplantation Proceedings 67, 67 (1997); Gutmann (2006), p. 4; Norba (2009), p. 55; cf. Oniscu/Forsythe, Vol. 38 Annals Academy of Medicine 365, 367 (2009); Ugowski (1998), p. 2.

[2] Beckmann, in Broelsch (ed.) (2006), p. 9; Deutsch/Spickhoff (2006), p. 433.

[3] Pühler et al., Vol. 25 Medizinrecht 584, 585 (2007).

[4] Commission oft he European Communities (2007), p. 2; Pühler et al., Vol. 25 Medizinrecht 584, 585 (2007).

[5] Matesanz (ed.), Council of Europe (2011); EULOD WP 2, Living Organ Donation Practices in Europe, p. 5.

[6] Bundestag printed paper 17/9030 (2012), p. 3.

[7] Gutmann/Schroth (2002), p. 1; Norba (2009), p. 24.

[8] Cf. Bundestag printed paper 17/9030 (2012), p. 4; cf. Bundeszentrale für gesundheitliche Aufklärung (2010); cf. Hauptmann/O´Connor, Vol. 336 The New England Journal of Medicine 422, 425 (1997); cf. Kirste, Vol. 81 Der Chirurg 778, 778 (2010); cf. Morgan/Deedat/Kenten, in Weimar/Bos/Busschbach (ed.) (2008), p. 99 f.; cf. Robert Koch-Institut (2003), p. 23.

L. Lopp, *Regulations Regarding Living Organ Donation in Europe*,
DOI 10.1007/978-3-642-33799-4_1, © Springer-Verlag Berlin Heidelberg 2013

Table 1 Living kidney donation (LKD) pmp

Country	LKD in 1992	LKD in 1999	LKD in 2008	LKD in 2009
EU				
Austria	1.6	4.9		8.3
Belgium/Luxemburg	1.1	2.5	Belgium: 4.2	
Bulgaria		2.0	1.6	2.1
Cyprus		48.3	49.0	
Czech Republic		1.7	2.8	2.6
Denmark	8.6	7.8	13.5	16.3
Estonia		0.7	2.2	3.0
Finland	3.4	0.8	1.7	1.1
France	0.8	1.3	3.5	
Germany	1.2 (West Germany)	4.6	6.9	7.3
Greece	6.5	8.5	4.6	3.0
Hungary		0.9	2.4	2.4
Italy		1.4	2.2	2.3
Latvia			0.4	2.6
Lithuania		3.8	1.5	2.6
Malta			12.5	12.5
Netherlands	5.5	8.4	25.2	25.3
Poland		0.3	0.5	0.6
Portugal	0.0	0.9	4.7	6.0
Romania			5.3	5.1
Slovakia		0.4	3.6	
Slovenia				0.5
Spain	0.4	0.4	3.4	5.0
Sweden	10.3	11.8	14.8	17.7
United Kingdom/Ireland	1.6	4.3	UK: 15.3 Ireland: 1.2	UK: 16.0 Ireland: 0.0
Additionally				
Moldova			0.6	
Norway	17.1	18.5	20.5	21.6
Switzerland	3.8	9.0	15.3	13.0

sufficiently reduced by only using organs donated by deceased persons,[9] LOD is a valuable supplement.[10] LOD, hence, seems to be a useful and desirable option, making it a worthwhile focus of concentration.

The (increasing) importance of LOD can be confirmed by the fact that the amount of LODs has risen considerably since the beginning of the nineties in several countries. Table 1 shows the development of living kidney donation per

[9] Cf. Ghods, Vol. 3 Iranian Journal of Kidney Diseases 183, 183 (2009); cf. Price (2010), p. 285.
[10] Cf. Neft, Vol. 20 Neue Zeitschrift für Sozialrecht 566, 566 (2011).

million people (pmp) in the countries selected for this study.[11] The countries that show an increase in the amount of LOD since 1992 (or since the earliest year that the publication of numbers was provided for that particular country) are italicised.

LOD is rather common in most countries considered; it can even be considered an established standard therapy today.[12] It became clear, however, that the rates of LOD differ significantly between the countries, showing that not all countries actually tap the full potential.[13] Not only does the medical development have an influence on the amount of LODs, the legal situation has an influence as well.[14] Legal regulations, while not automatically increasing the amount of LOD, are one important factor among several in influencing LOD numbers. This will be analysed in detail by examining the arguments in favour of and against restricting LOD by law. Such analysis will confirm the statement of some legal experts that LOD regimes create in part an *artificial scarcity*.[15]

By presenting the national transplant laws in the countries considered, possibilities for LOD regulation will be demonstrated. This study will show the extent of similarities and differences in existing national laws. It can be assumed that some regulations are *better* than others. The final aim, therefore, is to figure out the best possible way to regulate LOD in order to make a contribution to solve the problem of organ shortage while adequately adressing the ethical, legal, and political side-constraints of the subject.

A legal comparison of the national transplant laws makes sense to find out which regulations would be best to address the issues of LOD in need of regulation. This is also in accord with the European Unions *Action plan on Organ Donation and Transplantation* which strives "to promote the exchange of best practices of living donation programmes (Priority Action 3)." A legal comparison is, furthermore, a first step to make a harmonisation of the regulations for LOD possible. Since the Member States of the European Union are continually growing more and more together, the field of organ transplantation is as well; such harmonisation is desireable.[16]

---

[11] The numbers for 1992 are from Jakobsen, in Price/Akveld (ed.) (1997), p. 5. The numbers for 1999 are from Matesanz/Miranda, in Gutmann/Schroth (2002), p. 256 f. The numbers for 2009 and 2010 are from Transplant Procurement Management (2011). The criterion for choosing the countries is connected to the object of investigation (Schnitzer (1961), p. 105; Zweigert/Kötz (1996), p. 40 ff), and the intent of the investigation can be an indication of the selection of the included legal systems (Constantinesco (1972), p. 51; Ebert (1978), p. 38). Since this thesis focuses on the European Union, logically all European Union-Member States are considered. Furthermore, Moldova, Norway and Switzerland are included.

[12] Price (2010), p. 196; cf. Wagner/Fateh-Moghadam, Vol. 56 Soziale Welt 73, 74 (2005).

[13] Cf. EULOD WP 2 (2012), DOW: Deliverable 4, p. 4.

[14] Gutmann/Schroth (2002), p. 42.

[15] Cf. Evans, Vol. 15 Journal of Medical Ethics 17, 19 (1989); cf. Fateh-Moghadam, (2011); cf. Radcliffe Richards, in Weimar/Bos/Busschbach (ed.) (2011), p. 41.

[16] Cf. Gutmann/Schroth (2002), p. 41; cf. Prechern-Hauptmann, in Höfling (ed.) (2008), p. 97.

Not only have the Member States of the European Union become active in the field of (living) organ donation, but the European Union has also already carried out several actions in this area. It even passed a binding directive in this field: *Directive 2010/53/EU of the European Parliament and of the Council of 7 July 2010 on Standards of Quality and Safety of Human Organs Intended for Transplantation.* This *Directive* focusses on aspects of quality and safety of LOD (and post-mortem organ donation), but does not (completely) stipulate the requirements for LOD.

Does the European Union have the competence to regulate the requirements for LOD? This will be negated for hard law in chapter "Possibilities of Harmonisation". However, the option to pass soft law remains. The European Union is definitely amenable for such non-legally binding tools, which was confirmed by the *White Paper on Governance*. The *White Paper on Governance* was published by the European Commission in 2001 and promoted using such soft methods.[17] The option of the European Union to pass soft law to *regulate* the requirements for LOD, as being preferable towards no further measures, will thus be considered.

All in all, this thesis considers the national transplant laws, but also actions of the European Union that have already been done in this field and whether it has even further competences. The aim is to analyse the arguments in favour of and against legal restrictions of LOD to establish a best practice proposal in the end.

This thesis starts with an overview of LOD in chapter "Overview About Living Organ Donation". The overview includes an introduction of LOD and defines the relevant terminology. Chapter "Comparative Analysis of European Transplant Laws Regarding Living Organ Donation" compares the national transplant laws from the countries considered. First, it explains the method the comparison of laws will follow, and then actually compares the issues of LOD in need of regulation. It especially focuses on how LOD is restricted in the countries considered. Chapter "Analysis of the Normative Arguments That Dominate the Policy Arena About Necessity and Legitimacy of Legal Restrictions in Living Donor Transplantation" analyses the normative arguments that dominate the policy arena about the necessity and legitimacy of legal restrictions in LOD. The arguments in favour of legal restrictions will be compared with those against such restrictions. LOD has not only been recognised by the national laws, but by the European Union as well, which is active in the field of LOD. The actions the European Union has carried out so far are therefore presented in chapter "Actions of the European Union So Far". Whether the European Union has further possibilities to harmonise the legislation for LOD will be analysed in chapter "Possibilities of Harmonisation". Chapter "Best Practise Proposal for Living Organ Donation in the European Union" will focus on the possible content of a unified regulation for LOD and establish a best practise proposal. In the end, a summary of the entire dissertation will be made.

[17] Stefan, Vol. 14 European Law Journal 753, 760 (2008).

Overview About Living Organ Donation

Before analysing the legal regulations of LOD in the European transplant laws, LOD as such will be introduced. (1) I will start with presenting the underlying problem of organ shortage. (2) Afterwards, I will give the basic facts about LOD. (3) Last, I will clarify the used terminology.

I. Problem of Organ Shortage

The problem of organ shortage has already been presented. Transplantable organs are scarce, because the demand for organs is higher than their supply.[1] (1) Several options seem possible to solve this problem. (2) LOD in particular has a great impact on the amount of donor organs.

1) Possibilities to Serve the Problem of Organ Shortage

Cronin makes the following statement: "Prima facie meeting this demand is good. Why would it not be? Saving life is not only, not in itself immoral, it is also one of the most wonderful things that one individual can do for another."[2]

To solve the problem of organ shortage, several options exist. Those will be explained in the following.

[1] Coppen (2010), p. 14; Hoyer, Vol. 21 Pediatric Nephrology 1364, 1364 (2006); Keller, Vol. XXXII Stetson Law Review 855, 868 (2003); cf. Neuberger/Price, Vol. 327 British Medical Journal 676, 676 (2003); Steinbrook, Vol. 353 New England Journal of Medicine 441, 441 (2005); WHO Guiding Principles on Human Cell, Tissue and Organ Transplantation, Preamble, 1.

[2] Cronin, in Weimar/Bos/Busschbach (ed.) (2008), p. 100.

L. Lopp, *Regulations Regarding Living Organ Donation in Europe*,
DOI 10.1007/978-3-642-33799-4_2, © Springer-Verlag Berlin Heidelberg 2013

First, a machine could artificially carry out the task of the human organ. For example, renal failure can be treated with dialysis,[3] which is an alternative mechanism for the cleansing of the bloodstream.[4] The problem with dialysis is its negative side effects. First, the patient's quality of life is impaired due to several hospital stays. This severely limits his free time activities and could require him to search for other employment.[5] Second, the long-term function of the transplanted kidney is worsened the longer the dialysis is carried out. Therefore, transplantation before even starting the dialysis treatment (preemptive transplantation) is favoured.[6] This is usually only possible with LOD, because waiting time is typically required before receiving an organ donated by a deceased person.[7] Hence, LOD not only enhances the quality of life of the recipient, but even prolongs life or is life-saving. Third, patients that are treated with dialysis have a higher risk for several diseases in comparison to the patients who received a transplantation before dialysis became necessary (e. g., heart attack, high blood pressure, anaemia).[8] The negative aspects of dialysis are intensified in the treatment of diabetes. The life expectancy of a diabetic is exceptionally connected to the amount of time he has to undergo dialysis treatment while waiting for a kidney transplantation.[9] Fourth, the costs incurred for a kidney transplantation are lower than the costs incurred for dialysis.[10] Consequently, a machine artificially carrying out the tasks of the organ is no ideal solution.[11] In addition, several organs, for example the liver, cannot be replaced by artificial machines for a longer period of time.[12] The only life-saving option in

---

[3] Schutzeichel (2002), p. 61.

[4] Cf. Bundestag printed paper 15/5050 (2005), p. 88; Pschyrembel (1998), p. 345.

[5] Swiss Dispatch of 12. September 2001 on a Federal Law on Transplantation of Organs, Tissues and Cells, p. 148.

[6] Bos, in Weimar/Bos/Busschbach (ed.) (2008), p. 23; Daar et al., Vol. 11 Transplantation Review 95, 102 f. (1997); Godlee, Vol. 343 British Medical Journal (2011); Kasiske et al., Vol. 7 Journal of the American Society of Nephrology 2288, 2295 (1996); Kirste, in Rittner/Paul (ed.) (2005), p. 80; Roodnat et al., Vol. 9 American Journal of Transplantation 2150, 2154 (2009).

[7] Cf. Akkina et al., Vol. 57 American Journal of Kidney Diseases 152, 152 (2011); cf. Hamza et al., Vol. 45 Der Urologe 60, 63 (2006); Health Council of the Netherlands (2003); Kirste, Vol. 81 Der Chirurg 778, 780 (2010); cf. O'Hara et al., Vol. 17 Progress in Transplantation 180, 182 (2007); Swiss Dispatch of 12. September 2001 on a Federal Law on Transplantation of Organs, Tissues and Cells, p. 97.

[8] Swiss Dispatch of 12. September 2001 on a Federal Law on Transplantation of Organs, Tissues and Cells, p. 97.

[9] Passim Wolfe et al., 341, New England Journal of Medicine 1725–1730 (1999).

[10] Bundestag printed paper 15/5050 (2005), p. 14; Mueller/Case/Hook, Vol. 22 Transplantation Review 200, 201 (2008); O'Hara et al., Vol. 17 Progress in Transplantation 180, 180 (2007); Price (2010), p. 21; cf. Robert Koch-Institut, (2003), p. 26; Swiss Dispatch of 12. September 2001 on a Federal Law on Transplantation of Organs, Tissues and Cells, p. 58; Weigend/Zielinska, Vol. 14 Medizinrecht 445, 449 (1996).

[11] Kühn (1998), p. 24 f.; Norba (2009), p. 33.

[12] Bundestag printed paper 15/5050 (2005), p. 13; cf. Dahlke et al., Vol. 46 Psychosomatics 58, 58 (2005); Erim/Malagó/Valentin-Gamazo/Senf/Broelsch, Vol. 35 Transplantation Proceedings 909, 909 (2003); Matas et al., Vol. 343 The New England Journal of Medicine 433, 436 (2000); Mühlbacher (2009), p. 3.

such events is transplantation.[13] Therefore, artificial machines cannot be considered at present as a real substitution for an organ donation.

Second, organ shortages have lead to the attempt to transplant organs from animals into human beings (so-called Xenotransplantation).[14] This procedure is still in the experimental phase and has not yet been performed successfully.[15] Therefore, this cannot be seen as an alternative to organ transplantation.

Thus, organ transplantation is the best option in cases of organ failure. However, to meet the demand of organ shortage, the rate of organ donation must increase.[16] A more extensive use of post-mortem organ donation and the further development of LOD are the two possibilities to increase the amount of donor organs. It has already been explained that it cannot be assumed that the waiting time for donor organs could be sufficiently reduced by using organs donated by deceased persons.[17] LOD is, consequently, a useful and desirable option, making it worthy of consideration.

2) *Impact of Living Organ Donation*

Increasing the use of LOD is not depicted here as being the only and best option to solve the problem of organ shortage. It should still be made an enormous effort to increase the amount of deceased donation.[18] Because of the existing organ shortage, the extended use of LOD is nevertheless often the only possibility for a patient that suffers from organ failure.[19] Restricting LOD by law also means disabling several patients from receiving a healing donor organ. Many claim that such a restriction artificially increases the existing shortage of donor organs even more.[20] Because of the shortage of donor organs, the legal restrictions on LOD have to be justified and any unjustified barriers to LOD should be removed.[21] The German Federal Constitutional Court, for example, addressed the problem by referring to the German basic

[13] Bock (1999), p. 53; Norba (2009), p. 33; Levinsky, Vol. 343 The New England Journal of Medicine 430, 430 (2000); Pichlmayr, in Toellner (ed.) (1991), p. 22; cf. Reason (1) Directive 2010/53/EU of the European Parliament and of the Council of 7 July 2010 on standards of quality and safety of human organs intended for transplantation.

[14] Hofer (2006), p. 24; Keller, Vol. XXXII Stetson Law Review 855, p. 871 (2003); Norba (2009), p. 31; Schutzeichel (2002), p. 273; Ugowski (1998), p. 9.

[15] Hofer (2006), p. 24; Keller, Vol. XXXII Stetson Law Review 855, p. 873 (2003); Kühn (1998), p. 25; Nickel/Schmidt-Preisigke/Sengler (2001), p. 5.

[16] Winter, in Broelsch (ed.) (2006), p. 53.

[17] Cf. Ghods, Vol. 3 Iranian Journal of Kidney Diseases 183, 183 (2009); Price (2010), p. 285.

[18] Kirste, in Rittner/Paul (ed.) (2005), p. 79; Stangl, in Rittner/Paul (ed.) (2005), p. 32.

[19] Frei, Commission printed paper 15/141 (2004), p. 1; cf. Jakobsen, in Price/Akveld (ed.) (1997), p. 4; Land, in Oduncu/Schroth/Vossenkuhl (ed.) (2003), p. 211; cf. Price (2010), p. 20; cf. Roodnat et al., Vol. 10 American Journal of Transplantation 821, 821 (2010).

[20] Fateh-Moghadam (2011), p. 1 ff.

[21] Ibid.

rights. Art. 2 II German *Constitution* guarantees that "[e]very person shall have the right to life and physical integrity." This basic right is violated in cases where a governmental regulation denies a therapy that is medically possible and would be life prolonging or would at least reduce the suffering considerably.[22] The *Charter of Fundamental Rights of the European Union* (*CFREU*), which reflects common values of the Member States,[23] also regulates that "[e]veryone has the right to life"[24] and that "[e]veryone has the right to respect for his or her physical and mental integrity."[25] The *CFREU* does not mention LOD in particular, but these two fundamental rights definitely have to be kept in mind in any examination of LOD.

The following statement by the European Commission, which has been considered the executive branch of the EU,[26] attests to the need for more LODs: "The use of living donors is an increasing alternative given the failure to meet the growing need for organs with cadaver donation."[27] And the Council of the European Union mentions "[t]he promotion of living donation" as one of the "[n]ew alternatives for expanding the donor pool."[28]

It has already been shown that the amount of LODs has risen considerably since the beginning of the nineties in several of the countries considered. Since 2001, the United States of America has even had more living kidney donors than deceased donors.[29]

On the one hand, LOD is an established standard therapy.[30] On the other hand, LOD is very problematic and the subject of a controversy that cannot be ignored. LOD involves two patients instead of only one[31] — one of them being a healthy person.[32] Thus, the ethical principle of nonmaleficence, which asserts the obligation one has to avoid inflicting harm on another, is concerned.[33] To justify this, the principle of autonomy is needed. According to the principle of autonomy, every person has a right of self-determination. He can make his own decisions, which

[22] German Federal Constitutional Court [Bundesverfassungsgericht], 16 March, 1982, Case No. 1 BvR 938/81, in BVerfGE 60, 123, 132; German Federal Constitutional Court [Bundesverfassungsgericht], 11 August, 1999, Case No. 1 BvR 218/98, in Vol. 46 Neue Juristische Wochenschrift 3399, 3400 (1999).

[23] Cf. Calliess, in Ehlers (ed.) (2007), p. 534; Cf. Walter, in Ehlers (ed.) (2007), p. 16.

[24] Art. 2 (1) Charter of Fundamental Rights of the European Union.

[25] Ibid., Art. 3 (1).

[26] Marmor/Freeman/Okma, in Marmor/Freeman/Okma (ed.) (2009), p. 8.

[27] European Commission (2006).

[28] Commission of the European Communities (2008).

[29] Gutmann (2006), p. 3.

[30] Price (2010), p. 196; cf. Wagner/Fateh-Moghadam, Vol. 56 Soziale Welt 73, 74 (2005).

[31] Swiss Dispatch of 12. September 2001 on a Federal Law on Transplantation of Organs, Tissues and Cells, p. 30.

[32] Schreiber (2004), p. 18; Schutzeichel (2002), p. 100; Wagner/Fateh-Moghadam, Vol. 56 Soziale Welt 73, 77 (2005).

[33] Beauchamp/Childress (2001), p. 113.

includes having the right to make a decision about his own body.[34] Consequently, any legal rule on LOD that restrains autonomy infringes on the freedom of both the donor and recipient.[35]

Even if the principle of nonmaleficence and the principle of autonomy are considered to be universal, both maxims are not absolute. If there is a conflict between these two principles, both must be weighed before reaching a judgement in individual cases.[36] As a consequence of this tension, LOD is in need of regulation.[37]

II. Basic Facts About Living Organ Donation

Organ transplantation is defined as the removal of an organ from a donor's body to transplant into a recipient's body; the donor's organ replaces the recipient's failing or absent organ.[38] (1) The donor can either donate the organ while alive or post-mortem. These two existing procedures for organ transplantation will be deliminated. (2) Second, a brief overview about the historical development of LOD will be given. (3) Third, it will be presented which organs can be donated by a living person.

1) Delimitation of Living Organ Donation and Post-Mortem Organ Donation

Organs can be removed either from living or deceased persons. Thus, living and deceased donors must be distinguished.[39] They are classified according to the moment of their death. In compliance with the brain death criterion, a human being is pronounced dead if he exhibits an irreversible end of all brain activity.[40] The criterion of brain death is widely established and is enforced in virtually all

---

[34] Bundestag printed paper 15/5050 (2005), p. 34; Fischer, in Ahrens et al. (ed.) (1999), p. 545; Forkel, Vol. 23 JURA 73, 78 (2001).

[35] Gutmann, Vol. 15 Medizinrecht 147, 147 (1997); Nickel/Schmidt-Preisigke/Sengler (2001), p. 89.

[36] Beauchamp/Childress (2001), p. 5, p. 18 f.; Gutmann et al., Terasaki (ed.) (1995), p. 356; cf. Lamb, in Price/Akveld (ed.) (1997), p. 43 f.; cf. Price (2000), p. 254.

[37] This area of tension will be explained further in chapter "Analysis of the Normative Arguments That Dominate the Policy Arena About Necessity and Legitimacy of Legal Restrictions in Living Donor Transplantation".

[38] Coppen (2010), p. 10; Joint Council of Europe/United Nations Study (2009); Schreiber (2004), p. 19; Ugowski (1998), p. 8; Wollenek/Wolner, in Brandstetter/Kopetzki (ed.) (1987), p. 10 f.

[39] Kalchschmid (1997), p. 35.

[40] Kienapfel (1990), p. 3 at 11; Mayrhofer, in Brandstetter/Kopetzki (ed.) (1987), p. 55.

countries considered.[41] Therefore, LOD takes place when the donor does not experience brain death, or the irreversible end of all brain activity.[42]

2) Historical Overview

LOD already has a rather long history. The possibility of replacing a sick organ with an organ donated by a living person has been available since the 1950s. The first living kidney transplantation with long-term success took place in 1954 between two identical twins.[43] In 1959, the first kidney-homotransplantation[44] was performed,[45] and, in 1962, the first living kidney transplantation between people who were not genetically related took place.[46] In the meantime, transplantation medicine, and therefore LOD as well, evolved into a common treatment.[47] This was exemplified in Table 1 presented in the chapter "Introduction." The first successful living liver donation was not performed until the late 1980s for the benefit of a child and in the early 1990s for adults.[48] The increasing importance of LOD should not hide the fact that organ demand still far exceeds the amount of organs being donated.[49]

---

[41] See Belgium, France, Greece, Italy, the Netherlands, Poland and Switzerland (Swiss Dispatch of 12. September 2001 on a Federal Law on Transplantation of Organs, Tissues and Cells, p. 87); cf. Oniscu/Forsythe, Vol. 38 Annals Academy of Medicine 365, 366 (2009); cf. Nickel/Schmidt-Preisigke/Sengler (2001), p. 7.

[42] Schreiber (2004), p. 52.

[43] Gruessner/Benedetti, in Gruessner/Benedetti (ed.) (2008), p. xvii; Hauptmann/O´Connor, Vol. 336 The New England Journal of Medicine 422, 423 (1997); Jowsey/Schneekloth, Vol. 22 Transplantation Reviews 192, 192 (2008); Keller, Vol. XXXII Stetson Law Review 855, 865 f. (2003); Losse, in Toellner (ed.) (1991), p. 4; Ross/Thistlethwaite, Vol. 122 American Academy of Pediatrics 454, 454 (2008); Schreiber (2004), p. 16; Spital, Vol. 38 American Journal of Kidney Diseases 189, 189 (2001); Zenker, in Kaufmann et al. (ed.) (1979), p. 481.

[44] In this type of organ transplantation, organs from different individuals from the same species are transplanted (Bock (1999), p. 46; Kühn (1998), p. 19 f.; Norba (2009), p. 34; Schreiber (2004), p. 19 f.).

[45] Vogt/Karbaum, in Toellner (ed.) (1991), p. 12.

[46] Schutzeichel (2002), p. 74.

[47] Bock (1999), p. 29; de Klerk (2010), p. 14; Kalchschmid (1997), p. 3; Swiss Dispatch of 12. September 2001 on a Federal Law on Transplantation of Organs, Tissues and Cells, p. 34.

[48] Erim et. al., Vol. 81 Der Chirurg 820, 820 (2010).

[49] Teubner (2006), p. 3.

3) Possible Organs for Living Organ Donation

Not all organs are of use for LOD. To determine which organs could be used for LOD, one must differentiate (1) the medical and (2) the legal situation.

a) Medical Situation

Medically, the liver, the kidney, the lobes of the lung, segments of the pancreas and the small intestine can be donated by a living person.[50]

The liver could be subject to a LOD, because, from a medical point of view, it can regenerate.[51] This means that, after donating part of the liver, the remaining liver normally regenerates quickly and completely.[52] Living liver donation can be considered as a routine matter.[53]

The kidney is a paired organ and therefore one kidney can be donated by a living person.[54] It is the most frequent type of LOD.[55]

Lobes of the lung and segments of the pancreas and of the small intestine can also be donated while alive.[56] Living lung donation cannot be considered routine, though.[57] Because lung diseases usually affect both lungs, two lobes are essential for one to obtain enough volume and function. Two living donors are usually involved,[58] in contrast to an entire lung from a deceased person.[59] In addition, the surgery has a rather big impact on the donor. Her lung function is permanently impaired and she is no longer able to engage in intensive training, sports, or heavy

[50] Bundestag printed paper 15/5050 (2005), p. 8.

[51] Gutmann, in Schroth et al. (ed.) (2005), Sec. 8 at 31; cf. Norba (2009), p. 57; Siegmund-Schultze (1999), p. 110.

[52] Gutmann, in Schroth et al. (ed.) (2005), Sec. 8 at 31; Norba (2009), p. 57.

[53] Cotler et al., Vol. 9 Liver Transplantation 637, 637 (2003); Neumann/Neuhaus/Schmeding, Vol. 81 Der Chirurg 804, 812 (2010); Swiss Dispatch of 12. September 2001 on a Federal Law on Transplantation of Organs, Tissues and Cells, p. 38.

[54] Bundestag printed paper 15/5050 (2005), p. 9; Erim et al., Vol. 81 Der Chirurg 820, 820 (2010); Land, in Oduncu/Schroth/Vossenkuhl (ed.) (2003), p. 212; Schutzeichel (2002), p. 65.

[55] Childress/Liverman (2006), p. 266; Esser (2000), p. 7; Steinberg, Vol. 145 Annals of Internal Medicine 197, 197 (2006); Swiss Dispatch of 12. September 2001 on a Federal Law on Transplantation of Organs, Tissues and Cells, p. 34.

[56] Bundestag printed paper 15/5050 (2005), p. 8; Joint Council of Europe/United Nations Study (2009); Siegmund-Schultze (1999), p. 98.

[57] Dahlke et al., Vol. 46 Psychosomatics 58, 61 (2005); Swiss Dispatch of 12. September 2001 on a Federal Law on Transplantation of Organs, Tissues and Cells, p. 39.

[58] Bowdish et al., Vol. 79 The Annals of Thoracic Surgery 418, 418 (2005); Bundestag printed paper 15/5050 (2005), p. 12; Hodson, Vol. 26 Journal of Medical Ethics 419, 419 (2000); Information from M. Bos; Norba (2009), p. 57; Price (2010), p. 204.

[59] Bowdish et al., Vol. 79 The Annals of Thoracic Surgery 418, 418 (2005).

work.[60] Furthermore, the surgery itself is rather risky for the donor.[61] The Netherlands, for example, reacted to this danger by abolishing this procedure for the most part. Instead, unilateral and bilateral lung transplants are performed by using lungs from post-mortem donors.[62] LOD with segments of the pancreas and of the small intestine are medically possible, but seldom take place.[63]

This review of the medical practice has shown that living kidney and living liver donation are the most common types of LOD. These two types will therefore be the focus in this doctoral thesis.

b) Legal Situation

The national transplant laws often differentiate between regenerative organs and non-regenerative organs. With regard to living kidney and living liver donation, this leads one to assume that the kidney is classified as a non-regenerative organ, while the liver is categorised as a regenerative organ. However, the distinction between regenerative and non-regenerative organs is primarily used to distinguish between bone marrow, which is regenerative, and kidneys, which are non-regenerative.[64] This is, for example, the case in Finland.[65] This thesis only focuses on solid organs; bone marrow donation is excluded from consideration.

A separate assessment of the liver is therefore necessary. As already stated, from a medical point of view, the liver is a regenerative organ. That does not automatically have an impact on its legal classification, though. In Germany, for example, the distinction between regenerative and non-regenerative organs used to be relevant with regard to the donor-recipient relationship, since only the LOD of non-regenerative organs was restricted. Whether this restriction applied to living liver donation was highly controversial. Several legal experts argued that the wording of the restriction only applied to non-regenerative organs. They claimed that the restriction did not apply to living liver donation because the liver, from a medical point of view, can regenerate.[66] In contrast, the parliamentary Enquete-Commission designated the liver as non-regenerative, causing it to be included in the restriction. They argued that the remaining part of the liver grows again, but the structure differs in comparison to the former liver. Therefore, according to the Enquete-Commission, the liver is a

[60] Information from M. Bos; Price (2010), p. 204.

[61] Hodson, Vol. 26 Journal of Medical Ethics 419, 420 (2000).

[62] Information from M. Bos.

[63] Bundestag printed paper 15/5050 (2005), p. 12.

[64] Cf. Gutmann, in Schroth et al. (ed.) (2005), Sec. 8 at 31.

[65] Information from K. Salmela.

[66] Esser, in Höfling (ed.) (2003), p. 217; Gutmann (2006), p. 21; Gutmann, in Schroth et al. (ed.) (2005), Sec. 8 at 31.

The dissenting view holds that the liver is also non-regenerative, since the removed parts of the liver do not regenerate with regard to the liver of the donor (Nickel/Schmidt-Preisigke/Sengler (2001), p. 99).

non-regenerative organ.[67] Meanwhile, this debate has become obsolete since the German *Act on the donation, removal and transplantation of organs* was amended in 2007. Because of this amendment, the restriction of the donor-recipient relationship currently applies to living kidney donation, the donation of a part of the liver and living donation of other non-regenerative organs.[68] Therefore, in Germany, the liver is viewed as a regenerative organ,[69] but is still treated with the same strictness as non-regenerative organs. The Swedish *Transplant Act* also contains different regulations for regenerative and non-regenerative organs. However, the *Guidance of the provisions* clarifies that provisions which contain regulations for non-regenerative organs apply to the partial transplantation of the liver.[70] This is similar in Finland, where "non-regenerative organs" means solid organs, including the kidney, but also part of the liver.[71] The situation is similar in the Netherlands as well, where the liver, in general, is not considered as regenerative.[72] In England, an organ is defined as "a differentiated and vital part of the human body, formed by different tissues, that maintains its structure, vascularisation and capacity to develop physiological functions with an important level of autonomy."[73] Regenerative and non-regenerative organs are not differentiated; rather, parts of organs are generally included. Consequently, living liver donation is included in the restrictions concerning LOD.[74]

In conclusion countries with less strict rules for regenerative organs do not apply those rules to living liver donation and other countries do not even differentiate between regenerative and non-regenerative organs. The distinction is relevant, however, if one is to distinguish the donation of solid organs (which includes living liver donation and living kidney donation) and bone marrow donation. This thesis only deals with the LOD of solid organs, not with bone marrow or blood donation, though. As just seen, the legal distinction between regenerative and non-regenerative organs is, from a legal perspective, irrelevant for the LOD of solid organs. Therefore, living kidney donation and living liver donation are treated rather equally from a legal perspective, making a legal distinction between regenerative and non-regenerative organs for the LOD of solid organs absurd.

[67] Bundestag printed paper 15/5050 (2005), p. 21.

[68] Cf. Fateh-Moghadam (2008), p. 259; cf. Norba (2009), p. 216 f.

[69] Gutmann (2006), p. 21.

[70] The Ministry of Health and Social Affairs Sweden 1997, The Swedish Transplant Act, Guidance on the provisions, Sec. 7.

[71] Information from K. Salmela. (In Finland, no living liver donation has happened so far.)

[72] Information from M. Bos.

[73] Reg. 2 (6) Human Tissue Act Regulations 2006.

[74] Fateh-Moghadam (2008), p. 281.

III. Terminology

The used terminology needs to be clarified. While doing this, the exercised types of LOD are outlined as well. The terminology for the different types of LOD itself is quite problematic,[75] but will not be discussed any further here. Rather, the following terms will be taken as a basis. A rough distinction is made between specified LOD and unspecified LOD. The term specified LOD comprehends all LODs with the intent to help a specific recipient. This includes direct LOD, but indirect LOD as well. Direct LOD means that a person donates directly to the intended recipient, for example, a parent's donation to his child or a child's donation to his parent. Indirect LOD means that a person donates to help a specific recipient, but the donation is only indirect. The organ from the related donor is not directly transplanted into the recipient. The donation still allows the recipient to receive an organ, because he receives an organ from a stranger in return for the donation of the relative which is given to another person as well. This type of LOD includes cross-over LOD, unbalanced living paired exchange, living paired cascade exchange, pool donation and list-paired exchange. The opposite of specified LOD is unspecified LOD. In such a case, a person donates an organ to an anonymous recipient.[76] This is LOD to a stranger. Directed altruistic LOD can either be classified as indirect LOD or as unspecified LOD. Another type of LOD, namely, unspecified non-directed donation catalysing cascade exchanges, connects indirect LOD and unspecified LOD. As seen, several types of LOD exist.[77]

[75] Hilhorst et al., in Weimar/Bos/Busschbach (ed.) (2008), p. 380.

[76] With respect to the used terminology see Dor et al., 2011 Transplantation 1 ff.

[77] The different types of LOD will be explained and analysed below.

Comparative Analysis of European Transplant Laws Regarding Living Organ Donation

As seen in Table 1 in the chapter "Introduction," the rates of LOD are very different in the countries considered. It can be assumed that various reasons contribute to these differences, the legal situation being one of those.[1] The different national transplant laws will hence be compared. (1) How the comparison of laws takes place will be presented, before (2) the issues of LOD in need of regulation are compared.

I. Comparison of Laws

Comparative law is the comparison of different legal systems.[2]

The legal regulations of organ donation — including of LOD — are not uniform in the European Union Member States. This discrepancy in the national transplant laws leads to several problems. (1) Hence, reasons why a comparison of laws is necessary are presented. (2) The approach of the following comparison of laws is described. (3) The countries that are included in the comparison have already been listed in Table 1 in the "Introduction."[3] In addition, the national sources of law that are considered will be listed and (4) relevant European and international documents.

[1] Cf. Fateh-Moghadam (2011), p. 1; cf. Radcliffe Richards, in Weimar/Bos/Busschbach (ed.) (2011), p. 41.

[2] Constantinesco (1972), p. 107 f.; Zweigert/Kötz (1996), p. 2.

[3] P. 2.

L. Lopp, *Regulations Regarding Living Organ Donation in Europe*,
DOI 10.1007/978-3-642-33799-4_3, © Springer-Verlag Berlin Heidelberg 2013

1) Reasons for a Comparison of Laws

First, a comparison of laws can be viewed as a necessary step towards a unification of different laws.[4] The Member States of the European Union continually grow more and more together. This goes hand in hand with the increasing level of cooperation between them. *Klein* correctly states: "None of us can escape the bombardment of information about what is happening in other countries."[5] This also applies to the field of organ transplantation. Strengthening the states' cooperation in this subject matter and finding common solutions for the problems connected to this issue is recommended.[6] This is only possible after the similarities and differences have been brought to light through a comparison of the national transplant laws.[7] This is accomplished in this chapter. The final aim of this thesis is to formulate a *best practise proposal for LOD*. This will take place in the drafting style of the Common Frame of Reference for European Private Law, thus by "expressing our ideas, of summarising what we found in the existing national laws [. . .], and of testing whether it is possible to draft at least *one* coherent set of rules [. . .]."[8] Therefore, a comparison of laws serves as a contribution to the unification of laws.[9]

Second, if all countries regulate organ donation differently, they will have differences in quality and safety requirements.[10] If the laws were adjusted, though, a similar high level of protection could be guaranteed, without restraining the development of LOD.[11] According to the *Directive 2010/53/EU of the European Parliament and of the Council of 7 July 2010 on standards of quality and safety of human organs intended for transplantation (Directive 2010/53/EU)*, this "would help to reassure the public that organs procured in another Member State carry the same basic quality and safety guarantees as those obtained in their own country."[12]

---

[4] Cf. Beck/Burchhard/Fateh-Moghadam, in Beck/Burchhard/Fateh-Moghadam (2011), p. 5; Zweigert/Kötz (1996), p. 16.

[5] Klein, in Marmor/Freeman/Okma (ed.) (2009), p. 305.

[6] Cf. Gutmann/Schroth (2002), p. 41; cf. Prechern-Hauptmann, in Höfling (ed.) (2008), p. 97. The following statement has been made with respect to health care in general, but can be applied to LOD as well "Increased pressure for policy change in health care and, with that, the inclination to look abroad for promising solutions of domestic problems" (Marmor/Freeman/Okma, in Marmor/Freeman/Okma (ed.) (2009), p. 1 f.).

[7] This is in accord with Ehlers who states that knowledge of the foreign law is necessary for a legal comparison (Ehlers, in Großfeld et al. (ed.) (2006), p. 38).

[8] von Bar (2008), p. 4.

[9] Cf. Marmor/Freeman/Okma, Vol. 7 Journal of Comparative Policy Analysis 331, 333 (2005); Zweigert/Kötz (1996), p. 16.

[10] Reason (5) Directive 2010/53/EU of the European Parliament and of the Council of 7 July 2010 on standards of quality and safety of human organs intended for transplantation.

[11] Guillod/Perrenoud, in Gutmann et al. (ed.) (2004), p. 168.

[12] Reason (6) Directive 2010/53/EU of the European Parliament and of the Council of 7 July 2010 on standards of quality and safety of human organs intended for transplantation.

Third, the danger of organ tourism is a problem resulting from non-uniform legislation.[13] If the regulations on organ transplantation are not assimilated, and if the problem of organ shortage cannot be solved, then we can assume that many will use the opportunity to benefit from the legal rules of other countries.[14] People suffering from organ failure may find that the only possibility for them to receive a transplantation they are not able to get in their home country is to travel abroad.[15] It is feared, for example, that patients will search for living donors, who consent to "donate" organs for a certain amount of money. This procedure would then be attempted in a country that does not regulate LOD or that at least does not require any specific donor-recipient relationship.[16] Furthermore, organ tourism is problematic because it is closely connected to the risk of the commercialisation of organs.[17] The legal prohibition of organ trade remains an important goal for the legislators in all the countries considered.

A comparison of national transplant laws is thus urgent and beneficial.

2) Procedure of the Comparison

How should this comparison be approached? One very unique aspect of this comparison is that it does not only include two legal systems like comparative law usually does, but contains 30 legal systems. Following the Introduction of the *Principles of European Contract Law*, not one national legal system will be the basis of the comparison; it will rather be "paid attention to all the systems."[18] Therefore, the best way to make this broad comparison possible has to be worked out.[19]

Zweigert and *Kötz*, the authors of the most influential German speaking introductory textbook for comparative law,[20] consider the principle of functionality as the most appropriate technique for a legal comparison.[21] The principle of functionality has also been used by the Commission on European Contract Law when it established the *Principles of European Contract Law*.[22] This approach assumes that

[13] Cf. Hohmann (2003), p. 192; cf. Price (2010), p. 17.

[14] Cf. Gutmann (2004), p. 24 f.

[15] Guillod/Perrenoud, in Gutmann et al. (ed.) (2004), p. 164 f.

[16] Hohmann (2003), p. 192.

[17] Guillod/Perrenoud, in Gutmann et al. (ed.) (2004), p. 165.

[18] Commission on European Contract Law (2006). See also von Bar/Zimmermann (2002), p. XXVIII.

[19] Cf. Ebert (1978), p. 142.

[20] Fateh-Moghadam, in Beck/Burchhard/Fateh-Moghadam (2011), p. 44.

[21] Fateh-Moghadam, in Beck/Burchhard/Fateh-Moghadam (2011), p. 44; Michaels (2005), p. 2; Zweigert/Kötz (1996), p. 33 ff.

[22] Meyer (2007), p. 98.

certain problems connected to one issue—here LOD—arise in every country.[23] In other words, a presumption of similarity exists—*praesumtio similitudinis*.[24]

There are seven central issues associated with LOD that must be addressed. (1) First, the different regulations with regard to the donor and recipient's consent and disclosure. For example, the formal requirements and the aspect of voluntariness. Furthermore, the concept known as informed consent exists in most of the countries[25] and is regarded in detail in a few countries. (2) Second, I address the required suitability of the transplantation. Both medical suitability and the normative ceilings of risk-benefit-ratios are included in the analysis. (3) Third, I describe the special classes of potential donors, namely minors and mentally incapacitated adults, and to what extent they are allowed to become living organ donors. (4) Fourth, the amount of restriction on the donor-recipient relationship with respect to competent adults is addressed. I roughly present the general debate about whether a certain relationship between donor and recipient should be required before describing the national regulations addressing this aspect. Afterwards, I explicate what forms of LODs are allowed. (5) Fifth, I illustrate which countries consider LOD as a subsidiary option only. I consider the relationship between LOD and post-mortem organ donation, and the relationship between LOD and alternative therapies. (6) Sixth, I examine several procedural issues, especially whether any kind of commission is involved in the process of LOD. (7) Seventh, the social security regulations for donors in selected countries are presented.

This doctoral thesis observes each issue separately and denotes it with a neutral term that does not originate in any particular legal system to abstract the issues.[26] Those generic terms must be supranational, autonomous and at least partly universal.[27] This is consistent with the *Draft of a Common Frame of Reference*, which states: "The terminology should be as suitable as possible for use across a large number of translations. It should therefore try to avoid legalese and technicalities drawn from any one legal system."[28] In the next step, this thesis scrutinises the national laws and organises the legal regulations for the issues into groups according to the kind of solution that is proposed, so that they can be analysed and compared.[29] Finally, this thesis compares how the different legal systems solve the same, or at least a similar, problem.[30]

---

[23] Zweigert/Kötz (1996), p. 34 ff.

[24] Ibid., p. 39.

[25] Cf. Gutmann/Schroth (2002), p. 41.

[26] Braga, in Lüke/Ress/Will (ed.) (1983), p. 104 ff.; Ebert (1987), p. 29; Zweigert/Kötz (1996), p. 10, p. 34 f.; Zweigert, Vol. 16 Rabels Zeitschrift für ausländisches und internationales Privatrecht 387, 395 (1951).

[27] Ebert (1978), p. 148.

[28] von Bar et al. (ed.), Draft Common Frame of Reference (2009), p. 29.

[29] Cf. Zweigert/Kötz (1996), p. 44.

[30] Ebert (1978), p. 28.

It cannot be denied, though, that the principle of functionality has its limits. Consequently, an operative functionalism is proposed. Proponents of an operative functionalism criticise the presented principle of functionality by asserting that the law does not merely solve social problems; it also creates several problems that are bound to legal requirements. Therefore, the considered problems are always enmeshed in legal structures, which is unrecognised by the principle of functionality.[31] Secondly, the principle of functionality does not consider the different dogmatic structures. Consequently, the most interesting aspects of comparative law might remain concealed.[32] Thirdly, many criticise the starting point of the principle of functionality, the assumption of similarity.[33] According to the supporters of operative functionalism, this confounds similar solutions with functional equivalent solutions.[34] This critique leads us to suggest an approach based on operative functionalism. The main element of this approach is that the problem and solution both depend on the structure. Thus, both the solution and the problem are considered; operative functionalism is reflexive.[35]

The criticism of the principle of functionality cannot be dismissed. This is especially true with regard to the presumption of similarity because, as stated initially, the national laws considered have similarities as well as differences. For example, most countries accept the donor's consent, yet there is no consensus with regard to the details. Nevertheless, to determine how a comparison of laws should be carried out, the purpose of the intended comparison has to be considered. As shown above, the differences in the national laws regulating LOD lead to several problems. Therefore, the final goal of this thesis is to harmonise the different regulations, thus formulating a best practise proposal for LOD. The same issues for a LOD regulation basically arise in every country considered. The different solutions must first be observed before any judgments can be made. Structuring the comparison in this way most effectively presents the different regulations for the seven central issues connected to LOD. Therefore, in this thesis, the principle of functionality is applied. This, however, is no general judgement about the two possible approaches, but applies to this text only.

[31] Fateh-Moghadam (2008), p. 13 ff.; Fateh-Moghadam, in Beck/Burchhard/Fateh-Moghadam (ed.) (2011), p. 52 ff.

[32] Fateh-Moghadam (2008), p. 13 ff.; Fateh-Moghadam, in Beck/Burchhard/Fateh-Moghadam (ed.) (2011), p. 52 ff.

[33] Cf. Michaels (2005), p. 31 ff.

[34] Fateh-Moghadam (2008), p. 13 ff.; Fateh-Moghadam, in Beck/Burchhard/Fateh-Moghadam (2011), p. 52 ff.

[35] Fateh-Moghadam (2008), p. 13 ff.; Fateh-Moghadam, in Beck/Burchhard/Fateh-Moghadam (2011), p. 52 ff.

3) Relevant National Sources of Law Included in the Comparison

In a comparison of legal systems, all national sources of law of the countries that have already been listed[36] have to be considered. In doing so, everything that completely or at least partly concerns the legal frame is respected.[37] Almost all the countries regulate LOD by law, so the relevant law(s) are examined. One must keep in mind that the most relevant laws, namely the national transplant laws, are embedded in the whole legal system. Other laws that influence the legal and practical situation have to be taken into account as well, particularly, the national laws that regulate the national system of consent and the national Criminal Codes. In addition to the legal situation, the actual medical practice is, in certain cases, also observed.

Austria is an exceptional case. It has not yet regulated LOD,[38] only post-mortem organ donation. Austria established the opt-out model for post-mortem donation.[39] That means that anyone who has not refused post-mortem donation while being alive automatically becomes an organ donor after death. Consequently, many organs from deceased persons are available. That makes LOD less relevant.[40] However, LOD is also carried out and mentioned by several laws (Sec. 120 II *ASV* and Sec. 26 I e *KAG*), so it can be assumed that it is generally permitted.[41] While determining the requirements to justify LOD, general civil and criminal law must be applied.[42] In addition, the *Position Paper* from the Austria Health Limited Liability Company[43] published in 2004 is important. The *Position Paper* is not binding law, only *soft law*. It must nevertheless be regarded because it reflects common medical practice.[44]

---

[36] See Table 1 in the chapter "Introduction", p. 2.

[37] Ebert (1978), p. 150; Sacco (2001), at 23; Zweigert/Kötz (1996), p. 10, p. 34.

[38] Aigner, Vol. 9 Annals of Transplantation 11, 12 (2004); Bruckmueller/Schumann, in Roxin/Schroth (ed.) (2010), p. 844; Commission of the European Communities (2008); Häckel (1997); Kalchschmid (1997), p. 4; Kert, Falcone Project, Austria, p. 2, p. 22.

[39] Kert, Falcone Project, Austria, p. 11; Neyer/Scheuring, 2004 Österreichische Pflegezeitschrift 12, 12.

[40] Information from K. Bruckmueller.

[41] Kopetzki, in Brandstetter/Kopetzki (ed.) (1987), p. 88.

[42] Cf. Aigner, Vol. 9 Annals of Transplantation 11, 12 (2004); information from K. Bruckmueller; Kopetzki (1988), p. 250. It is sometimes stated that the regulations about post-mortem donation are applied analogous. This would be illegal in terms of criminal law, though. In detail, organ trafficking with organs of living people would then be punishable analogous to the criminal act of organ trafficking of organs of deceased, even though this is not clearly regulated by law. Therefore, this criminal analogy should be avoided (Information from K. Bruckmueller).

[43] The Health Limited Liability Company was established in 1991 and the Austrian Ministry of Health entrusted this company to accomplish several tasks connected to organ donation (Prechern-Hauptmann, in Höfling (ed.) (2008), p. 85).

[44] Information from K. Bruckmueller.

Ireland also lacks the central legislation or statutory framework regulating organ donation.[45] It has only established a *Guide to Professional Conduct and Ethics for Registered Medical Practitioners*. The Medical Council's intent for establishing the *Guide* was to clarify several specific areas, for example, questions connected to the aspect of consent.[46] The effects of the *Guide* are similar to those of the Austrian *Position Paper*: the *Guide* has no legal force, but physicians accept it insofar as it is applicable.[47] A failure to follow the *Guide* may lead to disciplinary action against the physician, leaving physicians to take the *Guide* very seriously.[48]

Unfortunately, this thesis cannot address the legal situations of all countries that were planned to be included in an equal manner due to the lack of input of some countries. The legal situation of LOD in Malta is unclear. Malta did not have any legislation for organ transplantation in 2004.[49] There is still no information available regarding any legislation that may have been passed since 2004. In Cyprus, LOD was not regulated by law in 1999,[50] and no current information about its actual legal situation could be assembled. LOD is a common medical practice in Cyprus, though, since it had the highest rate of LODs in Europe in 2009.[51] The same problem exists in Luxembourg. But, at least no LODs have been carried out there during the last few years.[52]

4) European and International Level

Not only the countries recognised the increasing importance of LOD, but the European Union and the World Health Organization as well. A very extensive action of the European Union started in 2007 with a *Communication from the Commission to the European Parliament and the Council—Organ Donation and Transplantation: Policy Actions at EU Level* (*Communication from the Commission*).[53] The result of this was to intend first, ensuring quality and safety, second, increasing organ availability and third, avoiding organ trafficking.[54] Hence, in addition to presenting the national legal situations of LOD, I examine the relevant European approaches. The view of the WHO is also mentioned where relevant.

---

[45] Bacik, Falcone Project, Ireland, p. 14.

[46] Guide to Professional Conduct and Ethics for Registered Medical Practitioners Introduction.

[47] Information from D. Madden.

[48] Ead.

[49] Information from D. Price; cf. Nickel/Schmidt-Preisigke/Sengler (2001), p. 6.

[50] Lilie, in Fischer/Lilie (ed.) (1999), p. 136.

[51] Gesundheit Österreich GmbH (2009).

[52] Ibid.

[53] This document will be regarded in detail in chapter "Actions of the European Union So Far".

[54] Council of the European Union (2007), p. 45.

With regard to Europe as a whole, the *Convention for the Protection of Human Rights and Dignity of the Human Being with regard to the Application of Biology and Medicine* (*Convention on Human Rights and Biomedicine*) and its *Additional Protocol to the Convention on Human Rights and Biomedicine concerning Transplantation of Organs and Tissues of Human Origin* (*Additional Protocol on Transplantation of Organs*) are examined. Both were established by the Council of Europe, an international organisation that was founded in 1949, which covers the entire European continent by now.[55] It is independent of the European Union and its organs.[56] However, since it covers the entire European continent,[57] its documents will also be considered in this comparison. One of its aims is to achieve greater unity between its members, especially through harmonising legislations on subjects that are of common interest.[58] Neither the *Convention on Human Rights and Biomedicine* nor the *Additional Protocol on Transplantation of Organs* apply to all Member States of the Council of Europe. Pursuant to Art. 33.3 of the *Convention on Human Rights and Biomedicine*, it comes into effect "on the first day of the month following the expiration of a period of 3 months after the date of the deposit of its instrument of ratification, acceptance or approval." It only "provide[s] a common framework for the protection of human rights and human dignity in both longstanding and developing areas concerning the application of biology and medicine."[59] It, however, has legal meaning in the countries that have ratified it. At the moment, with regard to the countries considered, Bulgaria, the Czech Republic, Denmark, Estonia, Finland, Greece, Latvia, Lithuania, Moldavia, Portugal, Rumania, Slovakia, Spain and Switzerland have ratified the *Convention on Human Rights and Biomedicine*. With respect to those countries, there could be discrepancies between the regulations of the *Convention on Human Rights and Biomedicine* and the national laws. However, this will not be taken into consideration. The *Additional Protocol on Transplantation of Organs* has been ratified by Bulgaria, Estonia, Finland, Hungary, Moldova, Slovenia and Switzerland. The cases where national law contradicts the *Additional Protocol on Transplantation of Organs*, representing an inconsistency, will also not be taken into account.

The two documents that have additionally been adopted by the Council of Europe — *Recommendation on Liver Transplantation from Living Donors* and *Resolution on Harmonisation of Legislations of Member States Relating to Removal, Grafting and Transplantation of Human Substances* — are also considered.

---

[55] Council of Europe (2011).

[56] Cf. Council of Europe (2011); cf. Haratsch/König/Pechstein (2010), p. 17.

[57] Hendriks, in Weimar/Bos/Busschbach (ed.) (2011).

[58] Resolution (78) 29 on Harmonisation of Legislations of Member States Relating to Removal, Grafting and Transplantation of Human Substances; see also Walter, in Ehlers (ed.) (2007), p. 1.

[59] Secretary General of the Council of Europe (1996), No. 7.

In 2004, the European Union already considered the transplantation of human tissues and cells to be "a great strongly expanding field of medicine offering great opportunities for the treatment of as yet incurable diseases."[60] Consequently, they passed the *Directive 2004/23/EC of the European Parliament and of the Council on Setting Standards of Quality and Safety for the Donation, Procurement, Testing, Processing, Preservation, Storage and Distribution of Human Tissues and Cells.* It only applies to tissues and cells, not to organ donation.[61] It is therefore not directly relevant for LOD, but the *Directive 2004/23/EC*, which has a direct effect on the national laws, is a good example of how the European Union has taken steps towards a harmonisation of the national laws of the Member States. The implementation of the *Directive 2004/23/EC* into each country's national law could create a situation in which the national implementations might not correspond with each other.[62]

The most recent European document that concerns LOD, the *Directive 2010/53/EU* will also be taken into account. The aim of the *Directive 2010/53/EU* is to lay "down rules to ensure standards of quality and safety for human organs [...] intended for transplantation to the human body, in order to ensure a high level of human health protection."[63] Due to the *Directive 2010/53/EU*, one of the most essential subject matters of public health — organ transplantation, which often determines life or death — has come to the fore of European Union law.[64] It is the first formal document passed by the European Union that addresses the donation and transplantation of organs.[65]

The *CFREU* does not particularly focus on LOD, but needs to be considered as well, since organ donation concerns fundamental rights. The *CFREU* came into effect when the *Treaty of Lisbon* did,[66] consequently, on 1 December 2009.[67] According to Art. 6 I *Treaty on European Union* (*TEU*), the *CFREU* is legally binding.[68] As stated in the Preamble of the *CFREU*, its main purpose is "to strengthen the protection of fundamental rights [...] by making those rights more visible in a Charter." Thus, the *CFREU* was created to embody the values of the

---

60 Directive 2004/23/EC of the European Parliament and of the Council on Setting Standards of Quality and Safety for the Donation, Procurement, Testing, Processing, Preservation, Storage and Distribution of Human Tissues and Cells, (1).

61 Ibid., (7) & (8).

62 Cf. Fateh-Moghadam (2008), p. 21.

63 Article 1 Directive 2010/53/EU of the European Parliament and of the Council of 7 July 2010 on standards of quality and safety of human organs intended for transplantation.

64 Wollenschläger, in Müller-Graff/Schmahl/Skouris (ed.) (2011), p. 447 f.

65 Joint Working Party of The British Transplantation Society and The Renal Association (2011), p. 21.

66 Jarass (ed.) (2010), p. V; Meyer, in Meyer (ed.) (2011), p.5.

67 Jarass (ed.) (2010), p. 5.

68 Jarass (ed.) (2010), p. 4. The Treaty establishing a Constitution for Europe was rejected, though, and the Charter has not become part of the Treaties of the European Union (Jarass (ed.) (2010), p. 4).

European citizens and to document those by establishing a common codex.[69] With regard to the content of the *CFREU*, it "reaffirms [...] the rights as they result, in particular, from the constitutional traditions and international obligations common to the Member States, the European Convention for the Protection of Human Rights and Fundamental Freedoms, the Social Charters adopted by the Union and by the Council of Europe and by the case-law of the Court of Justice of the European Union and of the European Court of Human Rights."

In addition to the European documents, the WHO is also quite pertinent. In the *WHO Guiding Principles on Human Cell, Tissue and Organ Transplantation,* the aim is to provide an ethical and acceptable framework for the acquisition and transplantation of organs. The courts have the jurisdiction to determine how the principles are to be implemented.[70] With respect to the type of effect the principles have, the WHO noted that they "have greatly influenced professional codes and practices as well as legislation around the world."[71] Initiatives of the WHO are not binding, but hold political meaning.

Thus, beyond the consideration of national regulations, the European and the international levels are covered where relevant.

In the following, a detailed account of the central issues of LOD that are in need of regulation will be presented, mainly describing how each country currently regulates these issues. The concept of informed consent in particular is discussed in detail for selected countries.

II. The Issues of Living Organ Donation in Need of Regulation

As already stated, there are basically seven issues of LOD in need of regulation.[72] In addition to the transplant laws, I analyse the regulations that are embedded into the general national legal systems which specifically apply to organ donation. Of particular relevance to this dissertation is how one gives valid consent in the national legal system that he is in. (1) The general opinion of selected countries towards the concept widely known as informed consent will first be described, (2) followed by a discussion of the seven issues of LOD.

1) Concept of Informed Consent

This section describes what the national laws require with regard to consent and disclosure and provides the general meaning and relevant background information for the doctrine widely known as informed consent for selected countries.

---

[69] Burr, in Tettinger/Stern (ed.) (2006), p. 213.

[70] WHO Guiding Principles on Human Cell, Tissue and Organ Transplantation, Preamble, 4.

[71] Ibid., Preamble, 3.

[72] P. 18.

As will be shown below, the legal systems considered basically agree that LOD requires the donor's consent and a reasonable amount of information for him. The countries do not share a common view with regard to whether consent is sufficient to justify LOD or if it is only necessary, but not sufficient. While addressing the issue of informed consent, the countries will consequently be divided into two groups: (1) The countries in which the patient's consent is not enough to justify LOD and (2) the countries that consider consent as necessary and sufficient to legitimate a LOD.[73]

a) Countries in Which the Patient's Consent Is Necessary, But Not Sufficient

The legal guidelines in England, France, Italy and Spain state that mere consent is not sufficient to justify LOD; rather, an additional legitimising element is required.

Before describing the English situation, it needs to be explained how the concept of informed consent is not really found in English medical criminal law. That is, failure to inform the patient renders the consent ineffective only if the patient was not even informed in a general manner about the nature and the purpose of the operation or informed of the identity of the acting person.[74] As a consequence, the following discussion focuses mainly on consent, not on informed consent.

In England, LOD is *prima facie* illegal[75] and is considered common assault and battery.[76] The patient's consent plays a striking part, though, because the person's autonomy is respected, yet can also be weighed in view of other interests.[77] In general, for actions above the level of bodily harm, a further legitimating feature is required. All organ removal procedures are considered actions above the level of bodily harm. Consequently, the patient's consent is necessary for LOD, but not sufficient.[78] Those generally forbidden surgeries can only be justified by an additional objective element: good reason. That usually means public interest.[79]

For the field of medical operations, a medical exception for reasonable and proper treatment is accepted.[80] For non-therapeutic medical operations, an additional exception is necessary. So far, a classification of LOD has not yet taken place.[81] The *Human Tissue Act 2004* and the *Human Tissue Act Regulations 2006* only recognise the legitimacy of LOD implicitly by stating that the Human Tissue

---

[73] Guillod/Perrenoud, in Gutmann et al. (ed.) (2004), p. 160.

[74] Fateh-Moghadam, in Roxin/Schroth (ed.) (2010), p. 930.

[75] Price (2000), p. 243.

[76] Fateh-Moghadam, in Roxin/Schroth (ed.) (2010), p. 898 & p. 929.

[77] Fateh-Moghadam (2008), p. 135 f.

[78] Cf. Price (2000), p. 244 ff.

[79] Fateh-Moghadam (2008), p. 140.

[80] Fateh-Moghadam (2008), p. 140; Fateh-Moghadam, in Roxin/Schroth (ed.) (2010), p. 904.

[81] Fateh-Moghadam (2008), p. 141.

Authority can authorise LODs in exceptional cases.[82] In addition, the *Human Tissue Act 2004* standardises, in Sec. 33, a criminal prohibition with a reservation to grant permission.[83] This makes LOD "illegal unless [...]."[84] To justify LOD, several legal requirements have to be fulfilled. Therefore, the possibility of LOD is only implicitly accepted.[85]

Summing up, the patient's consent plays a striking role in England. In fact, it is even described as being the "underlying principle" of the *Human Tissue Act 2004*.[86] However, the patient's consent is not sufficient to justify LOD.

To understand this particular stance, one must look to the constitutional situation in England. Even though the general attitude of English constitutional law is rather liberal since "everybody is free to do anything, subject only to the provisions of the law,"[87] this declaration cannot be equated with subjective fundamental rights and does not protect the individual from explicit bans.[88] The subjective rights of individuals have not been set out in a complete constitution or in a code so far. Consequently, courts cannot challenge any legislation that restricts or authorises the restriction of civil liberties. In addition, the possibility of constitutional judicial review does not exist in England.[89] This may have changed since the *Human Rights Act 1998* came into effect, which incorporated the *European Human Rights Convention* into the English law. Art. 8 of the *European Human Rights Convention* describes a right to dispose of one's body.[90] This could be the legal start to challenge the constant, paternalistic jurisdiction of the House of Lords with regard to consent.[91] It grants the courts the possibility of protecting guaranteed rights. The courts are entitled to determine which legislative provisions do not comply with the *Convention*. Afterwards, the Government may order the amendment of legislation to conform to the *Convention*.[92] However, the influence of the concept of informed consent in English criminal law is still rather weak.[93]

In France, on the one hand, the *liberté individuelle* is legally protected. This fundamental right was initially recognised by the Constitutional Council of the French Republic in 1977. The French *Constitution* also implies this right by institutionalising the ordinary jurisdiction as keeper of individual freedom. This

[82] Fateh-Moghadam, in Roxin/Schroth (ed.) (2010), p. 912 f.

[83] Fateh-Moghadam (2008), p. 165.

[84] Price (2010), p. 209.

[85] Fateh-Moghadam (2008), p. 165.

[86] Winterton (2004).

[87] House of Lords, 13 October, 1988, Case No. 1 AC 109 (per Lord Goff of Chieveley).

[88] Fateh-Moghadam, in Roxin/Schroth (ed.) (2010), p. 903.

[89] Barendt (1998), p. 46 ff.

[90] Ibid.

[91] In the United Kingdom, paternalism is not considered as (legal) problem (Fateh-Moghadam, in Fateh-Moghadam/Sellmaier/Vossenkuhl (ed.) (2010), p. 26).

[92] Barendt (1998), p. 46 ff.

[93] Fateh-Moghadam, in Roxin/Schroth (ed.) (2010), p. 903 f.

fundamental right is phrased as a general clause. It also contains the right to dispose of one's own body.[94] On the other hand, one general principle that is also relevant for LOD is the "right for respect of the human body" (*le droit de tout être humain au respect de son corps*). This principle has two facets: the human body is inviolable (*le principle d'inviolabilité du corps humain*), and the human body is unavailable (*le principle d'indisponibilité du corps humain*).[95]

Requiring the patient's consent is a direct result of the "right for respect of the human body."[96] However, merely consenting is not sufficient to justify a LOD.[97] In addition, an authorisation by law (*autorisation de la loi*) is required.[98] This applies to cases where certain behaviour is usually punishable, but explicitly allowed under certain circumstances by another regulation.[99]

Consequently, consent is only one element among many. It can even be said that the *ordre public* is given priority compared to the patient's will.[100] In the past, no explicit authorisation by law existed for organ transplantation. Therefore, physicians who performed LODs were committing a punishable offence. Consequently, the French legislator had to establish an authorisation by law that allows physicians to perform LODs. The current authorisation to legally perform LODs is given in *Bioethics Law*.[101]

The right of self-determination plays a striking part in Spain. Art. 2 of the *Law on Self-determination of the Patient* requires the patient's consent with regard to every therapeutic treatment. Although LOD is not viewed as therapeutic treatment for the donor, it is still explicitly permitted as an exception. LOD can be justified if it is performed by a qualified person, if the donor gave his consent and if all other legal requirements are met.[102] If the requirements are not fulfilled, the result is equivalent to an offence of mutilation and therefore punished by the *Criminal Code*.[103] Therefore, mere consent is not sufficient to preclude the physician's liability for bodily harm, since an established legal exception has to be fulfilled additionally.[104]

In Italy, Art. 5 *Civil Code* determines that disposing of ones own body is forbidden if the act can cause a permanent disability of the person, if the act breaks the law or if it causes a public order offence.[105] Living kidney donation, however, is

[94] Bauer, p. 151 ff.

[95] Jung, Vol. 14 Medizinrecht 355, 357 (1996).

[96] Ibid.

[97] Cf. Gutmann/Schroth (2002), p. 50 f.

[98] Gutmann/Schroth (2002), p. 50 f.; Jung, Vol. 14 Medizinrecht 355, 357 (1996).

[99] Jung, Vol. 14 Medizinrecht 355, 357 (1996).

[100] Eser, in Weigend/Küpper (ed.) (1999), p. 472.

[101] Gutmann/Schroth (2002) , p. 50 f.

[102] Casabona, in Price/Akveld (ed.) (1997), p. 140.

[103] Ibid., p. 145.

[104] Cf. Gutmann/Schroth (2002), p. 51.

[105] Information from I. Marino.

allowed by law under specific preconditions. That is considered as an exception to Art. 5 *Civil Code*.[106]

As a consequence of this, the process of giving consent to LOD can be considered rather strict. The declaration to donate a kidney or part of the liver while being alive has to be given in front of the district court judge by the donor or the domicile of the clinical centre of transplantation. If the judge is convinced that consent is given voluntarily and based on the donors own interests, she executes the written component of the declaration.[107] The judge must first rule out the possibility that the organ donor is not being coerced, or that the organ is not being donated for financial gain.[108]

Thus, in the countries just described, consent is not sufficient to justify LOD. Consent is essential and plays a striking part, yet an additional validating element is required for justification.

b) Countries in Which the Patient's Consent Is Sufficient

In contrast to the countries just presented, in Germany, Austria, Portugal, Switzerland and the Netherlands, consent is decisive. It is sufficient to justify LOD.

In Germany, the "right of personality" is protected by Art. 2 I in connection with Art. 1 I of the *Constitution*.[109] It guarantees the personal sphere of life and the preservation of its basic conditions.[110] The protection contains different areas.[111] It also contains the "right to dispose of one's own body."[112] To take the "right of personality" into consideration is an essential part of the physicians area of responsibility.[113] Without this, the rule of the *Federal Medical Practitioners Act* that the physician serves the health of the individual person and of the whole people is not

[106] Art. 1 Legislation on Donation, Retrieval and Transplantation of Organs and Tissues.

[107] Gutmann/Schroth (2002), p. 51.

[108] Information from P. Bruzzone.

[109] von Arnauld, Vol. 40 Zeitschrift für Urheber- und Medienrecht 286, 286 (1996); di Fabio, in Maunz/Dürig (ed.) (2010), Art. 2 at 204; German Federal Constitutional Court [Bundesverfassungsgericht], 13 May, 1986, Case No. 1 BvR 1542/84, in Vol. 39 Neue Juristische Wochenschrift 1859, 1859 (1986); Jarass, Vol. 42 Neue Juristische Wochenschrift 857, 857 (1989); Jarass, in Jarass/Pieroth (ed.) (2011), Art. 2 at 1.

[110] German Federal Constitutional Court [Bundesverfassungsgericht], 21 November, 2007, Case No. 1 BvR 1620/04, in BVerfGE 121, 69, 90; German Federal Constitutional Court [Bundesverfassungsgericht], 3 June, 1980, Case No. 1 BvR 185/77, in BVerfGE 54, 148, 153; Jarass, Vol. 42 Neue Juristische Wochenschrift 857, 859 (1989).

[111] Jarass, in Jarass/Pieroth (ed.) (2011), Art. 2 at 39.

[112] Cf. German Federal Constitutional Court [Bundesverfassungsgericht], 14 September, 1983, Case No. 2 BvR 1952/93, in BVerfGE 89, 119, 130; Di Fabio, in Maunz/Dürig (ed.) (2010), Art. 2 at 204.

[113] German Federal Constitutional Court [Bundesverfassungsgericht], 25 July, 1979, Case No. 2 BvR 878/74, in Vol. 32 Neue Juristische Wochenschrift 1925, 1930 (1979).

fulfilled.[114] As a result, the patient is free to either allow or decline therapeutic treatments according to his will.[115] If the physician acts lege artis while performing a therapeutic treatment, but the patient did not give valid consent, the physician commits bodily harm.[116] A legislative reform of this was considered.[117] The intention of this was to ensure that any indicated medical treatment that was actually performed lege artis would not be considered as bodily harm.[118] With respect to the physician's duty to disclosure, this alternative ruled that the consent would be valid only if the person consenting has been informed about the type, the scope and the consequences of the treatment that have relevance for the decision of an understanding person.[119] In the area of civil law, this goal, however, is already reached by the "right of personality."[120] With regard to the criminal liability of the physician who acts with the patient's consent, the principle *volenti non fit iniuria* is applied. Consequently, consent is valid in principle and is, apart from the concrete danger of death, not limited.[121]

The patient's consent as a result of the right of self-determination basically has three functions. First, an intervention on a person's physical integrity must not be carried out against or without the person's will,[122] which means that organ removal is only allowed with the valid consent of the donor.[123] Second, the person concerned has the power to abandon his objects of legal protection based on his own convictions.[124] Applying this principle in the context of LOD leads to the result that nobody, except for the donor, is allowed to judge under what conditions he should be allowed to sacrifice his physical integrity with the aim of boosting someone else's state of health.[125] Third, from a physician's perspective, the patient's consent guarantees that the physician cannot be held legally liable for

[114] Ibid.

[115] Deutsch/Spickhoff (2006), p. 165; Wenzel (2007), p. 248.

[116] German Federal Constitutional Court [Bundesverfassungsgericht], 25 July, 1979, Case No. 2 BvR 878/74, in Vol. 32 Neue Juristische Wochenschrift 1925, 1929 f. (1979). A valid consent of the person concerned does generally usually remove an infringment of a fundamental right (Jarass, in Jarass/Pieroth (ed.) (2011), Art. 2 at 54).

[117] German Federal Constitutional Court [Bundesverfassungsgericht], 25 July, 1979, Case No. 2 BvR 878/74, in Vol. 32 Neue Juristische Wochenschrift 1925, 1930 (1979).

[118] Ibid.

[119] Ibid.

[120] German Federal Constitutional Court [Bundesverfassungsgericht], 25 July, 1979, Case No. 2 BvR 878/74, in Vol. 32 Neue Juristische Wochenschrift 1925, 1930 (1979). See also German Federal Constitutional Court [Bundesverfassungsgericht], 14 February, 1973, Order No. 1 BvR 112/65, in Vol. 26 Neue Juristische Wochenschrift 1221–1226 (1973).

[121] Fateh-Moghadam (2008), p. 132. For more details see p. 40.

[122] Supreme Court of the German Reich [Reichsgericht], RGSt. 25, 375 ff.

[123] Fateh-Moghadam (2008), p. 75.

[124] Rönnau, in Laufhütte/Rissing-van Saan/Tiedemann (ed.) (2006), before Sec. 32 at 146.

[125] Fateh-Moghadam (2008), p. 79.

the consequences of the medical operation since a medical operation based on the patient's consent is basically free from sanctions.[126]

LOD can therefore be justified by consent and does not depend on authorisation by some other authority.

In Austria, a surgeon that removes an organ from a living person principally commits a bodily injury.[127] Pursuant to Sec. 90 I *Criminal Code*, bodily injuries can be justified by the patient's consent.[128] This also applies to the organ removal for transplantation[129] and can be explained by the patient's autonomy. In Austria, the patient's autonomy is protected by the *Convention for the Protection of Human Rights and Fundamental Freedoms*. The *Convention* has had constitutional status in Austria since 1964.[130] Therefore, all rights of the *Convention* can be claimed at the Constitutional Court just like national legislation.[131]

As a result, the right of the person concerned to decide what happens to his own body is protected,[132] and every capable patient can decide whether a medical operation should be performed or not.[133] If an organ removal is performed without the donor's consent or even against his will, it is in conflict with law and no exceptions are granted.[134] Again, since consent is sufficient, no further consent by other authorities is necessary.

In Switzerland, the *Constitution* protects the right of self-determination.[135] It contains the right to organise essential parts of one's life by oneself. This includes the right to dispose of one's own body.[136] With respect to medical operations, the patient's right of self-determination is protected by the statutory duty to inform the patient about the surgery.[137] If the patient was not properly informed, the act is considered an illegal intervention into the patient's personal rights regardless of the patient's formal consent. Consequently, the physician can be held liable.[138] If the patient is informed, it is his responsibility to legitimate medical operations, whether they are therapeutic or non-therapeutic, by giving consent. With regard to LOD, the

---

[126] Eser, Vol. 97 Zeitschrift für die gesamte Strafrechtswissenschaft 1, 23 f. (1985); cf. Wenzel (2007), p. 266.

[127] Bruckmueller/Schumann, in: Roxin/Schroth (ed.) (2010), p. 844; Kert, Falcone Project, Austria, p. 2 & 8; cf. Kienapfel (1990), p. 193 f.; Neyer/Scheuring, 2004 Österreichische Pflegezeitschrift 12, 14.

[128] Bruckmueller/Schumann, in: Roxin/Schroth (ed.) (2010), p. 838 f.

[129] Cf. Kert, Falcone Project, Austria, p. 2; Schabel (2007), p. 92.

[130] Grabenwarter (2009), p. 16; Schabel (2007), p. 14.

[131] Grabenwarter (2009), p. 16.

[132] Cf. Berka (1999), p. 260; Grabenwarter (2009), p. 199; Schabel (2007), p.14.

[133] Schabel (2007), p. 15.

[134] Kert, Falcone Project, Austria, p. 3.

[135] Art. 10 II Swiss Constitution.

[136] Schweizer, in Ehrenzeller et al. (ed.) (2002), p. 160.

[137] Hausheer, in Fischer/Lilie (ed.) (1999), p. 321.

[138] Ibid., p. 327.

criteria established by the *Federal Act on the Transplantation of Organs, Tissues and Cells* are stricter. In this case, consent is only sufficient if no serious risks exist, and if the recipient is unable to be healed with an alternative therapeutic treatment.[139]

In the Netherlands, the right of integrity of the person and the human body is protected by the Dutch *Constitution.* As a consequence, a person has the right of self-determination. He can consent to an organ donation, but medical treatments for LOD are not directly mentioned in the *Constitution.*[140]

The patient's right to be informed,[141] as well as the fact that an examination or treatment requires the patient's consent,[142] are laid down in the *Civil Code.* Informed consent is also specifically mentioned in the *Law on Medical Treatment Agreement.* It stipulates that a mutual agreement between the physician and the patient is necessary to treat the patient for a specific condition. The patient's consent is only valid if the patient was sufficiently informed.[143] This also applies to the particular case of LOD.

c) Interim Result

Summing up, it becomes clear that the concept of informed consent is widely accepted, but not implemented equally. It is not enough to merely consider what the countries require in detail with regard to consent and disclosure. The legal framework in general also has to be considered, especially the national constitutions and the national criminal laws, in order to fully grasp the concept of informed consent and its implications. Only this type of consideration brings to light the peculiar starting point of each country and how that starting point uniquely affects the implementation of ethical principles, such as informed consent. Whether consent by the donor is sufficient to legalise LOD or an additional authorisation or validation is required is particularly impacted by the existing structures of national laws.

2) Requirements with Regard to Informed Consent

After describing the underlying concept of informed consent for selected countries, the concept with regard to LOD will be analysed. In the context of LOD, informed consent describes the consent of the persons concerned after being sufficiently

---

[139] Information from M. Monnier.

[140] Information from M. Bos.

[141] Art. 7:448 Civil Code.

[142] Ibid., Art. 7:450.

[143] Information from M. Bos.

informed.[144] Informed consent therefore consists of two parts: the consent itself and the disclosure of the person consenting.

The purpose of informed consent is to protect the patient's autonomy.[145] This is only possible if the person concerned understands the medical treatment and authorises it.[146] Informed consent, thus, puts "patients in charge of their own care."[147]

The Irish *Guide to Professional Conduct and Ethics for Registered Medical Practitioners* nicely characterises the purpose of informed consent: "The ethical and legal rationale behind this is to respect the patient's autonomy and their right to control their own life. The basic idea of autonomy is that everyone's actions and decisions are their own. Therefore, the patient has the right to decide what happens to their own body."[148]

The principle of informed consent can be found in almost all countries considered. It is also set down in Art. 3 II (a) *CFREU*. The requirement of informed consent can therefore be seen as a common value and the requirement's inclusion in the *CFREU* has given it constitutional status.[149] A comparison of the national laws, however, reveals that the legal frameworks are not completely uniform. In fact, the principle of informed consent is legalised in different ways and various detailed preconditions are requested.

a) Consent

As stated in the beginning of this section, the concept of informed consent consists of two parts, the consent itself and the disclosure of the person consenting. Hence, those two parts will be examined independently of each other, starting with the consent itself.

None of the legal systems considered define an obligation for LOD.[150] Rather, all countries require the potential donor to consent to the organ removal.[151] The

---

[144] Gutmann/Schroth (2002), p. 44.

[145] German Federal Constitutional Court [Bundesverfassungsgericht], 25 July, 1979, Case No. 2 BvR 878/74, in Vol. 32 Neue Juristische Wochenschrift 1925, 1930 (1979); Schneewind/ Sedlmayer, in Schroth et al. (ed.) (2006), p. 16; Spital, Vol. 38 American Journal of Kidney Diseases 189, 189 (2001); cf. Valapour, Vol. 22 Transplantation Reviews 196, 196 (2008); cf. Westerhäll, in Taupitz (ed.) (2000), p. 883.

[146] Spickhoff, Vol. 14 Zeitschrift für Europäisches Privatrecht 400, 407 (2006); cf. Westerhäll, in Taupitz (ed.) (2000), p. 883.

[147] Eyal, in Zalta (ed.) (2011).

[148] Principle 33.1 Irish Guide to Professional Conduct and Ethics for Registered Medical Practitioners.

[149] Höfling, in Tettinger/Stern (ed.) (2006), Art. 3 at 17.

[150] Gutmann/Schroth (2002), p. 44.

[151] For Germany see Murswiek, in Sachs (ed.) (2011), Art. 2 at 207.

intended purposes of requiring consent are to protect the donor and to safeguard the donor's self-determination.[152]

The Dutch legal situation can serve as a case in point. Art. 11 of the Dutch *Constitution* states, "Everyone shall have the right to inviolability of his person, without prejudice to restrictions laid down by or pursuant to Act of Parliament." As a consequence, it is considered a violation of the right to physical integrity if organs from one's body are removed. Even though this is deduced from the Dutch *Constitution*, it is not only relevant in the Netherlands, but can be considered as a generally recognised principle. To prevent a violation of the right to physical integrity, consent for organ donation is required.[153]

As a result, it can be noted that the donor's consent is required to justify LOD in all countries considered. A closer look at the specific requirements, however, reveals that noteworthy differences do exist. The diverse embodiment of the specific requirements in the national laws will be analysed next. (1) I examine whether consent is named as an explicit requirement in the national transplant laws. (2) I consider the point of time (when the consent must be submitted/obtained) of the required consent. (3) I analyse whether the consenting person must have the capacity to do so and (4) in how far voluntariness of the consent is required. (5) The formal requirements that the national transplant laws contain are compared. (6) I also consider the donor's possibility to withdraw his consent. (7) I analyse the potential limitations to give valid consent and (8) whether the recipient has to give consent as well.

aa) Consent as Explicit Requirement

The first slight difference between the countries is that some countries state the requirement for consent explicitly in their transplant law, even though some legal experts have noted that it should be taken for granted.[154] In contrast, in other countries this requirement only becomes clear due to the context and general legal principles.

[152] Cf. Deutsch, in Spickhoff (ed.) (2011), p. 3; Price, in Gutmann et al. (ed.) (2004), p. 146.

[153] Coppen (2010), p. 13 f.

[154] Gutmann/Gerok, in Collins et al. (ed.) (1997), p. 317; Gutmann/Schroth (2002), p. 48.

Belgium,[155] Bulgaria,[156] Estonia,[157] Finland,[158] Germany,[159] Norway,[160] Slovakia[161] and the United Kingdom[162] explicitly state in their national transplant laws that consent is required to justify LOD.

In Austria, Hungary, Italy, Latvia, Lithuania, Moldova, the Netherlands, Poland, Portugal, Spain and Switzerland, in contrast, the requirement of a valid consent is not explicitly stated in the national transplant law. It can be found in other places of the legal system or can be concluded based on the context, because the requirements for a valid consent are listed.

In the Netherlands, for example, Art. 11 of the *Constitution* can be examined again, since it guarantees the right to inviolability of her person. Consequently, removing organs from one's body is a violation of the right to physical integrity. Therefore, the person's consent is necessary to justify this.[163]

In summary, there is a consensus among the countries considered regarding the fact that the consent of the potential donor is necessary to justify LOD. Only the way of clarifying this differs.

bb) Time of Consent

With respect to the point in time of the consent, several countries explicitly declare that consent has to be given prior to the removal. It can reasonably be assumed, though, that this has to be the case anyway, regardless of whether it has been stated explicitly or not.

The German *Act on the donation, removal and transplantation of organs* does not explicitly state that consent has to be given prior to the organ removal. A reading of the relevant rule shows, though, that its context leads to this conclusion. The donor needs to be informed that his consent is a requirement for the organ removal.[164] Requirements normally need to be fulfilled before the respective act is performed, which leads us to assume that this applies to consent for LOD as well. However, we still do not have very detailed guidelines with respect to the time of the consent. "Prior to the removal" could mean one month prior to it, but it could also mean immediately before the operation is carried out. As already clarified, the requirement of valid consent is not completely isolated, but it does need to be

---

[155] Art. 5 Law on Retrieval and Transplantation of Organs.

[156] Art. 24 (1) Law of the Transplantation of Organs, Tissues and Cells.

[157] Sec. 9 (1), 1) Transplantation of Organs and Tissues Act.

[158] Sec. 2, I Act on the Medical Use of Human Organs and Tissues.

[159] Sec. 8 (1) No. 1 b) Act on the donation, removal and transplantation of organs.

[160] Sec. 1 Act relating to transplantation, hospital autopsis and the donation of bodies etc.

[161] Sec. 36 (2) Act on Healthcare, Healthcare-Related Services and on the Amendment and Supplementing of Certain Laws.

[162] Schedule 1, Part 1, 7 Human Tissue Act 2004.

[163] Coppen (2010), p. 14.

[164] Sec. 8 (2) Act on the donation, removal and transplantation of organs.

connected to the duty to disclosure, because the person concerned can give valid consent only after being sufficiently informed.[165] Thus, a person cannot give his consent until he has been informed, and if his being informed is delayed, then the surgery will be delayed as well. The German Federal Supreme Court ruled that the persons involved in normal ambulatory operations may be informed the day of the operation.[166] This ruling does not apply to LOD, though, because it is a stationary operation.[167] Rather, the person concerned needs to be informed in good time to be able to weigh the pros and cons of the operation to make use of his freedom of choice and to therefore reserve his "right of personality."[168] As a consequence, informing the donor on the day of the LOD is not sufficient[169] and consent must also be given earlier.

cc) Capacity to Consent

A significant amount of countries explicitly state that the person concerned must have the capacity to give consent. They do so with varying degrees of detail. Other national transplant laws, however, do not state this requirement explicitly, as it normally is a general legal principle, applicable in all cases, not only in transplantation.[170]

Belgium,[171] the Czech Republic,[172] Germany,[173] Greece,[174] Hungary,[175] Lithuania,[176] the Netherlands,[177] Poland,[178] Slovakia,[179] Spain[180] and

[165] Cf. Spickhoff (2004), p. 60.

[166] German Federal Supreme Court [Bundesgerichtshof], July 14, 1994, Case No. VI ZR 178/93, in Vol. 57 Neue Juristische Wochenschrift 3010, 3010 (1994).

[167] Cf. German Federal Supreme Court [Bundesgerichtshof], March 25, 2003, Case No. VI ZR 131/02, in Vol. 57 Neue Juristische Wochenschrift 2012, 2012 (1994).

[168] German Federal Supreme Court [Bundesgerichtshof], January 8, 1985, Case No. VI ZR 15/83, in Vol. 57 Neue Juristische Wochenschrift 1399, 1399 f. (1985); German Federal Supreme Court [Bundesgerichtshof], July 14, 1994, Case No. VI ZR 178/93, in Vol. 57 Neue Juristische Wochenschrift 3010, 3010 (1994); German Federal Supreme Court [Bundesgerichtshof], March 25, 2003, Case No. VI ZR 131/02, in Vol. 57 Neue Juristische Wochenschrift 2012, 2013 (1994).

[169] German Federal Supreme Court [Bundesgerichtshof], March 25, 2003, Case No. VI ZR 131/02, in Vol. 57 Neue Juristische Wochenschrift 2012, 2012 (1994).

[170] Cf. O'Neill, Vol. 29 Journal of Medical Ethics 4, 4 (2003).

[171] Art. 9 Law on Retrieval and Transplantation of Organs.

[172] Art. 3 (1) c) Transplantation Act on Donation, Removal and Transplantation of Organs and Tissue and on the Amendment of some Acts.

[173] Sec. 8 I No. 1 a Act on the donation, removal and transplantation of organs.

[174] Canellopoulou-Bottis, Vol. 7 European Journal of Health Law 427, 433 (2000).

[175] Sec. 206 (1) Act on Health.

[176] Art. 10, 2.Law on Donation and Transplantation of Human Tissues, Cells and Organs.

[177] Sec. 3, 1 Organ Donation Act.

[178] Art. 12, 1., 7) The Cell, Tissue and Organ Recovery, Storage and Transplantation Act.

[179] Sec. 36 (2) Act on Healthcare, Healthcare-Related Services and on the Amendment and Supplementing of Certain Laws.

[180] Art. 4, c. Law Regulating Organ Recovery and Transplantation.

Switzerland[181] explicitly state in their national transplant laws that the living donor must have the capacity to give consent. They do not provide any further details, though.

In Hungary, the *Act on Health* only mentions that "[o]rgans and tissues, with the exception of Subsection (5), may only be donated by a person with legal capacity."[182] The general law can be used for further information. Art. 15/A of the Hungarian *Civil Code* holds that legal statements made by incompetent persons shall be null and void. In contrast, those who are competent are entitled to sign contracts and to make other legal statements.[183] Hungary has three levels of capacity.[184] First, people with full capacity are mentally healthy people 18 years or older. Second, people with limited capacity are adolescents who are between 14 and 18 years old and some mentally ill adults. Third, people with no capacity to act are children who are younger than 14 years and some mentally ill adults. People with limited capacity can make a joint decision together with their legal proxy.[185] If a person has no legal capacity, her legal proxy provides consent on behalf of her. With respect to medical decisions, the practical situation is quite heterogeneous in Hungary, where a mental disorder does not necessarily exclude the ability to understand the situation. In fact, the process of informed consent is more individualised and tailored specially for the patient´s condition and personality.[186] Even if the patient is considered as having no capacity to make decisions about her medical care, she is still not completely ignored. She would receive a certain amount of information before her treatment.[187] This is, because the right to receive information on the one hand, and the right of self-determination on the other hand, are stipulated as two separate rights in the Hungarian *Act on Health.*[188]

In Switzerland, the general law can also be quoted for further information. Art. 16 *Civil Code* regulates that "[a] person has capacity to consent within the meaning of the law if he or she does not lack the ability to act rationally by virtue of being under age or because of mental illness, mental incapacity, inebriation or similar circumstances." A person has the capacity to consent if she has both a moment of intellectual understanding and a moment of volitional understanding. With respect to LOD, this means that the living donor must be able to assess the consequences of the intervention as well as the outcomes and risks.[189] Whether a person is considered to have the capacity to consent is judged individually for each case, but with

[181] Art. 12, a) Federal Act on the Transplantation of Organs, Tissues and Cells.

[182] Sec. 206 (1) Act on Health.

[183] Sándor (2003), p. 83.

[184] Information from J. Sándor.

[185] Ibid.

[186] Sándor (2003), p. 83.

[187] Ibid., p. 84.

[188] Information from J. Sándor.

[189] Swiss Dispatch of 12. September 2001 on a Federal Law on Transplantation of Organs, Tissues and Cells, p. 144.

regard to adolescents older than 15 years it can usually be assumed that they are capable of consenting.[190] This also applies to mental-health problems, where there is no general criterion for exclusion and each case is judged individually. One must assess, whether the denial to donate is more stressful for the donor than the donation would be.[191]

None of the countries mentioned defines what preconditions a potential donor needs to fulfil in order to be considered capable of giving consent.

In Germany, one has the capacity to consent only if she has the ability to evaluate the surgery with regard to its significance, risks and consequences.[192] This is a general principle, though, and does not solely apply to LOD. It can be concluded that a mentally handicapped person will usually not have the capacity to consent. Contrary cases are possible as well, though. If the surgeon is not sure whether the mentally handicapped potential donor has the capacity to consent, she has to enlist a psychiatrist's aid. If there are further doubts, the capacity to consent has to be negated in order to protect the mentally handicapped person.[193]

The laws in the United Kingdom that specifically apply to organ donation do not specify how to identify whether a person has the capacity to consent. Therefore, the Common Law in general has to be regarded. It assumes that an adult has the competence to take care of everyday concerns herself. This includes the capacity to consent to medical treatments.[194] The Common Law principles, with regard to the capacity to consent, have been put in concrete form and have been partly modified in the *Mental Capacity Act 2005*.[195] There are five central principles established in Sec. 1 *Mental Capacity Act 2005*.[196] The first principle states, "A person must be assumed to have capacity unless it is established that he lacks capacity." Sec. 2 *Mental Capacity Act 2005* defines in detail the requirements under which a person lacks the capacity to consent. The United Kingdom established the *Mental Capacity Act 2005* to evaluate whether a person lacks the capacity to consent. With regard to LOD, the *Mental Capacity Act 2005* shows that a person who wants to donate can only be incapacitated after his capacity to consent has been questioned by medical results.[197] This diagnostic proof can only take place by a medical expert, mainly by a psychiatrist.[198] With regard to a minor's capacity to

---

190 Hofer (2006), p. 111, 113.

191 The Swiss Academies of Arts and Sciences, Live donations of solid organs, Medical-ethical guidelines and recommendations, p. 14.

192 Eser/Sternberg-Lieben, in Schönke/Schröder (ed.) (2001), Sec. 223 at. 38; German Federal Supreme Court [Bundesgerichtshof], BGHSt 23, 1, 4; Fischer (ed.) (2011), before Sec. 32 at. 3c; Kindhäuser (ed.) (2010), Sec. 228 at. 8.

193 Lilie/Krüger, Vol. 81 Der Chirurg 787, 788 (2010).

194 Re T (Adult: Refusal of Treatment) [1993] Fam. 95, p. 112.

195 Fateh-Moghadam (2008) p. 198.

196 Ibid.

197 Ibid., p. 200.

198 Ibid.

consent, no rigid age-limit exists. In fact, in each case, whether the minor has the "sufficient understanding and intelligence to enable her to understand fully what was proposed" must be reviewed on an individual basis.[199] It is questionable, however, if this principle can be applied to LOD as well. As a result, no reason can be found for restricting this principle to curative treatment.[200] Therefore, whether minors are capable of consenting or not has to be decided for each case individually.[201]

Ireland's *Guide to Professional Conduct and Ethics for Registered Medical Practitioners* also clarifies that it is presumed that every adult patient has the capacity to make decisions about his own health care.[202] An adult is considered to lack the capacity to consent if he is unable to understand, retain, use or weigh the information he has been given to make the relevant decision, or if he is unable to communicate his decision. A judgement that a patient lacks the capacity to make a particular decision does not imply that he is unable to make other decisions or will be unable to make this or other decisions in the future.[203] In addition, the Irish *Guide to Professional Conduct and Ethics for Registered Medical Practitioners* defines how to investigate whether a patient possesses the capacity to give valid consent. Namely, a functional approach should be taken when assessing an individual's capacity. One is evaluated based on his ability to make the relevant choice given his level of understanding and retention of the information he has been given. He is also evaluated based on his ability to apply the information to his own personal circumstances and come to a decision.[204] Thus, in Ireland, the capacity to consent is also required and moreover, there are also regulations regarding how the individual answers the question.

The countries not yet mentioned explicitly require the consenting donor to have the capacity to consent in their transplant laws. However, as stated in the beginning of this section, consent can naturally never be valid without the consenting person having the capacity to do so. Hence, given the validity of this conclusion, all countries require the capacity to consent, regardless of whether it is stated explicitly in the transplant law, if it is merely a generally applied principle or if it is stated in some other manner in the national law.

---

[199] Gillick v. West Norfolk A.H.A. (1986) at (3).

[200] Fateh-Moghadam (2008), p. 200 f.

[201] Ibid., p. 204 ff.

[202] Principle 34.1 Irish Guide to Professional Conduct and Ethics for Registered Medical Practitioners.

[203] Ibid., Principle 34.4.

[204] Ibid., Principle 34.3.

dd) Voluntariness of Consent

The danger of force or coercion[205] being enacted towards the donor is an immanent problem of LOD.[206] Therefore, the issue of voluntariness needs to be discussed. (1) Whether voluntariness of consent as a mandatory requirement is explicitly stated in the national laws differs. Many of the national laws contain a regulation to that effect, not all though. (2) In addition, this section analyses how far the national laws contain procedural safeguards to ensure voluntariness.

(1) Voluntariness of Consent as Material Requirement

The Czech,[207] Dutch,[208] Estonian,[209] Irish,[210] Italy,[211] Moldovan,[212] Polish,[213] Portuguese,[214] Swiss[215] and the United Kingdom[216] transplant laws explicitly require that the consent is given voluntarily. With respect to the countries just listed, the precise phrasing differs. One must consent freely or voluntarily, or one must have an altruistic purpose.

In the United Kingdom, the donor must be informed of his right to be free of any kind of coercion or threat against him or anyone else (for example, family or friends), and that consent given under any such pressure will not be validated by the IA.[217] The donor also has to be informed that it is an offence to seek or to receive payment or any other reward for providing an organ or part of an organ for

---

[205] Philisophers roughly define coercion "as a threat to make someone seriously worse of than she is or should be, unless she consents." According to the prevailing medical view "the threshold for condoning conduct as too coercive [is set] far lower than this definition would suggest" (Eyal, in Zalta (ed.) (2011).

[206] Grenvik, in Lawin/Lenhart/Peter (ed.) (1990), p. 154.

[207] Art. 7 (4) Transplantation Act on Donation, Removal and Transplantation of Organs and Tissues and on the Amendment of some Acts.

[208] Sec. 3, 2 Organ Donation Act.

[209] Sec. 6 (1) Transplantation of Organs and Tissues Act.

[210] Principle 35.1 Irish Guide to Professional Conduct and Ethics for Registered Medical Practitioners.

[211] Art. 2 Legislation on Donation, Retrieval and Transplantation of Organs and Tissues.

[212] Art. 15 (2) Law on the Transplant of Human Organs, Tissues and Cells.

[213] Art. 12 1. 7 The Cell, Tissue and Organ Recovery, Storage and Transplantation Act.

[214] Art. 8, 1 Law on Harvesting and Transplanting Human Organs and Tissues.

[215] Art. 12, b) Federal Act on the Transplantation of Organs, Tissues and Cells.

[216] No. 30 Code of Practice 1 & No. 72 Code of Practice 2.

[217] IA = Independent Assessor. Independent Assessors act on behalf of the Human Tissue Authority. They have to interview the potential living organ donors. That is to ensure that the potential donor understands the risks, that no reward has been offered to him and that he was not pressurised to the intended LOD (Human Tissue Authority (2010).

transplantation, and that this offence is subject to significant penalties,[218] namely, the risk of a fine or up to three years imprisonment.[219]

The rule in Switzerland is similar. The donor has to be informed of the voluntariness and gratuitousness of the donation and that donating for money is punishable.[220]

The Belgium transplant law does not only require the consent to be free, but also that it happened consciously[221] and with an altruistic purpose.[222]

The Spanish regulation is similar.[223] It adds that organs must not be removed if there is suspicion that the consent was motivated by economical or other reasons, such as social or psychological influences.[224]

The Greek law in general requires the consent to be the outcome of a freely made decision.[225]

The national laws that have not been brought up do not contain a rule explicitly stating that consent has to be voluntarily. As mentioned above, none of the countries considered contains a duty to act as a living organ donor. Consequently, the donors consent must be voluntary regardless of whether this is specifically clarified by law or not.

Therefore, the initial hypothesis that all of the considered countries require the consent to be voluntary is confirmed.

(2) Procedural Safeguarding for Voluntariness

Beyond the theoretical legal situation, it is important to consider how the regulation for consent is actually put into practice. Only a few countries have a special mechanism to ensure that the regulation is complied with. England,[226] Germany,[227] Hungary,[228] Portugal,[229] Scotland[230] and Switzerland[231] have created commissions to ensure that the consent is voluntary and that the organ is not the object of organ trade.

---

[218] No. 90 Code of Practice 2.

[219] Ibid., No. 46.

[220] Art. 9, 2 b) Transplantation Ordinance.

[221] Art. 8 Sec. 1 Law on Retrieval and Transplantation of Organs.

[222] Ibid., Art. 9.

[223] Art. 9, 1. c. Royal Decree 2070/1999.

[224] Ibid., Art. 9, 2.

[225] Sándor/Drakopoulou/Bárd (2009), p. 8.

[226] No. 11 (3) (a) Human Tissue Act 2004 Regulations 2006.

[227] Sec. 8 III 2 Act on the donation, removal and transplantation of organs.

[228] Sec. 206 (3) Act on Health.

[229] Art. 6, 3 Law on Harvesting and Transplanting Human Organs and Tissues.

[230] Reg. 2 (5) (a) Human Organ and Tissue Live Transplants (Scotland) Regulations 2006.

[231] Art. 10, 1 Transplantation Ordinance.

In Germany, the commission has to give an expert opinion on whether there are real, substantiated grounds to assume that the potential donor's consent has not been given freely or that the organ is the object of prohibited trade.[232] The regulations in England,[233] Scotland[234] and Switzerland[235] work in a similar fashion.

In Hungary, LOD may be possible in exceptional cases, even if the donor is not "a) a lineal kin of the recipient, b) a sibling of a lineal kin of the recipient, c) a sibling of the recipient, d) a lineal kin of a sibling of the recipient." In those cases, however, LOD is only permitted after a hospital ethics committee has established that the donation has taken place without consideration, force, duress, coercion or deception.[236]

ee) Formal Requirements

Another important aspect of consent is whether the consent has to comply with certain formal requirements. This differs significantly throughout the countries considered. The following issues are dealt with: (1) Does the consent have to be in writing? (2) Does a witness have to be present? (3) Does a public authority have to approve the consent? (4) What miscellaneous requirements are additionally required in a few national laws?

(1) In Writing

A significant amount of countries require the consent to be in writing or to be recorded in a document, but not all countries demand this precondition.

Consent has to be in writing in Belgium,[237] Bulgaria,[238] the Czech Republic,[239] Estonia,[240] Finland,[241] Germany,[242] Hungary,[243] Latvia,[244] Lithuania,[245]

---

[232] Sec. 8 III 2 Act on the donation, removal and transplantation of organs.

[233] No. 11 (3) (a) Human Tissue Act 2004 Regulations 2006.

[234] Reg. 2 (5) (a) The Human Organ and Tissue Live Transplants (Scotland) Regulations 2006.

[235] Art. 10 Transplantation Ordinance.

[236] Sec. 206 (3) Act on Health.

[237] Art. 8 Sec. 2 1 Law on Retrieval and Transplantation of Organs.

[238] Art. 24 (1) Law of the Transplantation of Organs, Tissues and Cells.

[239] Art. 7 (4) Art.4 Transplantation Act on Donation, Removal and Transplantation of Organs and Tissues and on the Amendment of some Acts.

[240] Sec. 6 (3) Transplantation of Organs and Tissues Act.

[241] Sec. 3 I 1 Act on the Medical Use of Human Organs and Tissues.

[242] Sec. 8 II 4 Act on the donation, removal and transplantation of organs.

[243] Sec. 209 (1) Act on Health.

[244] Sec. 13 Law on the Protection of the Body of Deceased Human Beings and the Use of Human Tissues and Organs in Medicine.

[245] Art. 10, 2. Law on Donation and Transplantation of Human Tissues, Cells and Organs.

Moldova,[246] the Netherlands,[247] Norway,[248] Poland,[249] Slovakia,[250] Spain[251] and Switzerland.[252]

Among the countries that require consent to be in writing, the details about the content of the declaration differ. The declaration has to be signed by the person consenting in Belgium,[253] the Czech Republic,[254] Moldova[255] and the Netherlands.[256] In Spain, the document must be signed by the donor, the transplanting doctor and all further attendees.[257] In the Netherlands, the document must also be dated.[258] The provision in Germany requires the person providing the information to sign the declaration; in addition, the physicians attending the process of informing the donor and the donor both have to sign.[259] The document must also contain detailed information of the health insurance coverage, should potential health problems from the medical procedures arise.[260] With regard to this aspect, the Swiss regulation is similar, since the donor must be educated on his insurance coverage and the reimbursement of expenses.[261]

The Ministry of Health in Lithuania has prepared a standard form for the written consent[262]; the donor can fill in the information about himself, especially his name, surname, personal ID, what he wants to donate and to whom he wants to donate. Afterwards, the doctor of the health care agency checks over the form and sends it to the National Transplant Bureau.[263]

---

[246] Art. 15 (2) Law on the Transplant of Human Organs, Tissues and Cells.

[247] Sec. 6 Organ Donation Act.

[248] Sec. 1 Act relating to transplantation, hospital autopsis and the donation of bodies etc.

[249] Art. 12, 1., 7) The Cell, Tissue and Organ Recovery, Storage and Transplantation Act.

[250] Sec. 36 (2) Act on Healthcare, Healthcare-Related Services and on the Amendment and Supplementing of Certain Laws.

[251] Art. 4, c. Law Regulating Organ Recovery and Transplantation.

[252] Art. 12, b) Federal Act on the Transplantation of Organs, Tissues and Cells.

[253] Art. 8 Sec. 2 Law on Retrieval and Transplantation of Organs.

[254] Art. 7 (4) Transplantation Act on Donation, Removal and Transplantation of Organs and Tissues and on the Amendment of some Acts.

[255] Art. 15 (3) Law on the Transplant of Human Organs, Tissues and Cells.

[256] Sec. 6 Organ Donation Act.

[257] Art. 9, 4. Royal Decree 2070/1999.

[258] Sec. 6 Organ Donation Act.

[259] Sec. 8 II 4 Act on the donation, removal and transplantation of organs.

[260] Ibid., Sec. 8 II 5.

[261] Art. 9, 2 g) Transplantation Ordinance.

[262] Art. 10, 3. Law on Donation and Transplantation of Human Tissues, Cells and Organs.

[263] Information fromV. Anulyte.

The Swedish law only requires the consent to be in writing if the organ concerned is not regenerative,[264] or if the action can otherwise entail considerable harm or inconvenience to the donor.[265]

In contrast, there are also countries that do not require the consent to be in writing. In Austria, even though the consent has to be explicit and strictly personal,[266] written form is not required.[267] The Italian law does not explicitly require the consent of the donor to be in writing. The magistrate has to determine, though, that the conditions contemplated exist. If the magistrate determines that the conditions exist, he will proceed to draw up the associated declarations in writing.[268] The situation in the United Kingdom is similar. The *Human Tissue Act 2004* does not specify the format in which consent should be given or recorded. However, it is considered good practice to obtain written consent for significant procedures such as organ donation.[269] In Greece, the donor's consent can either be in writing or orally. In case consent is given orally, it "will be filed in a special Register of the medical unit where the transplant is scheduled." In addition, "[t]wo witnesses must sign the declaration."[270]

All in all, the majority of the considered countries demand the consent of the living organ donor to be in writing.

(2) Presence of a Witness

Belgium,[271] Poland[272] and Portugal[273] request a witness to be present while the donor consents, but the requirements regarding the witness differ. Belgium requires the attending witness to be an adult,[274] who then also has to sign the declaration.[275] In Poland, the consent must be granted before a physician.[276] In Portugal, the consent must take place in front of a doctor who is designated by the clinical

[264] As already stated, according to the Guidance on the provisions of the Swedish Transplant Act, the kidney, but also parts of the liver and the pancreas are considered as non-regenerative organs (Cf. The Ministry of Health and Social Affairs Sweden 1997, The Swedish Transplant Act, Guidance on the provisions, Sec. 6).

[265] Sec. 6 Transplant Act.

[266] Kopetzki, in Barta/Weber (ed.) (2001), p. 132.

[267] Ibid., p. 133.

[268] Art. 2 Legislation on Donation, Retrieval and Transplantation of Organs and Tissues.

[269] No. 58 Code of Practice 1.

[270] Canellopoulou-Bottis, Vol. 7 European Journal of Health Law 427, 433 (2000).

[271] Art. 8 Sec. 2, 1 Law on Retrieval and Transplantation of Organs.

[272] Art. 12, 1, 7 Cell, Tissue and Organ Recovery, Storage and Transplantation Act.

[273] Art. 8, 2 Law on Harvesting and Transplanting Human Organs and Tissues.

[274] Art. 8 Sec. 2, 1 Law on Retrieval and Transplantation of Organs.

[275] Ibid., Art. 8 Sec. 2, 2.

[276] Art. 12, 1, 7 Cell, Tissue and Organ Recovery, Storage and Transplantation Act.

director of the institution where the transplantation is going to happen. It has to be provided that she is not part of the transplant team.[277]

(3) Public Authority

A few of the countries require the consent to be approved by a public authority. The type of public authority involved, as well as any further, stated requirements, differ depending on the country considered.

In Bulgaria, written consent has to be certified by a notary[278] in the region of the location of the medical institution where the organ is to be removed.[279]

The Hungarian *Act on Health* requires the donor's consent to be incorporated in a public deed. The deed must contain the general requirements of consent. Beyond the general requirements, it must also contain the donor's declaration stating that the donation has taken place without force, duress, coercion or deception and that she gives consent to autopsy following her death.[280]

In Moldova, the government is involved. The consent needs to be drafted as an informed consent in a template approved by the government.[281]

In Portugal, the Entity of Verification and Admissibility of Collection for Transplantation controls the donation insofar as it is based on a valid free and informed consent.[282] However, if the donor is a foreigner without legal residence and donating a non-regenerative organ, the valid free and informed consent is insufficient. In fact, a court's authorisation is necessary.[283] The reason for this additional requirement is the fear of organ trafficking from persons coming from abroad. This legal rule is criticised by legal experts. Many argue that this regulation implies that the law has no confidence in the administrative control procedure envisaged by law for Portuguese citizens.[284]

In Spain, the potential donor's consent must take place in front of the public authority appointed in accordance with regulations.[285] Furthermore, the consent must take place in front of the judge of the locally responsible register office, but only after an explanation is given by the surgeon who will perform the removal. This is carried out in the presence of another physician who evaluates the donor's state of health and keeps him informed. In addition, the physician responsible for

[277] Art. 8, 2 Law on Harvesting and Transplanting Human Organs and Tissues.

[278] Art. 24 (1) Law of the Transplantation of Organs, Tissues and Cells.

[279] Ibid., Art. 24 (2).

[280] Sec. 209 (2) Act on Health.

[281] Art. 15 (4) Law on the Transplant of Human Organs, Tissues and Cells.

[282] Pereira, Vol. 18 European Journal of Health Law 1, 3 (2011).

[283] Art. 6 (6) Law on Harvesting and Transplanting Human Organs and Tissues.

[284] Pereira, Vol. 18 European Journal of Health Law 1, 3 (2011).

[285] Art. 4, c. Law Regulating Organ Recovery and Transplantation.

the transplantation and the person approving the removal for the centre (according to the declaration of authorisation) have to be present.[286]

Therefore, while it is not uncommon to involve an authority in the process of consent, it would also be an overstatement to consider it a generally required condition.

(4) Miscellaneous

Next to the categories just presented, some countries have additional, unique formal requirements that are hardly comparable to the other national laws.

(aa) Possibility of the Enactment of Additional Regulations Concerning the Formal Requirements In Belgium and Lithuania, the enactment of additional regulations concerning the formal requirements of consent are provided.

In Belgium, the King regulates how consent can be manifested.[287] He established the *Royal Decree of 30 October 1986*.[288] According to the *Royal Decree*, the following elements need to be included in the written consent: the name and age of the donor, and, if applicable, the name, age and capacity of the other person or persons giving consent; their signature and date of this signature; the name and age of the witness of age (e.g., not a minor); the signature of this witness and date of this signature and the name and place of the hospital where the consent is deposited. These formal requirements are obligatory, making the consent null and void if an element is missing.[289] In Lithuania, the Ministry of Health may establish the consent form[290] and has done so already.[291]

(bb) Person Who May Seek Consent In the United Kingdom, the healthcare professional has to seek the required consent. In addition, details about the qualifications are specified: The staff seeking consent should have a good understanding of the activities they are seeking consent for, and they should be in a position to answer questions that donors or their families may ask. Furthermore, healthcare professionals should obtain the support and guidance of their managers to develop the necessary skills for the implications of a donation and essential requirements of seeking consent.[292]

(cc) Space of Time for Donor to Reconsider His Decision The Spanish *Royal Decree 2070/1999* requests a space of time of at least 24 h between the signing of

[286] Art. 9, 4. Royal Decree 2070/1999.

[287] Art. 15 Law on Retrieval and Transplantation of Organs.

[288] Information from R. Wilmotte; Merchx, Falcone Project, Belgium, p. 6 f.

[289] Merchx, Falcone Project, Belgium, p. 6 f.

[290] Art. 10, 3.Law on Donation and Transplantation of Human Tissues, Cells and Organs.

[291] Information from V. Anulyte.

[292] No. 48 ff. Code of Practice 1.

the declaration and the removal of the organ.[293] The Swiss Federal Council has established a similar provision by determining that the donor must be granted with an appropriate amount of time to consider his decision.[294] How long this time for reflection has to be exactly cannot be determined in general. Consequently the term "appropriate" was chosen. Therefore, the "appropriate" period of time has to be decided by the physician who considers the concrete circumstances together with the patient.[295] The medical practice in Austria is similar. The donor must be informed early enough to come to a decision unhurriedly. He must have enough time to be clear in his mind about the decision. Beyond this, he must have the opportunity to ask further questions or to consult with relatives.[296] Providing time for reflection can be considered as being positive for the donor, who gets the chance to think over the facts calmly at home.[297]

The examples above demonstrate the great diversity with regard to the formal requirements of consent.

(5) Recapitulation

Preponderance in favour of the consent to be in writing becomes apparent. Apart from that, the countries do not conform with regard to the formal requirements concerning the consent.

ff) Withdrawal of Consent

Even though it is not a requirement for the consent to be valid, an important issue that cannot be ignored while presenting consent is whether a consent given once can be withdrawn. Almost all countries render the possibility for the potential donor to withdraw her consent. The purpose of this is to guarantee the autonomy of the donor, who can potentially change her mind.[298] It is also intended to limit or, at best, prevent the posterior regret of an organ removal.[299]

Only the Latvian, Polish, Swedish, Moldovan and Swiss transplant law do not *explicitly* contain this option.

All other considered countries provide the option for the potential donor to withdraw her consent at any time in their transplant law. Usually, they also clarify that this opportunity only applies before the surgery to remove the organ takes

[293] Art. 9, 5. Royal Decree 2070/1999.

[294] Art. 9, 3 Transplantation Ordinance.

[295] Information from M. Monnier.

[296] Kalchschmid (1997), p. 203.

[297] Information from M. Monnier.

[298] Bock (1999), p. 152; cf. Deutsch, in Spickhoff (ed.) (2011), p. 6.

[299] Bock (1999), p. 152.

place. For instance, Italy states that the "consent may be revoked at any time prior to the surgical operation."[300]

Several of these countries do not regulate anything in addition. A few national laws provide details with regard to the formal requirements of the withdrawal, though. Hungary[301] and Spain[302] clarify that the removal is permitted without any formal restrictions. The regulation in Germany is similar. It states that the withdrawal can happen in written form or orally.[303] Finland's regulation is more detailed. It emphasises that the potential donor is not required to state any reasons for her reset.[304] Spain highlights that the potential donor who withdraws her consent does not have to pay any compensation.[305]

England,[306] Finland,[307] Ireland,[308] Poland,[309] Scotland[310] and Switzerland[311] regulate the option to withdraw consent; it has to be part of the general disclosure of the potential donor. Polish law also requires the donor to be informed about the consequence of withdrawing the consent, namely a new organ must be located for the recipient.[312]

All in all, with regard to the donor's possibility to withdraw his consent, the considered countries hold a similar view.

gg) Limitation to Consent

Consent provides an opportunity for the patient to exercise her right of self-determination. Within a certain scope, the patient is completely free in deciding on whether she wants to give consent or if she denies it. This scope usually has certain barriers, though, namely (1) homicide upon request and (2) good morals.

---

[300] Art. 2 Legislation on Donation, Retrieval and Transplantation of Organs and Tissues.

[301] Sec. 209 (4) Act on Health.

[302] Art. 9, 5. Royal Decree 2070/1999.

[303] Sec. 8 II 6 Act on the donation, removal and transplantation of organs.

[304] Sec. 3 I 2 Act on the Medical Use of Human Organs and Tissues.

[305] Art. 9, 5. Royal Decree 2070/1999.

[306] No. 90, 3.Code of Practice 2.

[307] Sec. 3 II 1 Act on the Medical Use of Human Organs and Tissues.

[308] Appendix A Irish Guide to Professional Conduct and Ethics for Registered Medical Practitioners.

[309] Art. 12, 1., 8) The Cell, Tissue and Organ Recovery, Storage and Transplantation Act.

[310] Reg. 2 (7) Human Organ and Tissue Live Transplants (Scotland) Regulations 2006.

[311] Art. 9, 2 i) Transplantation Ordinance.

[312] Art. 12, 1., 8) The Cell, Tissue and Organ Recovery, Storage and Transplantation Act.

(1) Homicide Upon Request

A general standard of LOD is that it is not legal in cases where it ceases the donor's life or would be directly life-threatening.[313] This is due to the fact that most national criminal codes prohibit "homicide upon request." Beyond that general standard, the national laws differ with regard to the maximal amount of risks the donors can legally take.[314]

(2) Good Morals

Several countries declare consent that would be against good morals void. This means the aspect of morality also limits the scope in which the patient can give valid consent.

For example, the German *Criminal Code* states in Sec. 228, "Whoever commits bodily injury with the consent of the injured person only acts unlawfully if the act is, despite the consent, contrary to good morals." Therefore, an action that is against good morals is punishable, even if the donor gave her consent. When interpreting the term good morals, it has to be confined to its legal core.[315] That means that a violation of the value system of individual social groups or the criminal court in charge is not sufficient[316] and that a violation of good morals only takes place in very few, extreme cases.[317] According to the prevailing opinion, morality is violated if the bodily injury is viewed as unacceptable from the legal system. If the bodily injury is severe, it is seen as an offence to the legal system. The bodily injury that occurred and the danger for life and limb of the victim related thereto are taken into account by the judges. The decision regarding whether a violation of morality took place has to be based on the type and the severity of the outcome of the bodily injury and the degree of mortal danger.[318] A violation of morality exists if, based on an anticipatory, objective close examination of all decisive circumstances of the action, the person consenting is in concrete mortal danger

---

313 Cf. Gutmann/Schroth (2002), p. 49; Dickens/Fluss/King, in Chapman/Deierhoi/Wright (ed.) (1997), p. 98; Pfeiffer (2004), p. 45.

314 That will be described below, when the issue of "Safety and maximum extent of permissible risk taking by donors" is approached.

315 Fateh-Moghadam (2008), p. 133; German Federal Supreme Court [Bundesgerichtshof], 11 December, 2003, Case No. 3 StR 120/03, in Vol. 57 Neue Juristische Wochenschrift 1054, 1055 (2004); German Federal Supreme Court [Bundesgerichtshof], 26 May, 2004, Case No. 2 StR 505/03, in Vol. 57 Neue Juristische Wochenschrift 2458, 2458 (2004).

316 German Federal Supreme Court [Bundesgerichtshof], 11 December, 2003, Case No. 3 StR 120/03, in Vol. 57 Neue Juristische Wochenschrift 1054, 1056 (2004); German Federal Supreme Court [Bundesgerichtshof], 26 May, 2004, Case No. 2 StR 505/03, in Vol. 57 Neue Juristische Wochenschrift 2458, 2458 (2004).

317 Fateh-Moghadam (2008), p. 109.

318 German Federal Supreme Court [Bundesgerichtshof], 26 May, 2004, Case No. 2 StR 505/03, in Vol. 57 Neue Juristische Wochenschrift 2458, 2459 (2004).

because of the bodily injury.[319] When applying this to LOD, it becomes apparent that LOD is principally authorised due to the consent of the donor.[320] A limit, based on a violation of good morals, can then only be considered if an anticipatory and objective close examination of all decisive circumstances of the particular LOD shows that it places the donor in concrete, mortal danger.[321]

The regulation in the United Kingdom is different. "Public policy" is required to justify LOD, so that consent itself is not sufficient. In addition, a legally accepted good reason, or a public interest with regard to the infringement, must exist. Comparing this to the concept of "good morals" makes clear that both concepts address the limitations of the right of self-determination of one's own body. However, they have completely different, even antithetic, functions in criminal law, where the question regarding the right of disposal of one's own body is answered in a different manner in Germany than in the United Kingdom.[322] In German law, one has the right of disposal over his own body. Whereas in the United Kingdom, inviolableness of one's bodily integrity is guaranteed. Consequently, good morals restrict the power of disposal of a person, while public policy in the United Kingdom is an instrument in criminal law to establish an enlargement of freedom with regard to public interest.[323] Thus, consent is not limited by public policy as it is with regard to the concept of good morals.

The scope of the patient's consent is usually limited. This is evident in the established barriers of the countries that have been considered.

hh) Consent of the Recipient

Some countries also explicitly require the recipient to consent to the LOD. This is to assure her autonomy, even though medical ethics already states that due to the right of autonomy, the recipient has the right to decline an organ, and that this right has to be enforced in any case.[324] The Belgian situation can be used as a case in point. The *Law on Retrieval and Transplantation of Organs* does not provide any special regulation with regard to the consent of the recipient, but it must always be present anyway.[325] This is because, in addition to the law that specifically concerns the procurement and transplantation of organs, the general law applies. If one applies

[319] Fateh-Moghadam (2008), p. 133; German Federal Supreme Court [Bundesgerichtshof], 26 May, 2004, Case No. 2 StR 505/03, in Vol. 57 Neue Juristische Wochenschrift 2458, 2460 (2004).

[320] Fateh-Moghadam (2008), p. 134.

[321] Ibid, p. 135.

[322] Fateh-Moghadam (2008), p. 179 ff.; Fateh-Moghadam, in Beck/Burchhard/Fateh-Moghadam (ed.) (2011), p. 59 ff.

[323] Fateh-Moghadam (2008), p. 179 ff.; Fateh-Moghadam, in Beck/Burchhard/Fateh-Moghadam (ed.) (2011), p. 59 ff.

[324] Cf. Gutmann/Land, Vol. 384 Langenbeck's archives of surgery 515, 516 (2000).

[325] Merckx, Falcone Project, Belgium, p. 5.

the common rights, any medical operation is an invasion of one's physical integrity. The consent of the patient can justify this. Moreover, Art. 8 of the Belgian *Law of 22 August 2002* expressively states that any patient has the right to freely consent to any operation once the preliminary information has been achieved.[326] Several countries, in addition, mention exceptions to the requirement of "explicit consent" by the recipient.

(1) Explicit Consent of the Recipient

Bulgaria,[327] Estonia,[328] Hungary,[329] Italy,[330] Moldova,[331] Poland,[332] Portugal,[333] Slovenia[334] and Spain[335] explicitly require the recipient to consent. Bulgaria,[336] Estonia,[337] Slovenia[338] and Spain[339] also allow the legal representative to consent on behalf of the recipient.

What exactly the consent refers to differs, though. With regard to the decisive point of reference of the recipients consent, Bulgaria,[340] Hungary,[341] Italy,[342] Moldova,[343] Slovenia[344] and Spain[345] refer to the forthcoming transplantation procedure in general.[346] The Portuguese law is even more general by only requiring the recipients consent.[347] In Estonia, in contrast, the consent must be given with

---

[326] Information from R. Willmotte.

[327] Art. 32 No. 1 Law of the Transplantation of Organs, Tissues and Cells.

[328] Sec. 6 (1) and Sec. 7 (1) Transplantation of Organs and Tissues Act.

[329] Sec. 209 (6) Act on Health.

[330] Art. 4 Legislation on Donation, Retrieval and Transplantation of Organs and Tissues.

[331] Art. 16 (1) Law on the Transplant of Human Organs, Tissues and Cells.

[332] Art. 12, 1., 9) The Cell, Tissue and Organ Recovery, Storage and Transplantation Act.

[333] Art. 8, 1 Law on Harvesting and Transplanting Human Organs and Tissues.

[334] Art. 6 I 1 The Removal and Transplantation of Human Body Parts for the Purpose of Medical Treatment Act.

[335] Art. 6, c. and Art.15, 1. Royal Decree 2070/1999.

[336] Art. 32 No. 1 Law of the Transplantation of Organs, Tissues and Cells.

[337] Sec. 6 (1) and Sec. 7 (2) Transplantation of Organs and Tissues Act.

[338] Art. 6 II The Removal and Transplantation of Human Body Parts for the Purpose of Medical Treatment Act.

[339] Art. 6, c. and Art.15, 1. Royal Decree 2070/1999.

[340] Art. 32 No. 1 Law of the Transplantation of Organs, Tissues and Cells.

[341] Sec. 209 (6) Act on Health.

[342] Art. 4 Legislation on Donation, Retrieval and Transplantation of Organs and Tissues.

[343] Art. 16 (1) Law on the Transplant of Human Organs, Tissues and Cells.

[344] Art. 6 I 1 The Removal and Transplantation of Human Body Parts for the Purpose of Medical Treatment Act.

[345] Art. 6, c. and Art.15, 1. Royal Decree 2070/1999.

[346] Art. 32 No. 1 Law of the Transplantation of Organs, Tissues and Cells.

[347] Art. 8, 1 Law on Harvesting and Transplanting Human Organs and Tissues.

regard to the transplantation in general, and the implantation of the organ is permitted only after the recipient has given informed consent.[348] In Poland, the consent of the recipient must specifically refer to the implantation.[349]

Considering the formal requirements of the donors consent, it must be in writing in Estonia,[350] Hungary,[351] Moldova,[352] Slovenia[353] and Spain.[354] The formal requirements in Spain are even more extensive. The document proving the informed consent of the recipient must include the name of the health care centre, the date of the authorisation, the name of the recipient and, if necessary, the representative who consents to the transplantation. This document must be signed by the physician who informed the recipient, and by the recipient himself or by his representative.[355]

As seen, the recipient's consent is explicitly demanded by several laws. Finland's, Ireland's, Latvia's, Lithuania's, the Netherlands, Slovakia's, Sweden's, Switzerland's and the United Kingdom's transplant laws do not explicitly require the recipient's consent. However, due to the principle of autonomy, in normal cases, irrespective of the state of emergency, no one can get an organ implantation without consenting to this surgery. This applies to Austria as well.[356]

(2) Exceptions of the Requirement of "Explicit Consent" by the Recipient

Some countries mention exceptions to the requirement of explicit consent by the recipient. Explicit consent is commonly unnecessary if a delay in the performance of the procedure threatens the recipient's life or if it is impossible to obtain her consent. For instance, pursuant to the Moldovan transplant law, the transplant shall be made without the recipient's consent "if the recipient is not able to express his/her consent in writing due to some objective circumstances and connection with his/her family is not possible in due time, while delay would inevitably lead to the patient's death."[357] The exception included in the Czech law works in a similar fashion.[358]

---

[348] Sec. 6 (1) and Sec. 7 (1) Transplantation of Organs and Tissues Act.

[349] Art. 12, 1., 9) The Cell, Tissue and Organ Recovery, Storage and Transplantation Act.

[350] Sec. 6 (2) Transplantation of Organs and Tissues Act.

[351] Sec. 209 (6) Act on Health.

[352] Art. 16 (1) Law on the Transplant of Human Organs, Tissues and Cells.

[353] Art. 6 I 1 The Removal and Transplantation of Human Body Parts for the Purpose of Medical Treatment Act.

[354] Art. 6, c. and Art.15, 1. Royal Decree 2070/1999.

[355] Art. 15, 2. Royal Decree 2070/1999.

[356] Information from K. Bruckmueller.

[357] Art. 16 (3) Law on the Transplant of Human Organs, Tissues and Cells.

[358] Art. 7 (3) Transplantation Act on Donation, Removal and Transplantation of Organs and Tissue and on the Amendment of some Acts.

Table 1 Summary of the issues concerning consent

Considered aspect	Result after comparing the countries
Consent as explicit requirement	• About half of the countries considered explicitly mention the requirement of consent in their transplant laws, however, every country requires the donor's consent, even if it does not explicitly mention the requirement of consent in its transplant laws
Point of time of consent	• Common standard that consent must be given prior to the transplantation
Capacity to consent	• Required in all countries, but, only some countries explicitly demand the capacity to consent in their transplant law
Voluntariness of consent	• None of the countries considered contains a duty to act as living organ donor, consent must be voluntary, regardless of whether it is specifically required by the national transplant law or not
Formal requirements	• A significant amount of countries requires the consent to be in writing • A few countries request a witness to be present while the donor consents • Not uncommon to involve an authority in the process of consent, but this is also not a generally required condition
Withdrawal of consent	• Almost all countries render this possibility for the donor
Limitation to consent	• Common standard that LOD is not legal in case it ceases the donor's life or is directly life-threatening • Several countries declare consent that would be against good morals void
Consent of the recipient	• Some countries explicitly require this, but due to the principle of autonomy, no one can get an organ transplanted without consenting anyway

Even if the national transplant law does not explicitly state an exception, the general legal system usually allows a decision based on substituted or presumed consent due to the state of emergency in cases where therapeutic treatment is needed.[359] LOD is a therapeutic treatment for the organ recipient.[360] Consequently, the possibilities of consenting on behalf of the incapacitated patient and consent due to the state of emergency (one which the European legal systems have in common), are valid with regard to the recipient.[361]

ii) Summary of the Issues Concerning Consent

Overall, the considered countries consistently require both the donor and the recipient to give valid consent to justify LOD. However, after examining the national laws in detail, the similarities and differences in regulating this aspect became apparent. Table 1 summarises the findings made.

[359] Cf. Gutmann/Schroth (2002), p. 48; cf. Eyal, in Zalta (ed.) (2011).

[360] Cf. Gutmann/Schroth (2002), p. 48.

[361] Ibid.

b) Disclosure

In addition to the consent itself, the disclosure of the donor is also required for the informed consent to be valid.

The duty to provide disclosure means that the potential donor needs to be provided with adequate information about the medical operation. In most cases, the potential donor will not be medically educated.[362] In order to give him the possibility to reflect on his decision, he needs to be informed to such an extent that his consent is based on a comprehension of the process. Only then can he give legally valid consent, or in other words, informed consent.[363]

All countries require the potential donor to become sufficiently informed. This provision is either stated explicitly by law or is at least the result of a generally accepted legal principle.

With respect to the extent of the information, it has to be kept in mind that the organ removal is a non-therapeutic treatment for the donor. That leads to the hypothesis that the requirements for the amount of information are very high.[364] However, national provisions are not equal in depth, but vary significantly. (1) What degree of disclosure is required in the national laws? (2) Do the national laws explicitly require the donor to understand the provided information? (3) I furthermore present possible formal requirements of the education of the donor, and (4) I deal with the recipient, namely if he also has to get informed.

aa) Degree of Disclosure

Even though all countries have determined that the donor has to be informed, the precise regulations differ with regard to the content of the information.

Independent from the national laws, the *Consensus Statement of the Amsterdam Forum on the Care of the Live Kidney Donor* answers the question: What should the education of the kidney donor contain? The potential donor must be informed about the following issues: 1) The potential donor must be aware of the nature of the evaluation process. 2) The potential donor must be informed of his test results and the consequences of testing. This should include the possibility that conditions may be discovered that can impact future social status, which includes, e.g., healthcare and the insurability of the potential donor. 3) The potential donor must be aware of the risks involved with the organ removal, as assessed after the complete evaluation. This should include, among others, the risk of death, surgical morbidities, change in health and renal function, impact upon insurability/employability and

[362] Bock (1999), p. 144; Schreiber (2004), p. 76.

[363] German Federal Supreme Court [Bundesgerichtshof], 9 December, 1958, Case No. VI ZR 203/57, in Vol. 12 Neue Juristische Wochenschrift 811, 812 f. (1959); cf. Spickhoff (2004), p. 60.

[364] Bock (1999), p. 145 f.; Edelmann, Vol. 50 Versicherungsrecht 1065, 1067 (1999); Gutmann, Vol. 15 Medizinrecht 147, 151 (1997).

unintended effects upon family and social life. 4) The potential donor must be informed of individual healthcare provider and the social system responsibilities in the management of discovered conditions. 5) The potential donor must be aware of the situation of the recipient. Namely, the expected transplant outcomes for the recipient have to be part of the disclosure along with any specific recipient conditions which may have an impact upon the decision to donate the kidney. In addition, specific information about the recipient may be disclosed, but this must have the assent of the recipient. 6) The potential donor should also be informed of alternative renal replacement therapies available to the potential recipient.[365]

This recommendation is almost all-embracing and most of the national laws are not as extensive. The countries have different requirements concerning the amount and content of information that need to be provided to the donor. Therefore, several different preconditions required by the national laws will be discussed. I examine whether the national laws require informing the donor about (1) the possible consequences of the LOD, (2) the medical background, (3) the protection of the donor and (4) the impact on the recipient. (5) In addition, the miscellaneous requirements the national laws contain with regard to the content of the education of the donor will be considered.

(1) Disclosure About Consequences of the Living Organ Donation

A significant amount of the considered national laws contains the duty to inform the donor about possible consequences of the LOD. Of course, the possible consequences are usually more extensive than what the actual consequences will be in reality, but the donor has the right to know what might happen in the worst case.

The Belgian,[366] Moldovan[367] and Spanish[368] laws explicitly refer to the physical, mental, familial and social consequences that have to be disclosed.

The Dutch law only requests the supply of information regarding the consequences and risks for the donor's health and quality of life.[369]

The Swiss *Federal Act on the Transplantation of Organs, Tissues and Cells* only determines that the donor must have been informed comprehensively.[370] According to Art. 15, 1, the Federal Council shall specify the requirements the information must fulfil. This can be found in Art. 9 *Transplantation Ordinance*. Hence, the donor must be informed about the short- and long-term risks for her health.[371] In addition, the disclosure of the donor must contain possible psychological

[365] The Ethics Committee of the Transplantation Society, Vol. 78 Transplantation 491, 491 (2004).

[366] Art. 9, 2 Law on Retrieval and Transplantation of Organs.

[367] Art. 15 (3) Law on the Transplant of Human Organs, Tissues and Cells.

[368] Art. 4, b. and Art.9, 3. Royal Decree 2070/1999.

[369] Sec. 3, 2 Organ Donation Act.

[370] Art. 12 b) Federal Act on the Transplantation of Organs, Tissues and Cells.

[371] Art. 9, 2 c) Transplant Ordinance.

consequences and, going along with that, the possibility of receiving psychological care.[372]

The Swedish *Transplant Act* only states that the donor and, if necessary, the custodian or guardian has to be informed about the risks associated with the procedure.[373] To clarify the very broad term "risks," the *Guidance on the provisions* explains that the information has to focus on possible short-term and long-term consequences.[374]

The laws of England,[375] Hungary[376] and Ireland[377] contain the same demand. In particular, the appendix to the Irish *Guide to Professional Conduct and Ethics for Registered Medical Practitioners* requires one to supply information regarding any lifestyle changes which may be caused or required by the treatment.[378]

The German national law is also similar. The information has to contain possible or indirect repercussions and delayed sequelae of the intended organ removal for the donor's health.[379]

The regulation in the Czech Republic essentially contains the same request.[380] In addition, the physician performing the removal is obligated to repeat all of the information given immediately before the removal.[381]

In contrast, the Bulgarian,[382] Estonian,[383] Italian,[384] Latvian,[385] Norwegian,[386] Portuguese,[387] Scottish[388] and Slovakian[389] laws do not highlight specific consequences, they only require the disclosure about consequences and risks in general. The Scottish regulation also determines that any other relevant

[372] Ibid., Art. 9, 2 j).

[373] Sec. 10 Transplant Act.

[374] The Ministry of Health and Social Affairs Sweden 1997, The Swedish Transplant Act, Guidance on the provisions, Sec. 10.

[375] No. 90 Code of Practice 2.

[376] Sec. 209 Act on Health.

[377] Principle 35.2 Irish Guide to Professional Conduct and Ethics for Registered Medical Practitioners.

[378] Ibid., Appendix A.

[379] Sec. 8 II 1 Act on the donation, removal and transplantation of organs.

[380] Art. 7 (1) Transplantation Act on Donation, Removal and Transplantation of Organs and Tissue and on the Amendment of some Acts.

[381] Ibid., Art. 7 (3).

[382] Art. 24 (1) Law of the Transplantation of Organs, Tissues and Cells.

[383] Sec. 6 (2) Transplantation of Organs and Tissues Act.

[384] Art. 2 Legislation on Donation, Retrieval and Transplantation of Organs and Tissues.

[385] Sec. 13 Law on the Protection of the Body of Deceased Human Beings and the Use of Human Tissues and Organs in Medicine.

[386] Sec. 1 Act relating to transplantation, hospital autopsis and the donation of bodies etc.

[387] Art. 7 Law on Harvesting and Transplanting Human Organs and Tissues.

[388] Reg. 2 (7) Human Organ and Tissue Live Transplants (Scotland) Regulations 2006.

[389] Sec. 6 (1) Act on Healthcare, Healthcare-Related Services and on the Amendment and Supplementing of Certain Laws.

implications arising from the intended donation, including the effect on any children or dependent relatives of the donor, have to be discussed.[390] Thus, the Scottish duty concerning the information about consequences is extensive, while the phrasing of the regulation is very general. In Greek, the donor only has to be informed about possible risks.[391] The Lithuanian[392] and Polish[393] laws are generally comparable, as both refer to the possible effects of the organ donation on the donor's health. In Austria, it is also generally accepted that the donor has to be informed about all risks the medical operation involves.[394]

Therefore, almost all countries require the donor to become informed about possible consequences and risks.

Only Finland does not mention this requirement. In fact, the Finish *Act on the Medical Use of Human Organs and Tissues* only demands an explanation of the significance of the procedure for donors and recipients.[395] It does not provide any details about the interpretation of the exact meaning of "significance", making it unclear if the donor has to be informed about potential consequences or not.

(2) Medical Background

The donor will usually be a medical layperson. Hence, it can be presumed that he does not have any knowledge about the medical background of LOD. For that reason, many national laws, not nearly all though, require the information for the donor to contain a medical explanation, allowing him to learn about the medical procedure that takes place. The national laws that do so are very diverse with regard to their exact requirements and content.

The Portuguese law only mentions that the donor has to be informed about the treatment and its side effects.[396]

The Dutch,[397] Greek,[398] and Latvian[399] laws only demand the donor to be informed about the purpose and the nature of the intervention.

---

[390] Reg. 2 (7) Human Organ and Tissue Live Transplants (Scotland) Regulations 2006.

[391] Canellopoulou-Bottis, Vol. 7 European Journal of Health Law 427, 433 (2000).

[392] Art. 10. 9 Law on Donation and Transplantation of Human Tissues, Cells and Organs.

[393] 12, 1., 8) The Cell, Tissue and Organ Recovery, Storage and Transplantation Act.

[394] Aigner, Vol. 9 Annals of Transplantation 11, 12 (2004); Kert, Falcone Project, Austria, p. 5; Neyer/Scheuring, 2004 Österreichische Pflegezeitschrift 12, 14.

[395] Sec. 3 II 1 Act on the Medical Use of Human Organs and Tissues.

[396] Art. 7 Law on Harvesting and Transplanting Human Organs and Tissues.

[397] Sec. 3, 2 Organ Donation Act.

[398] Canellopoulou-Bottis, Vol. 7 European Journal of Health Law 427, 433 (2000).

[399] Sec. 13 Law on the Protection of the Body of Deceased Human Beings and the Use of Human Tissues and Organs in Medicine.

The Bulgarian,[400] Polish[401] and Scottish[402] laws generally require informing the donor about the medical procedure.

The Swedish *Transplant Act* only states that the donor has to be informed about the action, but it does not particularly refer to the medical background.[403] Again, the *Guidance on the provisions* renders further details. It clarifies that the information must include particulars of the nature of the action, for example, how it is performed and what material or organ is to be taken.[404]

The Estonian regulation requires the donor to be informed about "the purpose and the nature of the transplantation."[405] In addition, the donor has the right to receive information on her state of health, on the results of tests conducted on her organs and the suitability for treatment of the organs donated by her. She also has a right of confidentiality of identity.[406]

When informing the donor about the medical background, the German law is comparable with the Estonian rule. It requires the donor to be informed about the physical examination, and also of her right to be informed about the results. In addition, she has to be informed of the intent and type of intervention and also of her medical confidentiality.[407]

The Hungarian *Act on Health* also focuses on informing the donor about her state of health, including the medical evaluation.[408] In addition, the donor has to be informed about the recommended examinations and interventions[409] and about possible alternative procedures and methods.[410]

The laws in Norway[411] and the United Kingdom[412] require the donor to be informed about the nature of the medical procedure. Beyond that, in the United Kingdom, the donor should be provided with information about potential complications or side effects for the recipient, as such information on these factors could impact the life of the graft, or the recipient herself.[413]

Ireland includes a very detailed regulation in the appendix of its *Guide to Professional Conduct and Ethics for Registered Medical Practitioners.*

---

[400] Art. 24 (2) Law of the Transplantation of Organs, Tissues and Cells.

[401] 12, 1., 8) The Cell, Tissue and Organ Recovery, Storage and Transplantation Act.

[402] Reg. 2 (7) Human Organ and Tissue Live Transplants (Scotland) Regulations 2006.

[403] Sec. 10 Transplant Act.

[404] The Ministry of Health and Social Affairs Sweden 1997, The Swedish Transplant Act, Guidance on the provisions, Sec. 6.

[405] Sec. 6 (2) Transplantation of Organs and Tissues Act.

[406] Sec. 9^1 (2) Handling and Transplantation of Cells, Tissues and Organs Act.

[407] Sec. 8 II 1 Act on the donation, removal and transplantation of organs.

[408] Sec.13 (2) a) Act on Health.

[409] Ibid., Sec. 13 (2) b).

[410] Ibid., Sec. 13 (2) f).

[411] Sec. 1 Act relating to transplantation, hospital autopsis and the donation of bodies etc.

[412] Reg. 11(9) (a) Human Tissue Act Regulations 2006.

[413] No. 91 Code of Practice 2.

Accordingly, the donor has to be informed of the diagnosis and of any possible uncertainties about the diagnosis. Moreover, she has to be briefed about the purpose of the proposed investigation or treatment, and the details of the procedures or therapies involved, including methods of pain relief and the preparation for the procedure. She also has to be informed about possible experiences during or after the procedure, including common and serious side effects.[414]

The Swiss rule is detailed as well. The donor has to be informed of the purpose and the course of action of the preliminary diagnosis and the intervention.[415] She also has to be briefed on the necessity of a health examination[416] and the transplantation centre's duty to track the state of health of the donor.[417] In addition, the information has to contain the expected length of the hospital stay, the extent of a possible work incapacity after the procedure or other constraints for the donor.[418]

Since the donor will usually be a medical layperson, many countries require her to be informed of her medical background. Not all national laws contain this requirement, namely Austria, Belgium, Finland, Moldova, Spain and Switzerland do not explicitly require this.

(3) Protection of the Donor

Since LOD does not have any direct benefit for the donor's health, a certain kind of protection for him seems appropriate. However, only Bulgaria,[419] the Czech Republic[420] and Germany[421] require the donor to be informed about the rights and safety measures provided for him by law.

(4) Impact on the Recipient

One distinctive feature of LOD in comparison to other medical operations is that it does not only have an impact on one person, but on two persons, namely the donor and the recipient. This makes it a reciprocal duty to provide disclosure to both parties. (1) That first includes disclosing the impact of the LOD on the recipient to the donor.[422] This is taken into account by a few national laws and is understandable. It makes it possible for the donor to carry out a necessary subjective

414 Appendix A Irish Guide to Professional Conduct and Ethics for Registered Medical Practitioners.

415 Art. 9, 2 a) Transplantation Ordinance.

416 Ibid., Art. 9, 2 e).

417 Ibid., Art. 9, 2 f).

418 Ibid., Art. 9, 2 d).

419 Art. 24 (3) Law of the Transplantation of Organs, Tissues and Cells.

420 Art. 7 (1) Transplantation Act on Donation, Removal and Transplantation of Organs and Tissue and on the Amendment of some Acts.

421 Sec. 8 II 1 Act on the donation, removal and transplantation of organs.

422 Fateh-Moghadam (2008) p. 250.

risk-benefit equation according to his individual system of values.[423] (2) The second aspect of the reciprocal duty to disclosure is that the recipient is informed about the risks for the donor.[424] Whether the national laws consider this as well will be regarded below.

Austria,[425] Germany,[426] Hungary,[427] Ireland[428] and Spain[429] require the donor to be informed about the expected benefits for the recipient.

The Swiss *Transplantation Ordinance* does not explicitly refer to the expected benefits for the recipient. However, it contains the duty to brief the donor about expected advantages and possible disadvantages for the recipient as well as other possible therapies for the healing of the recipient.[430]

In Finland, the potential donor does not particularly have to be informed about the expected benefit for the recipient, but the significance of the procedure for the recipient must be explained.[431]

Even though LOD is a special case in the medical field, in the sense that two patients, as opposed to only one, are involved, it is noteworthy to consider that only a few countries consider this with regard to the content of the information for the donor.

(5) Miscellaneous

In addition to the aforementioned requirements, a few national laws require additional information. In most cases this means that information has to be adjusted to each individual case.

The specific requirements mentioned in the German *Act on the donation, removal and transplantation of organs* are mentioned above (e.g., the possible repercussions and delayed sequelae of the intended organ removal for the donor's health). In addition, the German *Act on the donation, removal and transplantation of organs* requires the donor to be informed about other facts which are evidently of importance to her with respect to the organ donation.[432] For example, it is important to know about the indirect repercussions, which include the danger of partial or complete incapacity to work and also the expected psychological consequences of a

---

[423] Ibid.

[424] Ibid.

[425] Bruckmueller/Schumann, in Roxin/Schroth (ed.) (2010), p. 845; Kert, Falcone Project, Austria, p. 5; cf. Kienapfel (1990), p. 193 at 30; Kopetzki (1988), p. 254.

[426] Sec. 8 II 1 Act on the donation, removal and transplantation of organs.

[427] Sec.13 (2), g) Act on Health.

[428] Appendix A Irish Guide to Professional Conduct and Ethics for Registered Medical Practitioners.

[429] Art. 4, b. an Art.9, 3. Royal Decree 2070/1999.

[430] Art. 9, 2 k) Transplantation Ordinance.

[431] Sec. 3 II 1 Act on the Medical Use of Human Organs and Tissues.

[432] Sec. 8 II 1 Act on the donation, removal and transplantation of organs.

successful (or possibly failed) transplantation, including the consequences for the relationship between the donor and the recipient.[433] The Community of Interest Living Kidney Donation claims, however, that the potential living kidney donor is not informed about potential risks in a sufficient was (e.g. loss of efficiency, fatigue).[434]

In England, the amount and the content of information for the donor also have to be personalised. The relevant information varies on a case-by-case basis and transplant teams should share information with donors following prior agreement with the potential recipient.[435]

In Hungary[436] and Ireland,[437] the patient should have a right to receive the information provided to him in a customised form. In addition, the patient has a right to pose additional questions during and after the information session.[438]

Thus, only a few countries explicitly require informing the donor in a customised way.

After observing all aspects that might have to be included in the disclosure of the donor, it becomes clear that all countries specify the exact content the donor has to be informed about. However, the countries considered have different opinions with regard to the relevant content; so the disclosure of the donor probably also differs in the medical practice in each country.

bb) Comprehension

Patients sometimes do not understand the information provided to them.[439] A few national laws have reacted to this predicament by explicitly requiring the donor to understand the information provided.

Hungary[440] and the United Kingdom[441] also focus on the language of the disclosure. In Hungary, it is ruled that, if necessary and if possible, the services of an interpreter or a sign language interpreter should be supplied for the provision of information. In addition, the patient should have a right to be informed with a "type" of language which is comprehensible for him considering his "age,

---

[433] Gutmann, in Schroth et al. (ed.) (2005), Sec. 8 at 43; Nickel/Schmidt-Preisigke/Sengler (2001), p. 109.

[434] Interessengemeinschaft Nierenlebendspende e.v. (2012).

[435] No. 91 Code of Practice 2.

[436] Sec. 13, (1) Act on Health.

[437] Principle 36.1 Irish Guide to Professional Conduct and Ethics for Registered Medical Practitioners.

[438] Sec. 13, (3) Act on Health; Principle 36.4 Irish Guide to Professional Conduct and Ethics for Registered Medical Practitioners.

[439] Eyal, in Zalta (ed.) (2011).

[440] Sec. 13 (8) Act on Health.

[441] No. 60 Code of Practice 1.

education, knowledge, state of mind and his wish expressed on the matter."[442] In the United Kingdom, particular consideration should be given to the needs of individuals and families, whose first language is not English.[443] The aspect of comprehension is also raised in a more general way, because it is explicitly required that the individual (or the person with parental responsibility) should understand the nature and purpose of what is proposed and should be able to make an informed decision.[444]

The Czech *Transplantation Act* also does not ignore the aspect that the donor has to understand the information. It requests that the information given has to be understandable.[445]

In Norway[446] and Sweden,[447] it is also required that the donor understands the meaning of the information and, in order to assure this, the physician has to verify it. This is the same in Austria.[448]

In the Netherlands, the party who is to remove the organ has to ensure that the donor is aware of the provisions of the *Organ Donation Act* regarding compensation for costs incurred.[449]

The regulation in Slovakia considering the comprehension of the person who is informed is rather detailed. The disclosure must be comprehensible and understandable without restraint. It must be in an "appropriate way to the intellectual, volitive and health condition of the instructed person."[450]

In Scotland, the report of the interview of the donor must ensure the capacity of the donor to understand the nature of the medical procedure and the risk involved. In addition, he must understand that his consent can be withdrawn at any time.[451]

Not all countries explicitly require the donor to understand the information provided.

cc) Formal Requirements

While all of the informational requirements discussed so far were content driven, there are also legal specifications regarding the form of the information provided to

[442] Sec. 13 (8) Act on Health.

[443] No. 60 Code of Practice 1.

[444] Ibid., No. 146.

[445] Art. 7 (1) Transplantation Act on Donation, Removal and Transplantation of Organs and Tissue and on the Amendment of some Acts.

[446] Sec. 1 Act relating to transplantation, hospital autopsis and the donation of bodies etc.

[447] Sec. 10 Transplant Act.

[448] Aigner, Vol. 9 Annals of Transplantation 11, 13 (2004).

[449] Sec. 3, 2 Organ Donation Act.

[450] Sec. 6 (2) Act on Healthcare, Healthcare-Related Services and on the Amendment and Supplementing of Certain Laws.

[451] Reg. 2 (7) Human Organ and Tissue Live Transplants (Scotland) Regulations 2006.

the donor. (1) Does the information have to be in writing? (2) Must the person entitled by law to inform the donor be specified? Both questions will be examined separately.

(1) Written Form

The Czech Republic,[452] Germany,[453] Hungary,[454] the Netherlands,[455] Poland,[456] Spain[457] and Switzerland[458] require the education of the potential donor to be in writing. In Hungary,[459] the Netherlands[460] and Switzerland,[461] the information must also be verbally communicated. In Germany, the person providing the information, the additional physician who is not involved in the transplantation process, and the donor have to sign the document that records the content of instruction.[462] The regulation in the Czech Republic is similar because all attending persons have to provide their dated signatures.[463] In Switzerland, those documents have to be stored for ten years.[464]

In Spain, all aspects that have to be included in the information must be certified in a medical certificate. These aspects include the state of health of the donor, the disclosed information, the donor's reason for motivation (freely expressed) and, if necessary, any evidence of external pressure being placed on him. Furthermore, this certificate has to include the names of other qualified people that worked together with the certified physician.[465]

In the Netherlands, it might not be enough to provide the information in writing. In fact, audiovisual media of the nature and purpose of the removal procedure and of the consequences and risks for the donor's health and quality of life have to be used if appropriate.[466] In the medical practice, some transplant centres use DVD's to inform the potential donor about the technique of the removal of a kidney. The

[452] Art. 7 (2) Transplantation Act on Donation, Removal and Transplantation of Organs and Tissue and on the Amendment of some Acts.

[453] Sec. 8 II 4 Act on the donation, removal and transplantation of organs.

[454] Sec 209 (1) Act on Health.

[455] Sec. 3, 2 Organ Donation Act.

[456] Art. 12, 1., 5) The Cell, Tissue and Organ Recovery, Storage and Transplantation Act.

[457] Art. 9, 3. Royal Decree 2070/1999.

[458] Art. 9, 1 Transplantation Ordinance.

[459] Sec. 209 (1) Act on Health.

[460] Sec. 3, 2 Organ Donation Act.

[461] Art. 9, 1 Transplantation Ordinance.

[462] Sec. 8 II 4 Act on the donation, removal and transplantation of organs.

[463] Art. 7 (2) Transplantation Act on Donation, Removal and Transplantation of Organs and Tissue and on the Amendment of some Acts.

[464] Art. 9, 4 Transplantation Ordinance.

[465] Art. 9, 3. Royal Decree 2070/1999.

[466] Sec. 3, 2 Organ Donation Act.

patient concerned can take the DVD home to watch it in private and to ask questions later in the hospital. DVD's are also used to inform the patient who suffers from renal failure about the different types of dialysis treatment. However, using audio-visual media is not an agreed standard.[467]

Several countries do not regulate the form of the information at all. One could assume that those countries do not have any specific formal requirements, such as the information has to be in writing etc. This, for example, is the case in the Belgian transplant law. It only states that the donor has to be informed in a clear and complete way.[468] No hints can be found regarding formal requirements. Overall, the majority of countries considered explicitly require the information of the donor to be in writing.

(2) Source(s) of Disclosure

Quite a lot of countries considered address who has to inform the donor. They have varying requirements regarding the person who has the duty to disclose the information.

The Belgian,[469] Moldovan,[470] Portuguese[471] and Swiss[472] laws require the doctor involved in the transplantation process to inform the donor. The Swiss law goes a step further by specifying that the doctor physically removing the organ has to provide the information.[473]

In the Netherlands, the person who informs the donor must also be involved in the transplantation process.[474] Namely, the donor team, which is separate from the transplant team, has to inform the donor. The rationale of this separation is to ensure that the donor has the freedom and the possibility to accept or decline becoming a donor. This is difficult, though, if the doctor who educates the donor is the same person who looks after and does the work-up for the recipient. The inclusion of two teams in the transplant process also enables one team to prepare the recipient while the other removes the kidney, shortening the cold ischemia time as much as possible.[475]

In Sweden, neither the transplanting physician nor the donor team have to inform the donor, but the one entitled to decide on the action.[476]

---

[467] Information from M. Bos.

[468] Art. 9 Law on Retrieval and Transplantation of Organs.

[469] Ibid., Art. 9 II.

[470] Art. 15 (3) Law on the Transplant of Human Organs, Tissues and Cells.

[471] Art. 7 Law on Harvesting and Transplanting Human Organs and Tissues.

[472] Art. 9, 1 Transplantation Ordinance.

[473] Ibid.

[474] Sec. 4 Organ Donation Act.

[475] Information from M. Bos.

[476] Sec. 10 Transplant Act.

The Bulgarian[477] and Hungarian[478] laws, in contrast, state that the potential donor has to be informed by a physician who does not participate in the process of organ transplantation.

Other countries demand more than only one specific person to gather the information of the donor.

The Polish law requires the transplantation physician as well as another physician, who is not part of the process, to inform the donor.[479]

The regulation in Germany is similar. It demands the attendance of two physicians, one who is not further specified and another one who may not be involved in the transplantation process.[480] The attendance of a physician who is not part of the transplantation process is stipulated to ensure that the information is carried out in an objective way.[481] In addition, in Germany, other expert persons must be present, where necessary.[482] This would normally be a person with specific expertise of the psychological questions and pursuant to insurance law.[483] The requests for the requirement of necessity to be fulfilled should not be too strict. With regard to this, some legal experts argue that psychological experts should attend every disclosure of the donor, since an appropriate psychological preliminary of the decision to donate is inevitably connected to proper information of the donor.[484]

In the Czech Republic, the physician evaluating the health of a living donor has to inform the donor. In addition, the consenting donor and the legal representative of a minor or the legal representative of a legally incompetent person may require another witness to be present while being informed. The informing physician should inform them about this possibility in advance.[485]

The Finnish law neither obliges the transplant surgeon to inform the donor, nor explicitly excludes him from informing the donor. In fact, "the donor's personal physician contributing to the decision to donate must personally make the explanation to the donor."[486]

In England and Scotland, the regulations do not address whether the transplant physician is allowed to inform the donor. Actually, in England the explanation must be carried out by a medical practitioner with appropriate qualifications.[487] In

[477] Art. 24 (3) Law of the Transplantation of Organs, Tissues and Cells.

[478] Sec. 209 Act on Health.

[479] Art. 12, 1., 5) The Cell, Tissue and Organ Recovery, Storage and Transplantation Act.

[480] Sec. 8 II 1, 3 Act on the donation, removal and transplantation of organs.

[481] Esser, in Höfling (ed.) (2003), p. 227; Weigend/Zielinska, Vol. 14 Medizinrecht 445, 448 (1996).

[482] Sec. 8 II 2 Act on the donation, removal and transplantation of organs.

[483] Gutmann, in Schroth et al. (ed.) (2005), Sec. 8 at 49.

[484] Esser, in Höfling (ed.) (2003), p. 227.

[485] Art. 7 (2) Transplantation Act on Donation, Removal and Transplantation of Organs and Tissue and on the Amendment of some Acts.

[486] Sec. 3 II 2 Act on the Medical Use of Human Organs and Tissues.

[487] No. 90 Code of Practice 2.

Scotland, it is important that "the person appears to the Scottish Ministers to be suitably qualified to conduct the interview." Furthermore, the interviewer cannot have any connection with the interviewee. She also is not allowed to be the person who provided the donor with the information about the nature of the medical procedure and the risk involved in the organ removal.[488] In addition, the name and the qualifications of the person who provided the information about the nature of the medical procedure and the risk involved in the organ removal are necessarily recorded.[489]

Several countries do not mention who has to inform the donor; however, it can be concluded in those cases where the informant is specified that he has to be a physician. What kind of physician this has to be, and whether any kind of "third person" has to be present as well, differs.

dd) Disclosure of the Recipient

As clarified above, the recipient always has to give valid consent, even if it is not explicitly stated in the national transplant law. Since the requirement of consent and the requirement of informing the person consenting are connected, another important aspect in the context of LOD is the duty to disclosure of the recipient.

Only the transplant laws in Estonia,[490] Hungary,[491] Moldova,[492] Poland,[493] Portugal,[494] Slovenia[495] and Spain[496] explicitly request for the recipient to be informed. Most of these countries demand the information to contain possible risks and consequences. Hungary[497] and Poland,[498] however, also stipulate that the recipient be informed about the possible risks and consequences the organ donation might have for the donor, thereby taking into account that LOD involves two patients. These countries have considered the second part of the reciprocal duty to disclosure described above, namely that the recipient should also be informed of the donor's risks.

Scotland also takes the recipient into consideration. The required report must address whether any evidence of an offer or reward exists. In addition, it focuses on

[488] Reg. 2 (8) Human Organ and Tissue Live Transplants (Scotland) Regulations 2006.

[489] Ibid. Reg. 2 (7).

[490] Sec. 6 (2) Transplantation of Organs and Tissues Act.

[491] Sec. 209 (5) Act on Health.

[492] Art. 16 (1) Law on the Transplant of Human Organs, Tissues and Cells.

[493] Art. 12, 1., 9) The Cell, Tissue and Organ Recovery, Storage and Transplantation Act.

[494] Art. 7 Law on Harvesting and Transplanting Human Organs and Tissues.

[495] Art. 6 I 2 The Removal and Transplantation of Human Body Parts for the Purpose of Medical Treatment Act.

[496] Art. 6 Law Regulating Organ Recovery and Transplantation.

[497] Sec. 209 (5) Act on Health.

[498] Art. 12, 1., 9) The Cell, Tissue and Organ Recovery, Storage and Transplantation Act.

cases of direct donation. In cases involving a direct donation, the relationship between donor and recipient has to be regarded and any evidence of duress or coercion affecting the decision of the recipient to receive an organ or part of it must be covered in the report. Moreover, any difficulties that occurred while communicating with the interviewed recipient have to be reported, accompanied by an explanation of how those difficulties were overcome.[499]

The other transplant laws do not explicitly declare that the recipient has to be informed. Nevertheless, it can be assumed that many of the legal systems considered require the recipient to be informed before the transplant process takes place. This can be justified given the transplantation of an organ is a medical operation, which is usually considered to be justifiable battery due to the informed consent of the person solely concerned.

For example, the German *Act on the donation, removal and transplantation of organs* does not require informing the recipient. However, this condition is considered to be self-evident. The organ is implanted into the recipient with the intent to cure him. However, the physician's duty for every therapeutic treatment is to supply the patient with information. This consequently applies to the organ implantation as well.[500]

Thus, a generally accepted guideline is for both donors and recipients to be informed. Additionally, many claim that the recipient should be informed about the risks for the donor, because harming the donor might lead to a severe psychological burden for him.[501]

ee) Summary of the Issues Concerning Disclosure

In summary, informing the organ donor and the recipient is a general standard (or general practice). The details regarding the donor's disclosure of the donor possess similarities and differences. This will be shown roughly in Table 2.

c) Interim Result

It can be concluded that the concept of informed consent requires both the donor and the recipient to consent validly after having been sufficiently informed prior to the procedure. Thus, the concept of informed consent is a mandatory requirement in all of the countries considered. However, the details are regulated rather differently.

---

[499] Reg. 2 (6) Human Organ and Tissue Live Transplants (Scotland) Regulations 2006.

[500] Cf. Esser, in Höfling (ed.) (2003), p. 230; cf. Nickel/Schmidt-Preisigke/Sengler (2001), p. 94.

[501] Fateh-Moghadam et al., Vol. 22 Medizinrecht 19, 30 (2004).

Table 2 Summary of the issues concerning disclosure

Considered aspect	Result after comparing the countries
Disclosure matters	• Almost all countries require the donor to be informed about possible consequences • Many countries require the information of the donor to contain a medical explanation • Only a few national transplant laws require the donor to become informed about the rights and safety measures provided for him by law • Only a few national transplant laws require the donor to become informed about the situation of the recipient
Comprehension	• Only a few national transplant laws explicitly require the donor to understand the information provided to him
Formal requirements	• Several countries require the disclosure of the donor to be in writing • Not all countries specify the person who has to inform the donor, but in the countries that do, the donor usually has to be informed by a physician; the kind of physician this has to be and whether he has to be involved in the transplant process or not differs
Disclosure of the recipient	• Generally accepted that the recipient is to be informed, but only a few national transplant laws explicitly require this

3) Suitability of the Transplantation

The intended purpose of organ removal is to transplant the removed organ into the recipient to cure his organ failure. This purpose can be reached only if the transplantation is medically suitable. Consequently, living organ transplantation only makes sense in cases in which the recipient could be cured.

(1) With regard to this aspect, I address the necessity of a medical evaluation. (2) Besides, I discuss whether countries require an anticipated benefit for the recipient. (3) I examine the potential donor and his suitability to act as a living organ donor and (4) potential donor protection and the maximum extent of permissible risks. (5) Last, I examine the risk-benefit equations required in a few national laws.

a) Medical Evaluation

In practice, the medical suitability of a transplantation can only be judged sufficiently after a medical evaluation.

Thus, Bulgaria,[502] the Czech Republic,[503] Estonia,[504] Lithuania[505] and the United Kingdom[506] require the donor and recipient to undergo a medical check-up prior to the process of organ transplantation.

[502] Art. 4 (3) Law of the Transplantation of Organs, Tissues and Cells.

[503] Art. 6 (1) Transplantation Act on Donation, Removal and Transplantation or Organs and Tissues and on the Amendment of some Acts.

[504] Sec. 9 (1), 2) Transplantation of Organs and Tissues Act.

[505] Art. 10, 8 Law on Donation and Transplantation of Human Tissues, Cells and Organs.

[506] No. 59 Code of Practice 2.

The Czech Republic regulates this in a rather detailed manner. Various medical evaluations and procedures have to be carried out to evaluate the donor's health capability to donate an organ. In order to protect the donor, the health and life risks have to be eliminated.[507] The health care facility that performs the organ removal is responsible for the medical evaluation of the donor.[508] The Ministry of Health has established details for the medical evaluation of the donor in *Announcement 437/ 2002*.[509] This *Announcement* also contains details about the scope of the medical evaluation.[510] The donor or his legal representative are also given the option to ask another physician, who is neither involved in the organ removal nor in the further transplantation procedure, to assess the health risks of the donor.[511]

Poland[512] and Spain[513] only mention the medical investigation of the donor's health.

In Europe, the *Additional Protocol on Transplantation of Organ* states that appropriate medical investigations and interventions shall be carried out before the organ removal to evaluate and to reduce physical and psychological risks to the health of the donor.[514] The most recent European document on organ donation, the *Directive 2010/53/EU* goes into a similar direction.[515]

Since the medical suitability of a transplantation can only be addressed after a medical evaluation, it could easily be assumed that it does not make a difference if this is explicitly required or not.

A recent survey on the frequency of psychological screenings in pre-LOD evaluation showed that pre-LOD psychological screenings are very common. About 66 % of the living kidney donation units that were included in the survey stated that a psychological screening is routine in the pre-LOD evaluation and even 80 % of the living liver transplantation units answered that they do so. Living kidney and liver transplant centres both stated that a psychological screening is also done in most other cases if problems are identified.[516]

---

507 Art. 6 (1) Transplantation Act on Donation, Removal and Transplantation or Organs and Tissues and on the Amendment of some Acts.

508 Ibid., Art. 6 (2).

509 Ibid., Art. 6 (5).

510 Norba (2009), p. 291.

511 Art. 6 (6) Transplantation Act on Donation, Removal and Transplantation or Organs and Tissues and on the Amendment of some Acts.

512 Art. 12, 1., 4) The Cell, Tissue and Organ Recovery, Storage and Transplantation Act.

513 Art. 9, 3. Royal Decree 2070/1999.

514 Art. 11 Additional Protocol on the Convention on Human Rights and Biomedicine, on Transplantation of Organs and Tissues of Human Origin.

515 Art. 15, (2) Directive 2010/53/EU of the European Parliament and of the Council of 7 July 2010 on standards of quality and safety of human organs intended for transplantation.

516 EULOD WP 2 (2012), DOW: Deliverable 4, p. 9 & p. 10.

b) Anticipated Benefit for the Recipient

Since the purpose of organ transplantation is to cure the recipient, several laws explicitly require the expectation that the organ donation will benefit the recipient.

This is the quintessence of the Czech,[517] Estonian,[518] German,[519] Hungarian,[520] Polish,[521] Slovakian[522] and Spanish[523] regulations with respect to this aspect.

Additionally, the Estonian law requires an improvement of the recipient's quality of life after transplantation.[524]

The German law requires the transplantation to save the recipient's life or to cure a serious disease, to prevent it from worsening or to alleviate the resulting complaints.[525]

The Spanish regulation is similar. The transplantation must be performed with the intention to advance the health or the general living conditions of the recipient.[526]

Hungary[527] and Poland[528] merely demand a justification of the transplantation for the recipient.

The national laws that require histocompatibility[529] between donor and recipient have the same intention (e.g., the Bulgarian *Law of the Transplantation of Organs, Tissues and Cells*).[530]

---

[517] Art. 3 (1) a) Transplantation Act on Donation, Removal and Transplantation of Organs and Tissues and on the Amendment of some Acts.

[518] Sec. 5, 3) Transplantation of Organs and Tissues Act.

[519] Sec. 8 I No. 2 Act on the donation, removal and transplantation of organs.

[520] Sec. 208 Act on Health.

[521] Art. 12. 1., 3) The Cell, Tissue and Organ Recovery, Storage and Transplantation Act.

[522] Sec. 36 (1) (b) Act on Healthcare, Healthcare-Related Services and on the Amendment and Supplementing of Certain Laws.

[523] Art. 9, 2. Royal Decree 2070/1999.

[524] Sec. 5, 3) Transplantation of Organs and Tissues Act.

[525] Sec. 8 I No. 2 Act on the donation, removal and transplantation of organs.

[526] Art. 4, 1. and Art.8, 1. e) Royal Decree 2070/1999.

[527] Sec. 208 Act on Health.

[528] Art. 12. 1., 3) The Cell, Tissue and Organ Recovery, Storage and Transplantation Act.

[529] Histocompatibility literally means that tissues are compatible (MedicineNet (2003)). Explained in a more medical way, it defines the degree of antigenic similarity between the tissues of different individuals (U.S. National Library of Medicine (1999)). If tissue is transplanted from one person to another, it will be rejected if the recipient's immune system recognises the transplant as foreign (Porth (2010), p. 354). In case of full histocompatibility between donor and recipient, the immune system of the recipient will not consider the transplant to be foreign and therefore will not reject it (MedicineNet (2003)). The higher the degree of similarity of tissues, the better the prospect of success is (Bock (1999), p. 48 f.). However, the organ recipient has to take immunosuppressive drugs her whole life anyway to suppress the rejection of the received organ (Coppen (2010), p. 11; German National Ethics Council (2007), p. 9; Robert Koch-Institut (2003), p. 28). Therefore, organ donation makes sense only if it is expected that the system of the recipient will accept the transplant. Consequently, this kind of matching of donor and recipient seems desirable.

[530] Art. 25 No. 2 Law of the Transplantation of Organs, Tissues and Cells.

Several national laws only require the transplantation to have a therapeutic purpose without any further reference to the potential benefits for the recipient. For example, the Moldovan *Law on the Transplant of Human Organs, Tissues and Cells* enumerates the basic transplant principles. One principle states that transplantation is performed for the therapeutic benefit of the recipient.[531] The Moldovan law does not provide any additional information with regard to the expected success of the transplantation, nor do the other laws that reference the required therapeutic purpose of the transplantation.

On the one hand, the main purpose of organ transplantation is to cure a sick person. On the other hand, living organ transplantation is only possible if a healthy person is ready to donate an organ. It is thus essential to scrutinise whether the intended transplantation is expected to have success. Even if the national transplant law does not explicitly state the requirement of a therapeutic purpose, most countries consider organ removal pursuant to general criminal law illegal if no therapeutic benefit is intended.[532] Therefore, it is generally agreed upon that a donation is only permissible if it is expected to benefit the recipient.

c) Suitability of the Donor

Several countries explicitly require the suitability of the donor. Generally, every healthy person can act as living organ donor, whereas the suitability for each individual case is examined separately.[533] A person is primarily qualified as a living organ donor when he possesses the physical requirements for an organ donation. Moreover, he must be mentally qualified.[534] Absolute criteria for exclusion as living organ donor are especially diseases like aids, hepatitis, malignancy and septicaemia.[535]

Estonia[536] and Germany[537] rule that a person can be declared suitable to act as a living organ donor if she is in the appropriate state of health. In Estonia, this declaration has to be made by a doctor through laboratory testing.[538] In addition, if necessary, a doctor has the right to carry out supplementary testing to ascertain the suitability of a donor.[539] Germany, in contrast, does not state how to approach

[531] Ibid., Art. 3 b.

[532] Gutmann/Schroth (2002), p. 50.

[533] Bock (1999), p. 46.

[534] Schreiber (2004), p. 92.

[535] Bock (1999), p. 47; Roodnat et al., Vol. 89 Transplantation 1263, 1265 (2010); Schreiber (2004), p. 97 f.

[536] Sec. 8^1 (1) Handling and Transplantation of Cells, Tissue and Organs Act.

[537] Sec. 8 I No. 1 c) Act on the donation, removal and transplantation of organs.

[538] Sec. 8^1 (1) Handling and Transplantation of Cells, Tissue and Organs Act.

[539] Ibid., Sec. 8^1 (5).

this question, but it can be assumed that the procedure is similar to the Estonian approach.

The Swiss regulation does not only impose an obligation on the attending doctor to verify the suitability of the donor,[540] but also contains a list of people who are generally excluded to act as donors.[541] Pursuant to this, "any person who has received a transplant of organs, tissues or cells of animal origin or transplant products obtained there from" cannot act as an organ donor. Further, any person "whose organs, tissues or cells could transmit pathogenic agents or otherwise harm the health of the recipient" is excluded as organ donor. Art. 31, 2, c *Federal Act on the Transplantation of Organs, Tissues and Cells* that states that the Swiss Federal Council shall specify "the cases in which organs, tissues or cells may be transplanted in spite of the test results being reactive" applies notwithstanding.[542] The Swiss Federal Council listed such cases in Annex 5 of the *Transplantation Ordinance*. And Art. 23, 1 of the *Transplantation Ordinance* codifies that a surgeon with sufficient experience or a person trained for this exercise supervised by a surgeon, has to judge whether the person considered is suitable to act as donor. Annex 5, 1 *Transplantation Ordinance* contains a list with requirements for the suitability of the donor.

In the Czech Republic, organ removal is prohibited by law if there is a reason to presume that this might seriously endanger the donor's life or health. It is also forbidden if one suspects that the donor suffers from a disease or condition that might endanger the recipient's life or health. This ban does not apply, if the risk posed to the recipient's health is outweighed by the transplantation being life-saving for him.[543] Cases which only slightly infringe on the recipient's rights are excluded from this rule.[544]

The Hungarian *Act on Health* refers to the organ itself; it requires the organ to be suitable for transplantation. In addition, the donor must meet the conditions for organ transplantation and there must not be a medical contraindication of the transplant.[545]

The Polish law authorises the Health Minister to establish a decree stating the health requirements a person has to meet in order to become a living organ donor. In addition, he has to work out a list of medical examinations and auxiliary diagnostic tests that every potential organ donor should undergo; and the contraindications to organ donation, considering the living organ donor's state of health.[546] The Health

[540] Art. 30, 1 Federal Act on the Transplantation of Organs, Tissues and Cells.

[541] Ibid., Art. 30, 2.

[542] Ibid.

[543] Art. 3 (3) Transplantation Act on Donation, Removal and Transplantation of Organs and Tissues and on the Amendment of some Acts.

[544] Norba (2009), p. 291.

[545] Sec. 208 Act on Health.

[546] Art. 12, 6 Cell, Tissue and Organ Recovery, Storage and Transplantation Act.

Minister has already established a regulation that contains rules about medical diagnosis before and after LOD.[547]

In the United Kingdom, the practitioners concerned must decide after a full medical assessment whether the potential donor is medically fit and suitable to actually act as living organ donor.[548]

Several national transplant laws mention the suitability of the living organ donor as a specific requirement. The explicit regulations reveal that the amount of details provided differs.

d) Safety and Maximum Extent of Permissible Risk Taking by Donors

LOD can only be performed legally if the donor validly consents thereto, but the mere consent of the donor does not mean that this justifies every LOD.

It is generally agreed upon that LOD is not legal in cases in which it leads to the death of the donor or would be directly life-threatening.[549] Hence, LOD can usually not be justified if it contains a certain amount of risk. The impact the donor's consent has with respect to LOD is limited. Therefore, the legality of medical procedures does not only depend on the purpose of the procedure and on the consent of the person concerned, but the extent of harm is considered as well.[550]

While national laws agree that LOD is forbidden if it is life-threatening or even ceases the donor's life, the national laws differ with regard to the maximal amount of risk the donor is legally allowed to take.

Despite the lack of a specific law regulating LOD in Austria, their legal situation will be used to exemplify the issue just described. For organ removal, valid consent can only be given with respect to battery or threat to physical safety, not to manslaughter, though, since Sec. 77 *Criminal Code* penalises "homicide upon request."[551] Consequently, consenting to the removal of a pivotal organ is not possible, because this would mean consenting to one's own manslaughter, what is against the unavailability of life.[552]

[547] Information from D. Lewandowska.

[548] No. 59 Code of Practice 2.

[549] Dickens/Fluss/King, in Chapman/Deierhoi/Wright (ed.) (1997), p. 98; cf. German Federal Supreme Court [Bundesgerichtshof], 26 May, 2004, Case No. 2 StR 505/03, in Vol. 57 Neue Juristische Wochenschrift 2458, 2450 (2004); cf. Gutmann/Schroth (2002), p. 49.

[550] Price (2000), p. 246.

[551] Kert, Falcone Project, Austria, p. 3; cf. Brandstetter, in Brandstetter/Kopetzki (1987), p. 106; cf. Kienapfel (1990), p. 26 at 19; Kopetzki (1988), p. 251.

[552] Häckel (1997); Kert, Falcone Project, Austria, p. 3; cf. Kienapfel (1990), p. 26, at 19.

aa) European Level

In Europe as a whole, the *Additional Protocol on Transplantation of Organs* states in Art. 11: "[t]he removal may not be carried out if there is a serious risk to the life or health of the donor." According to the corresponding explanatory report, the transplant medical team looking after the donor, or the body authorising the donation, should assess the risk. If they come to the conclusion that there is a significant risk of death or long term severe disability to the donor, the donation procedure should not be carried out. While assessing the risks for the donor, physical risks and psychological factors are decisive. Generally, the donor's interests obtain priority, but in certain cases the donor's risks are considered in light of the potential benefits the recipient may receive.[553] Thus, the explanatory report defines the meaning of a serious risk and clarifies who is capable of judging this question.

bb) National Laws

The national laws considered contain rather different regulations with regard to the risk the donor is allowed to take.

Bulgaria,[554] Greece,[555] Hungary,[556] Lithuania,[557] Norway,[558] Spain,[559] Sweden[560] and Switzerland[561] do not allow the organ removal if it implies that the donor's life will be in danger. Bulgarian[562] and Hungarian[563] laws clarify that, as a result, only one of a paired organ or a part of a self-storing organ can be donated by a living donor. Additionally, Spanish law states that the function of the organ removed must be compensated by the organism of the donor in an adequate and sufficiently safe way.[564]

In England, the Law Commission states in its *Consultation Paper Consent in the Criminal Law* that the existence of risks to the donor has led to the differentiation between the uses of regenerative tissue, non-regenerative tissue that is essential for

---

553 Art. 11 Additional Protocol to the Convention on Human Rights and Biomedicine concerning Transplantation of Organs and Tissues of Human Origin, Explanatory Report.

554 Art. 24 (1) Law of the Transplantation of Organs, Tissues and Cells.

555 Canellopoulou-Bottis, Vol. 7 European Journal of Health Law 427, 433 (2000).

556 Sec. 209 (4) Act on Health.

557 Art. 10, 9 Law on Donation and Transplantation of Human Tissues, Cells and Organs.

558 Sec. 1 Act relating to transplantation, hospital autopsis and the donation of bodies etc.

559 Art. 3 e. and Art.9, 1. b. Royal Decree 2070/1999.

560 Sec. 5 Transplant Act.

561 Art. 12, c) Federal Act on the Transplantation of Organs, Tissues and Cells.

562 Art. 25, No. 1, No. 2 Law of the Transplantation of Organs, Tissues and Cells.

563 Sec. 209 (4) Act on Health.

564 Art. 3 e. and Art.9, 1. b. Royal Decree 2070/1999.

life and other non-regenerative tissue. Consent can only justify the removal of regenerative tissue and non-regenerative tissue that is not essential for life.[565]

Finland,[566] Greece,[567] Hungary,[568] Lithuania,[569] Norway,[570] Portugal,[571] Slovakia,[572] Sweden[573] and Switzerland[574] do not only consider the donor's life as a limitation. The removal of an organ that causes the donor any major health hazard or serious harm is also prohibited.

The Swedish regulation states that "[b]iological material intended for transplantation or some other medical purpose may not be taken from a living person if it is to be feared that the action will seriously endanger the donor's life or health."[575] Taking a look at the development of the Swedish legal situation with regard to transplantation reveals that the first transplant law in Sweden was already established in 1975, but has been amended a few times since then.[576] The provision concerned is essentially identical to Sec. 6 of the 1975 *Transplantation Act*. Thus, endangering the donor's life or health can be considered one of the fundamental principles that applies in all situations, no matter how strong the reason of the potential donor may be for her wish to donate.[577]

The regulation in Latvia is stricter than the Swedish rule, because it only allows a minimum risk to the donor's health.[578]

The Estonian national law compares the allowed risk resulting from LOD with the risks from any other surgical operation. It has to be ascertained, by way of medical investigations, that the risk to the life or health of the donor involved in the organ removal is not higher than the risk implied by any other surgical operation of the same degree of complexity.[579]

The Polish regulation works similarly by ruling that the risk involved in the transplantation procedure is not allowed to exceed the expected range that is

---

565 Law Commission (2012), p. 111 f.

566 Sec. 2 II Act on the Medical Use of Human Organs and Tissues.

567 Canellopoulou-Bottis, Vol. 7 European Journal of Health Law 427, 433 (2000).

568 Sec. 209 (4) Act on Health.

569 Art. 10, 9 Law on Donation and Transplantation of Human Tissues, Cells and Organs.

570 Sec. 1 Act relating to transplantation, hospital autopsis and the donation of bodies etc.

571 Art. 6, 7 Law on Harvesting and Transplanting Human Organs and Tissues.

572 Sec. 36 (1) (a) Act on Healthcare, Healthcare-Related Services and on the Amendment and Supplementing of Certain Laws.

573 Sec. 5 Transplant Act.

574 Art. 12, c) Federal Act on the Transplantation of Organs, Tissues and Cells.

575 Sec. 5 Transplant Act.

576 The Ministry of Health and Social Affairs Sweden 1997, The Swedish Transplant Act, Introduction, p. 1.

577 The Ministry of Health and Social Affairs Sweden 1997, The Swedish Transplant Act, Guidance on the provisions, Sec. 5.

578 Sec. 13 Law on the Protection of the Body of Deceased Human Beings and the Use of Human Tissues and Organs in Medicine.

579 Sec. 9 (1), 2) Transplantation of Organs and Tissues Act.

admissible for these types of procedures. In addition, the proceeding must not substantially harm the donor's health.[580]

Germany stipulates that the donor is exposed to no foreseeable risk beyond that of the operation. By mentioning the risk of the operation, the general risk of an operation is meant.[581] Thus, any threat the surgical intervention poses to the donor is decisive. This depends on the kind and extent of the operation.[582] Hence, LOD is prohibited if the risk of the surgery for the donor is increased, for example, if the patient has a disease.[583] However, several legal experts consider this regulation to be problematic because a possible living kidney donor might not be allowed to donate due to the increased risk involved with the operation, even though her general risk is still lower than the risk for a healthy living donor of a lung lobe, whose donation has been approved.[584] The legislator's intention is to minimise the risks for the potential donor by eliminating every danger that exceeds the inevitable risk connected to the operation.[585] Furthermore, the donor's health is explicitly taken into account because the donor should not suffer any serious damage to her health other than the direct results of the organ removal.[586]

Summing up, almost all national transplant laws explicitly set boundaries for the maximum risk a living donor is allowed to take. The fact that most of the national criminal codes prohibit "homicide upon request" verifies the initial hypothesis that it is common for LOD to be illegal when it ceases the donor's life or would be directly life-threatening. The extent of other restrictions placed on the donor regarding the risks she can legally take differs in the countries considered.

e) Risk-Benefit Equation

Because LOD involves two people and because a certain risk is always taken by the donor to benefit the recipient, European documents and some national laws have implemented a risk-benefit equation.

aa) European Level

The Council of Europe *Resolution (78) 29 on Harmonisation of Member States Relating to Removal, Grafting and Transplantation of Human Substances*

---

[580] Art. 12, 1., 4) The Cell, Tissue and Organ Recovery, Storage and Transplantation Act.

[581] Bundestag printed paper 13/4355 (2996), p. 21; Nickel/Schmidt-Preisigke/Sengler (2001), p. 95.

[582] Roche (2003), p. 1367.

[583] Schreiber (2004), p. 92.

[584] Gutmann, Vol. 15 Medizinrecht 147, 152 (1997).

[585] Schreiber (2004), p. 92 f.

[586] Sec. 8 I Nr. 1 c) Act on the donation, removal and transplantation of organs.

(*Resolution (78) 29*) is relevant here. The increase of organ transplantation was already visible in 1978, leading the Council of Europe to desire a harmonisation of the legislation on this subject-matter. With respect to the requirement of a risk-benefit equation, Art. 5 of the Council of Europe *Resolution (78) 29* states "[w]here removal of substances presents a foreseeable substantial risk to the life or the health of the donor, a removal may only be permitted exceptionally when it is justified by the motivations of the donor, the family relationship with the recipient and the medical requirements of the case." Thus, a risk-benefit equation is deemed reasonable.

bb) National Laws

Only the Belgian,[587] the Dutch,[588] the Slovakian[589] and the Slovenian[590] laws define the risk-benefit equation. Furthermore, in Austria, a risk-benefit equation is common medical practice.

The Slovenian rule is very general. It states that "the risk may not be disproportionate to the benefit the recipient is expected to obtain."[591]

The Slovakian risk-benefit equation is also worded generally. It assesses the recipient's benefits according to the donor's detriment, in opposition to the Slovenian rule, which assesses the recipient's benefit to the donor's risk.[592]

The Belgian[593] and the Dutch[594] rules are more specific. They only apply to cases in which an organ removal could have serious/permanent consequences for the donor's health. In Belgium, the risk-benefit equation is also relevant in cases where a non-regenerative organ is being removed. In those cases, an organ can only be removed if the recipient's life is at risk. Furthermore, the Belgian law requires that the transplantation of an organ donated by a deceased does not produce an equally satisfactory result.[595] In contrast, the Dutch provision does not refer to organs from deceased persons, but does additionally require that no other equally

---

[587] Art. 6, Sec. 1 Law on Retrieval and Transplantation of Organs.

[588] Sec. 3, 3 Organ Donation Act.

[589] Sec. 36 (1) (c) Act on Healthcare, Healthcare-Related Services and on the Amendment and Supplementing of Certain Laws.

[590] Art. 7 The Removal and Transplantation of Human Body Parts for the Purpose of Medical Treatment Act.

[591] Ibid.

[592] Sec. 36 (1) (c) Act on Healthcare, Healthcare-Related Services and on the Amendment and Supplementing of Certain Laws.

[593] Art. 6, Sec. 1 Law on Retrieval and Transplantation of Organs.

[594] Sec. 3, 3 Organ Donation Act.

[595] Art. 6, Sec. 1 Law on Retrieval and Transplantation of Organs.

Table 3 Summary of the issues concerning suitability of the LOD

Considered aspect	Result after comparing the countries
Medical evaluation	• A few national transplant laws require a medical check-up of donor and recipient prior to the organ transplantation; but a medical evaluation is always necessary to answer questions connected to the issue of suitability of the transplantation
Anticipated benefit for the recipient	• It is common for a LOD to only be permissible if a benefit for the recipient can be reasonably expected
Suitability of the donor	• Several countries mention this as a specific requirement, but the details of this requirement differ
Safety and maximum extent of permissible risk taking by donors	• It is generally accepted that LOD is illegal if it ceases the donor's life or is life-threatening; beyond this agreed standard, the national laws differ with regard to the maximal amount of risks the donor may take
Risk-benefit equation	• Only a few national transplant laws require a risk-benefit equation

suitable treatment option exists. Having a look at the practical implementation reveals that living kidney donation is still possible.[596]

In Austria, the donor's risk is correlated with the recipient's benefit, making the donor's risk a justifiable proportion of the recipient's benefit.[597] Consequently, the transplantation must be suitable to save the recipient's life or to improve her state of health on the one hand, while there are no harmful effects for the donor beyond the general risk of the surgery on the other hand. With respect to regulation, the Austrian Federal Ministry of Labour, Health and Social Welfare clarifies that medicine cannot guarantee that the medical intervention prior to the transplantation will be successful. Therefore, LOD is also tolerable in cases with only a small prognosis to succeed.[598]

All in all, only a few national laws require the performance of a risk-benefit equation.

f) Summary of the Issues Concerning Suitability of the Living Organ Donation

Some relevant issues regarding the suitability of LOD are regulated similarly, whereas others are regulated differently. The result of the comparison is roughly presented in Table 3.

[596] Price (2000), p. 257.

[597] Aigner, Vol. 9 Annals of Transplantation 11, 12 (2004); Hohmann (2003), p. 157.

[598] Hohmann (2003), p. 157.

4) Minors and Mentally Incapacitated Adults

Countries usually have basic rules that apply to all potential donors. In addition, the national laws regularly contain special regulations for minors and mentally incapacitated adults. This distinction is mainly based on one's ability to give consent or the lack thereof.[599] Minors and mentally incapacitated adults will be the subject of the upcoming section.

The intent of these particular requirements for minors and mentally incapacitated adults is to protect them because legal transactions typically do not only entail rights, but also duties.[600] For example, Austria's Sec. 21 (1) *Civil Code* states that minors are under the special protection of the law. Several national laws contain the same rule for minors and for mentally incapacitated adults. Some, however, treat these two groups differently. The following discussion will make the distinction as well, dealing (1) with minors and (2) then with incapacitated adults. It will be investigated whether LOD in these circumstances is legal at all, and, if it is legal, then how is it restricted.

a) Minors

In the countries considered, there is a consensus insofar as no country applies exactly the same rules to minors as it does to competent adults. This unequal treatment is explained by the fact that minors might not be able to evaluate the situation properly.[601] In addition, from a legal perspective, for consent to be valid, the person concerned must have the capacity to consent. As a result, each country has some sort of special regulation for minors, either for legal treatment in general or in the context of LOD.

There is also a widespread consensus about the age limit of minors. In virtually all of the countries taken into consideration, anyone 18 years or older is legally considered an adult. The only exception is Scotland, where the "threshold" is 16.[602]

The majority age is typically not defined in the national transplant laws, but rather in the Civil Codes, making it valid for the whole legal system and not only for the field of organ transplantation. For example, Sec. 2 of the *Civil Code* in Germany regulates that "[m]ajority begins at the age of eighteen." The regulation in Hungary is slightly different. Art. 12 of the Hungarian *Civil Code* adds the exception "unless they are married" to the general law, which states that people who have not yet reached the age of 18 years are deemed minors. In Sweden, the *Code of Parenthood and Guardianship* rules that 18 is the age at which minors become legally

[599] Cf. Cheyette, Vol. 41 Boston College Law Review 465, 468 (2000).

[600] Cf. Cheyette, Vol. 41 Boston College Law Review 465, 468 (2000); Eisenhardt (2007), p. 94; cf. Spickhoff, Vol. 54 World Medical Journal 9, 9 (2008).

[601] Leipold (2008), p. 127.

[602] Sec. 53 (17) Human Tissue (Scotland) Act 2006.

competent. Interestingly, minors between 15 and 18 are considered separately because they will probably have attained a certain degree of maturity. The decision about organ transplantation is considered as being very difficult. Therefore, it is reasonable that the custodian, and not the young person, should decide whether the transplantation considered is going to be performed or not. If the minor concerned cannot be deemed capable of giving informed consent due to insufficient maturity, then the rules of the *Code of Parenthood and Guardianship* are probably sufficient to enable the custodian to forbid the action with legal effect.[603]

The question if minors are allowed to act as living organ donors is answered unequally in the national laws considered. (1) Several countries completely prohibit minors to become living organ donors, while (2) some permit minors to become living organ donors under specific circumstances.

aa) Minors Are Never Allowed to Act as Living Organ Donors

Several national laws contain a regulation that completely prohibits minors from becoming living organ donors. In favour of completely banning minors from becoming organ donors, many argue that the ban provides legal certainty and maximises the legal protection of minors. Opponents of the ban claim that there might be cases in which the organ removal of a minor is ethically justifiable. In addition, it is opined that a complete prohibition does not take the differences between more or less mature minors into account.[604]

Austria,[605] Bulgaria,[606] the Czech Republic,[607] Estonia,[608] Finland,[609] Germany,[610] Greece,[611] Hungary,[612] Italy,[613] Lithuania,[614] the Netherlands,[615]

---

[603] The Ministry of Health and Social Affairs Sweden 1997, The Swedish Transplant Act, Guidance on the provisions, Sec. 8.

[604] Gutmann/Schroth (2002), p. 55.

[605] Bruckmueller/Schumann, in Roxin/Schroth (ed.) (2010), p. 845.

[606] Art. 24 (5), 1 Law of the Transplantation of Organs, Tissues and Cells.

[607] Art. 4 Transplantation Act on Donation, Removal and Transplantation of Organs and Tissues and on the Amendment of some Acts.

[608] Sec. 10 Transplantation of Organs and Tissues Act.

[609] Sec. 5 Act on the Medical Use of Human Organs and Tissues.

[610] Sec. 8 I Nr. 1 a) Act on the donation, removal and transplantation of organs.

[611] Canellopoulou-Bottis, Vol. 7 European Journal of Health Law 427, 433 (2000).

[612] Sec. 206 (1) Act on Health.

[613] Art. 1 & 2 Living-donor kidney transplant.

[614] Art. 10 Law on Donation and Transplantation of Human Tissues, Cells and Organs.

[615] Sec. 5 Organ Donation Act.

Moldova,[616] Portugal,[617] Scotland,[618] Slovakia,[619] Spain[620] and Switzerland[621] prohibit minors from acting as living donors for solid organs. Selected regulations are presented and explained in the following text.

In Switzerland, living kidney donations are often carried out between family members. This leads to the danger that one could be pressured into becoming an organ donor by other family members, making it very difficult for her to decline the possible LOD. It is assumed that a mature person can usually handle such a situation better than an immature person who depends on the family to a greater degree.[622] In addition, in the *Medical-Ethical Guidelines and Recommendations about Live Donations of Solid Organs*, The Swiss Academies of Arts and Sciences refers to living organ donors who are of full age, but are still young adults. If such a young adult intends to donate an organ, her ability to judge has to be thoroughly reviewed.[623]

The arguments in Germany brought forward in favour of this ban are similar to the Swiss justification.[624]

In Scotland, a living minor can only legally donate an organ if it is the result of a domino transplantation;[625] the organ donated by a living minor cannot be for a single transplantation only.

In Austria, the question of whether the minor has the capacity to understand and the ability to judge the situation has to be assessed individually in each case. The organ removal is a non-therapeutic intervention and does not improve the minor's well-being.[626] Because of this it is argued that not even he himself can consent to

[616] Art. 15 (2) Law on the Transplant of Human Organs, Tissues and Cells.

[617] Art. 6, 4 Law on Harvesting and Transplanting Human Organs and Tissues.

[618] Sec. 17 Human Tissue (Scotland) Act 2006.

[619] Sec. 36 (2) Act on Healthcare, Healthcare-Related Services and on the Amendment and Supplementing of Certain Laws.

[620] Art. 4, a) and Art.9, 1. a) Royal Decree 2070/1999.

[621] Art. 12, a) Federal Act on the Transplantation of Organs, Tissues and Cells.

[622] Swiss Dispatch of 12. September 2001 on a Federal Law on Transplantation of Organs, Tissues and Cells, p. 145.

[623] The Swiss Academies of Arts and Sciences, Live donations of solid organs, Medical-ethical guidelines and recommendations, p. 12.

[624] Gutmann, in Schroth et al. (ed.) (2005), Sec. 8, at 7.

[625] Sec. 17 Human Tissue (Scotland) Act 2006. A transplantation is considered a domino transplantation if an organ that failed was removed and replaced by an organ donated, and the parts of the removed organ that are still healthy are transplanted into another person. Cardiac valves, usually donated as part of a liver removal, are an example of such a domino transplantation (Swiss Federal Office of Public Health (2011)). A second example is lung transplantations, where it can be necessary to transplant a heard along with the lung. The removed heard that is still healthy can then be donated to another person (Cf. Keller, Vol. XXXII Stetson Law Review 855, 870 (2003); Nickel/Schmidt-Preisigke/Sengler (2001), p. 90; Swiss Federal Office of Public Health (2011); Nickel/Schmidt-Preisigke/Sengler (2001), p. 90).

[626] Kalchschmid (1997), p. 199 f.; Kopetzki, in Brandstetter/Kopetzki (ed.) (1987), p. 89.

the organ removal, even if he is capable of judging the situation.[627] Some legal scientists argue that parents should not be able to give consent on behalf of their child, even if the LOD could save the life of their other child. They conclude that LOD by a minor is not possible.[628] In contrast, other legal experts permit an exception from this policy. Minors should be able to become organ donors in cases where the organ donation is for the benefit of a sibling or twin. They claim exceptions are possible for people who are only partially capable of consenting.[629] Since this type of LOD contains the danger of a conflict of interest for the parents of the potential donor and the recipient, many recommend that consent be given by an independent authority.[630]

All in all, several countries completely exclude minors from the possibility of acting as living organ donors, without specifying any exceptions related to special circumstances.

bb) Minors Are Allowed to Act as Living Organ Donors Under Specific Circumstances

Belgium, England, Norway and Sweden do not exclude minors a priori as living donors. However, they still are not on the same level as competent adults. Rather, these countries define the specific circumstances that have to be met to justify LOD by a minor.

The Belgium *Law on Retrieval and Transplantation of Organs* requires the removal to be free of any serious consequences for the donor, the removed organ must be regenerative and the organ donation must be for the benefit of a donor's sibling.[631] Since 2002 all three requirements have to be fulfilled cumulatively, since they are connected via "and," while the word "or" was used before.[632] This change enhanced the protection of minors. In Belgium, the term "regenerative organ" includes living donation of a part of the liver.[633] With regard to the necessary consent, the minor's exact age is decisive. If she is between 12 and 18 years old, she must give her consent. If she is not yet 12 years old, she must have the opportunity to give her opinion. However, if she is not able to express her will due to her underdeveloped mental capacity, her parents or her legal guardian must give explicit consent.[634]

[627] Kalchschmid (1997), p. 200.

[628] Bruckmueller/Schumann, in Roxin/Schroth (ed.) (2010), p. 845.

[629] Burgstaller/Schütz (2010), Sec. 90 at 130.

[630] Kalchschmid (1997), p. 200.

[631] Art. 7, Sec. 1 Law on Retrieval and Transplantation of Organs.

[632] Nys, The Belgian Law on retrieval and transplantation of organs, p. 6 (Power Point Presentation).

[633] Information from R. Wilmotte.

[634] Art. 7, Sec. 2 Law on Retrieval and Transplantation of Organs.

In Norway, the donor is normally at least 18 years old. If he is younger, he may donate "[w]hen special circumstances so indicate."[635] The minor concerned must consent. The guardian and the person with parental custody have to give their approval. In addition, the approval of the Directorate of Health Services is necessary.[636]

In England, each case must be judged individually to determine if a minor is capable of consenting or not.[637] If a minor is considered to have the capacity to consent, then LOD is not impossible, but would still be an exceptional case.[638] In case a minor is incapable of consenting, parental responsibility applies.[639] Nevertheless, for decisions about non-therapeutic operations — like LOD—an approval of the court is considered obligatory.[640] With regard to the material terms of reference, the best-interest-principle has to be applied.[641] While coming to a decision on whether a LOD would be in the best interest of the minor concerned, all the evidence available has to be taken into account. Not only medical interests, but also potential emotional, psychological and social factors should be considered.[642] Additional details about the examination, namely, whether an act is in the best interest of the person concerned, are stated in Sec. 4 *Mental Capacity Act 2005*. According to this regulation, the person's past and present wishes and feelings have to be regarded. One must also attempt to find out what the person concerned would do if he did have the ability to consent for himself. It is not clear, however, if an English court would actually accede to an intended organ removal even in cases where both the minor and the parents consented to it.[643] Thus, even if the legal framework in England does not generally rule out the possibility of LOD by minors, this procedure cannot be considered a common practice.

In Sweden, a minor can act as a living organ donor if the donor and the intended recipient are related and if it is not possible to obtain medically suitable biological material from some other person. In addition, the National Board of Health and Welfare has to give permission, which the custodian or guardian has to apply for. This application has to be approved by the physician entitled to decide whether permission should be conferred or not.[644] According to the *Guidance on the provisions*, the requirement that "medically suitable biological material cannot be taken from anybody else" means that there is no other donor who matches genetically or is equally suitable. This is also the case if there is indeed another medically

[635] Sec. 1 Act relating to transplantation, hospital autopsis and the donation of bodies etc.

[636] Ibid.

[637] Fateh-Moghadam (2008), p. 204 ff.

[638] Ibid., p. 206.

[639] Sec. 2 Children Act 1989.

[640] Fateh-Moghadam (2008), p. 207.

[641] Ibid.

[642] Appendix A, A9 Code of Practice 2.

[643] Information from D. Price.

[644] Sec. 8 Transplant Act.

suitable donor who refuses to donate the biological material.[645] "Related" requires the donor and recipient to be related in the formal sense — this also includes adopted children.[646] Consent has to be provided by a custodian or guardian. If the donation is intended from one sibling for transplantation to another, then a guardian — who has been appointed to the minor — has to consent. Since "the action may not be taken against the wishes of the donor," the minor has a veto in every case, even when she has absolutely no understanding of its meaning or consequences.[647] The National Board of Health and Welfare has to prove whether the prerequisites of the provision are satisfied. If the minor seeks to donate non-regenerative biological material, which applies to living kidney donation and to living liver donation, exceptional reasons are needed to grant permission.[648] Permission may be granted if the recipient may die if the transplantation does not take place and there is no other conceivable donor.[649] Statistics show that no donation of a solid organ by a minor has been approved yet, even though the possibility is granted by law.[650]

All in all, the national laws disagree with regard to the question of whether minors are allowed to act as living organ donors. The majority of laws completely prohibit minors from becoming living organ donors.

b) Mentally Incapacitated Adults

The countries considered usually have special regulations for mentally incapacitated adults. Those rules are often similar to those for minors, so references to the above-mentioned information are often made in the following discussion. In Bulgaria, for example, according to the *Law about the Person and the Family*, the regulations about minors always apply to mentally incapacitated adults as well in case no special rule exists.[651] Independently of the countries considered, the *WHO Guiding Principles on Human Cell, Tissue and Organ Transplantation* applies the same principles to both minors and mentally incapacitated adults. Guiding Principle 4 states: "[w]hat is applicable to minors also applies to any legally incompetent person."

Before dwelling on the requirements of organ donation for mentally incapacitated adults, it must be clarified what it means to be a mentally incapacitated adult. With respect to minors, the legal systems clearly determine a specific age at which people are considered to attain majority. Regarding mentally

---

645 Ibid.

646 Ibid.

647 Ibid.

648 Ibid.

649 The Ministry of Health and Social Affairs Sweden 1997, The Swedish Transplant Act, Guidance on the provisions, Sec. 8.

650 Information from C. Möller.

651 Information from C. Nikolova.

incapacitated adults, the situation is not as clear. The national transplant laws usually do not contain a definition for a "mentally incapacitated adult." Thus, the general national civil laws have to be observed. For example, Sec. 104 of the German *Civil Code* states that "[a] person is incapable of contracting if he is in a state of pathological mental disturbance, which prevents the free exercise of will, unless the state by its nature is a temporary one." In Austria, legal capacity means to have the ability to authorise and obligate oneself by legal transactions.[652]

It should also be mentioned that several countries differentiate not only between fully capable adults and those who are mentally incapable, but also between an established third category of adults, namely those whose loss of discretionary ability is only partial. For instance, Hungary has a special system with three levels of competence/capacity.[653]

The laws reveal that there is no general standard for addressing mentally incapacitated adults and whether or not they should be able to becoming living organ donors. Should they be excluded from this procedure categorically, or, if they should be allowed to act as living organ donors, then under what circumstances? (1) Several countries completely prohibit adults deemed incapacitated to act as living organ donors, while (2) other national laws allow incapacitated adults to act as living organ donors.

aa) Mentally Incapacitated Adults Are Never Allowed to Act as Living Organ Donors

Bulgaria,[654] the Czech Republic,[655] Estonia,[656] Finland,[657] Germany,[658] Hungary,[659] Italy,[660] Lithuania,[661] Moldova,[662] the Netherlands,[663] Portugal,[664]

---

652 Barta (2004).

653 P. 36.

654 Art. 24 (6) Law of the Transplantation of Organs, Tissues and Cells.

655 Art. 4 Transplantation Act on Donation, Removal and Transplantation of Organs and Tissues and on the Amendment of some Acts.

656 Sec. 10 (1) Transplantation of Organs and Tissues Act.

657 Sec. 5 I 1 Act on the Medical Use of Human Organs and Tissues.

658 Sec. 8 I, 1, No. 1 a) Act on the donation, removal and transplantation of organs.

659 Sec. 206 (1) Act on Health.

660 Art. 2 Living-donor kidney transplant.

661 Art. 10, 4 Law on Donation and Transplantation of Human Tissues, Cells and Organs.

662 Art. 15 (2) Law on the Transplant of Human Organs, Tissues and Cells.

663 Sec. 4 Organ Donation Act.

664 Art. 6, 5 Law on Harvesting and Transplanting Human Organs and Tissues.

Scotland,[665] Slovakia,[666] Spain[667] and Switzerland[668] explicitly state in their transplant laws that the living donor must be mentally competent. Therefore, all these countries do not only completely prohibit minors to act as living organ donors, but also mentally incapacitated adults.

Requiring a mentally competent donor means that both persons fully incapable and persons temporarily or partially incapable are excluded from acting as donors. Even though this seems to apply anyway, Lithuania explicitly prohibits the removing of organs "from fully or partially incapable donors."[669] The Spanish regulation produces a similar result, since "the donor must have full mental faculties." Thus, Spain also excludes persons whose loss of capability is only partial from becoming a living organ donor.[670]

The Czech Republic excludes fully incompetent persons from becoming living organ donors, as well as persons with limited legal competence. In addition, it also precludes "persons who have expressed their consent to the donation, but, with regards to their current medical condition, it can be reasonably presumed that despite full advice that they did not, or could not, assess all consequences relating to removal of regenerative tissue for their own health."[671]

In Hungary, the *Act on Health* requires the organ donor to have legal capacity. As stated above, civil law in general has three levels of competency.[672] 15/A *Civil Code* states that legal statements made by incompetent persons shall be null and void and that their guardian shall proceed on their behalf. This general legal rule leads to the same result as the regulation in the *Act on Health*, since both exclude adults that are mentally incompetent from consenting to a LOD. Prior to making a decision, the guardian hears the views and requests of the person in his custody. If that person is of sound mind, then the guardian abides by his requests if possible.[673] People who exhibit only a partial loss of discretionary ability and who are under guardianship are able to make their own legal statements in all matters where the court did not limit their competency in its ruling restricting legal competency. Health care rights, if the legal competency of persons is placed under guardianship, may be completely restricted by court order.[674] Thus, with regard to partially incapacitated potential donors, the Hungarian *Act on Health* and the Hungarian

[665] Sec. 17 (6) (a) (ii) Human Tissue (Scotland) Act 2006.

[666] Sec. 36 (2) Act on Healthcare, Healthcare-Related Services and on the Amendment and Supplementing of Certain Laws.

[667] Art. 4, b) and Art.9, 1. a) Royal Decree 2070/1999.

[668] Art. 12, a) Federal Act on the Transplantation of Organs, Tissues and Cells.

[669] Art. 10, 4 Law on Donation and Transplantation of Human Tissues, Cells and Organs.

[670] Art. 4, b) and Art.9, 1. a) Royal Decree 2070/1999.

[671] Art. 4 Transplantation Act on Donation, Removal and Transplantation of Organs and Tissues and on the Amendment of some Acts.

[672] P. 36.

[673] Sándor/Bárd (2009), Hungary, p. 9 f.

[674] Art. 14, sec. (6) Civil Code.

Civil Code achieve a similar result. In both cases, the potential donor cannot give valid consent.

The Human Tissue (Scotland) Act 2006 contains the same rule for mentally incapacitated adults as it does for minors. Both groups are only permitted to become organ donors if they are participating in a domino organ transplant operation.[675]

In summary, it is clear that several countries completely prohibit mentally incapacitated adults from donating organs while being alive. Yet, as the upcoming section will show, other countries have more lenient legal guidelines when it comes to the LOD of adults incapable of giving valid consent.

bb) Mentally Incapacitated Adults Are Allowed to Act as Living Organ Donors Under Specific Circumstances

In Austria, Belgium,[676] England[677] and Sweden,[678] mentally incapacitated adults are allowed to act as living organ donors under specific circumstances. For the most part, the same requirements that are applicable in the case of LOD by minors are also applicable in the case of incapacitated adults.

In Austria, general rules have to be applied when addressing the question of whether a mentally disabled person might become a living organ donor. The donor's capacity to understand and his ability to make good judgments should be individually assessed in every case.[679] If the potential donor has the capacity to understand and the ability to judge, she can consent for herself, whereas the custodian's consent is needed for a severe medical intervention, such as LOD. If the person concerned is unable to consent, the custodian can consent after being sufficiently informed, obtaining an authorisation by the court and if the granting the decision is for the potential donor's well-being. Since the organ removal does not enhance the well-being of the potential donor, the custodian cannot give valid consent in that case.[680] Thus, a mentally incapacitated adult is allowed to become a living organ donor if she has the capacity to understand and the ability to judge in the individual case concerned, and if the custodian also consents in that case. If the potential donor does not have this capacity, then the custodian is also unable to consent on her behalf, since the organ removal would not be in accordance with the well-being of the donor.

In Belgium, mentally incompetent adults have been able to legally become living organ donors since 2007.[681] The requirements that have to be met in order

[675] Sec. 17 (6) (a) (ii) Human Tissue (Scotland) Act 2006.

[676] Art. 6 Sec 2 Law on Retrieval and Transplantation of Organs.

[677] Appendix A-A2-A7 Code of Practice 2.

[678] Sec. 8 Transplant Act.

[679] Kalchschmid (1997), p. 201 f.

[680] Ibid.

[681] Information from R. Wilmotte.

to donate an organ are different than those for minors, though. The general rule for LOD by an adult states that a removal that puts the donor at risk, or that relates to a non-regenerative organ, is only legal if the recipient's life is at risk and transplanting an organ of a deceased does not produce an equally satisfactory result. This rule also applies to mentally incapacitated adults. In addition, the requirement to consent is considered with respect to incompetent adults. In this case, the adult's legal representative or the patient-designated representative may give consent,[682] and if there is none, or if he fails to act, the cohabiting spouse, the legally cohabiting partner or the actually cohabiting partner can exercise the rights of the incapacitated adult. If this person refuses or if there is no such person, an adult child, a parent or an adult sibling of the patient, can, in descending order, assess the rights of the adult concerned. If these people refuse or if no such persons exist, the health professional involved has to exercise the patient's interests, possibly after a multidisciplinary consultation. This also occurs if there is a conflict between two or more representatives of equal rank.[683] Interestingly, according to some legal experts in Belgium, the fact that non-regenerative organs may be removed from incompetent adults could put the Belgium law in conflict with international law. Specifically, Art. 20 of the *Convention on Human Rights and Biomedicine* only allows the removal of regenerative tissue from incompetent adults, and that is only allowed under very strict conditions.[684]

With respect to the Swedish regulation, the *Guidance on the provisions* clarifies that the term "mental disturbance" is used in legislation on compulsory psychiatric care, penal law, civil law and procedural law. In the transplantation field, the category also includes persons, whose mental disturbance consists solely of retardation. The mental disturbance and the capacity for granting consent have to be connected somehow. Thus, situations might arise in which a person suffering from a mental disturbance may be perfectly capable of understating the implications of consent. Therefore, the person who obtains consent, say the physician who has to decide on the action, has to independently clarify if the prospective donor is capable of understanding the implication of consent. If it is not clear without ambiguity that the person with a mental disturbance understands the implication of his consent, permission must be obtained from the National Board of Health and Welfare. Until 2010 no permissions were granted for mentally disturbed persons, though.[685] Thus, neither minors, nor mentally incapacitated adults have acted as living donor of a solid organ yet, even though it is legally allowed.

In England, Sec. 1 (5) *Mental Capacity Act 2005* states that "[a]n act done, or decision made, under this Act for or on behalf of a person who lacks capacity must be done, or made, in his best interests." Applying this to LOD means that it has to be

---

682 Art. 6, Sec. 2 Law on Retrieval and Transplantation of Organs.

683 Nys (2009), p. 103.

684 Nys, The Belgian Law on retrieval and transplantation of organs, p. 5 (Power Point Presentation).

685 Information from C. Möller.

evaluated for each case individually whether the particular LOD would be in the best interest of the mentally incapacitated adult. Details about the best-interest-principle have already been given in the examination of minors becoming living organ donors. To recapitulate, the conclusion was that a LOD is permitted if an objective examination shows that it would be in the best interest of the person concerned. The best interest is not limited to the best medical interest. It can also be answered affirmatively if the surgery involved risks and disadvantages for the person's health and takes place predominantly for the interest of a third person. Hence, LOD by mentally incapacitated adults is not impossible in general.[686]

The countries have thus no common standard for mentally incapacitated adults and their right to act as living organ donors.

c) Summary of Minors and Mentally Incapacitated Adults

All in all, several countries completely exclude minors as well as mentally incapacitated adults from becoming living organ donors. A few countries do permit minors to act as living organ donors. This applies to mentally incapacitated adults as well. All countries that allow this, however, only do so under special circumstances.

Figure 1 shows the countries that allow minors and/or mentally incapacitated adults to act as a living organ donor in proportion to those that do not allow that.

d) European Tendencies Concerning Minors and Mentally Incapacitated Adults

Recent European documents also address the issue of how to treat minors and mentally incapacitated adults with regard to living organ transplantation.

The *Convention on Human Rights and Biomedicine* states in Art. 20 that "[n]o organ or tissue removal may be carried out on a person who does not have the capacity to consent under Article 5." An exception to this strict ban is only provided for regenerative tissues, not for organs, though. The regulation in the *Additional Protocol on Transplantation of Organs* is identical.

The *Recommendation on Liver Transplantation from Living Donors* includes the same ban without any exceptions for living liver donation.[687]

Thus, these documents prefer to completely exclude minors as living organ donors.

---

[686] Fateh-Moghadam (2008), p. 212.

[687] Appendix iv Recommendation No. R (97) 16 on Liver Transplantation from Living Donors, Council of Europe, Committee of Ministers.

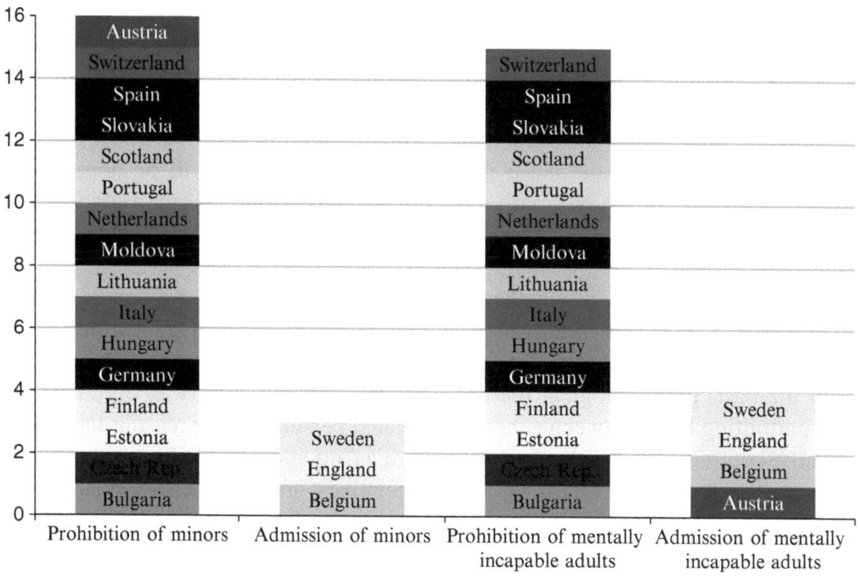

Fig. 1 Admission and prohibition of minors and mentally incapacitated

5) Restrictions on Living Organ Donation by Competent Adults

As mentioned in the last section that focussed on minors and mentally incapacitated adults, the general rules of organ transplantation established by the national laws apply to competent adults. The countries regulate the issues of LOD differently for adults, for example, the details about consent. In addition to the issues already described, the countries have different requirements, especially regarding the relationship between donor and recipient. (1) An overview about the debate on this aspect will be given, (2) before presenting the national rules.

a) Debate on Restriction of Donor-Recipient Relationship Required for Living Organ Donation

The national transplant laws considered do not have a common standard regarding the required donor-recipient relationship. This aspect is noteworthy, since regulations that confine the range of possible living organ donors or even prohibit certain persons from becoming living donors for particular recipients affect the total amount of LODs and even impede medically possible LODs in individual cases.[688]

Whether a certain relationship between donor and recipient is required, and what constitutes this relationship, is answered differently. The debate about the

[688] Cf. Gutmann/Schroth (2002), p. 60.

restriction of the donor-recipient relationship does basically contain the following two questions: Is a genetic relationship necessary or is an emotional relationship sufficient? Should no relationship at all be demanded? The arguments brought forth in favour of the different possibilities, as well as those opposing them, will be roughly presented below in a descriptive, neutral way. A more detailed presentation of the debate and a judging of the arguments will be done in the chapter "Analysis of the Normative Arguments That Dominate the Policy Arena About Necessity and Legitimacy of Legal Restrictions in Living Donor Transplantation."

A prominent debate is whether LOD should only be allowed in a genetic relationship. This means that a mere emotional relationship (e.g., partner, close friend)[689] between a potential donor and a potential recipient is insufficient.

In the past, immunological reasons were brought forward as an argument for only permitting LOD between genetically related persons. These reasons have been recently contradicted by the fact that there is no medical difference anymore between the results of LODs of genetic relatives and non-genetic-relatives.[690] That organs with low tissue compatibility can lead to very good outcomes is essentially the result of the invention of new pharmaceuticals. The differences of the functionality between organs donated by genetically related persons or by those with low tissue compatibility shrink increasingly.[691] This particularly applies to kidney donation, which is the most important type of organ donation in terms of it being the most often donated organ.[692]

In favour of permitting LOD between persons that are only emotionally related, many argue that there is no evidence showing the potential of coercion being higher than in cases of LOD between genetically related persons.[693]

In addition, the sociological changes that have been affecting our society are cited. Experts claim that a large amount of people do not live in "classic family connections" anymore, making it very difficult for these people to find a living organ donor if a genetic relation is required. Many believe that close partnerships without a formal marriage should not be evaluated as lesser than blood relationships.[694]

---

[689] Cronin, Vol. 343 British Medical Journal (2011); Thiruchelvam, Vol. 343 British Medical Journal (2011); Zuidema et al., in Weimar/Bos/Busschbach (ed.) (2008), p. 306.

[690] Cf. American Medical Association, Vol. 284 The Journal of the American Medical Association 2919, 2923 (2000); Bock (1999), p. 116; Gutmann/Schroth (2002), p. 60 f.; Hamza et al., Vol. 45 Der Urologe 60, 64 (2006); Matas et al., Vol. 343 The New England Journal of Medicine 433, 433 (2000); Ross et al., Vol 74 Transplantation 418, 419 (2002); Swiss Dispatch of 12. September 2001 on a Federal Law on Transplantation of Organs, Tissues and Cells, p. 98.

[691] Cf. Breyer/Kliemt, in Höfling (ed.) (2008), p. 31; Schutzeichel (2002), p. 76 f.

[692] Edelmann, Vol. 50 Versicherungsrecht 1065, 1066 (1999); Heuer/Conrads, Vol. 15 Medizinrecht 195, 201(1997); Gutmann/Schroth (2002), p. 60 f.

[693] Cf. Gutmann, Vol. 27 Zeitschrift für Rechtspolitik 111, 112 f. (1994).

[694] Cf. Gutmann, Vol. 27 Zeitschrift für Rechtspolitik 111, 113 (1994); Heuer/Conrads, Vol. 15 Medizinrecht 195, 201 (1997).

Another important discussion focuses on whether any kind of relationship — genetic or emotional — is necessary to donate an organ while alive, or whether LOD should be allowed without any pre-existing relationship between donor and recipient.

Many consider taking a healthy organ from a healthy person for the purpose of LOD to be against the principle of nonmaleficence.[695] Opponents claim, however, that forbidding unspecified LOD because it violates the principle of nonmaleficence is inconsistent because a healthy donor is concerned in an equal way if he donates for the benefit of a relative.[696]

In favour of restricting LOD to relatives, possible risks for the donor's health in general are mentioned.[697] Opponents reply that the risks for the donor's health are equally high if the donation is for the benefit of a completely unrelated recipient or if it is for the benefit of a relative.[698]

The discussion in Germany will be used as an example. The German Bundestag argues that a restriction of the donor-recipient relationship guarantees the voluntariness of the LOD and that a general liberalisation of the circle of donors entails the danger of organ trade.[699] Several academics disagree with the aspect that a relationship guarantees the voluntariness of LOD. They state that the donor might be pressurised within the family as well,[700] because family members might express a certain expectation towards the potential donor or his family status might be affected seriously if he refuses to donate. As a consequence, it is considered to be very difficult for the potential donor to disengage himself from this family pressure.[701] Other arguments against the German Bundestag claim the voluntariness and the absence of financial incentives can be assured for all types of LOD, for example, the donation or allocation of the organs by a central authority such as Eurotransplant could be anonymous, making the mentioned risks non-existent.[702] In addition, prosecuting organ trade as a criminal offence is generally mentioned as a safeguard to assure that the potential costs of organ trade outweigh the benefits.[703]

One argument by the German Federal Constitutional Court is similar to the German Bundestag's argument. According to the court, which is usually not very

---

[695] Cf. Bundestag printed paper 15/5050 (2005), p. 35.

[696] Cf. Fateh-Moghadam (2008), p. 265 f.

[697] Heuer/Conrads, Vol. 15 Medizinrecht 195, 201 (1999); Ross, Vol. 30 The Journal of Law, Medicine and Ethics 440, 440 (2002).

[698] Matas et al., in Gutmann et al. (ed.) (2004), p.195; Spital, in Gutmann et al. (ed.) (2004), p. 136.

[699] Bundestag printed paper 13/4355 (1996), p. 20.

[700] Bock (1999), p. 118; Erim et al., Vol. 35 Transplantation Proceedings 909, 909 (2003); Heuer/Conrads, Vol. 15 Medizinrecht 195, 201 (1997); Keller, Vol. XXXII Stetson Law Review 855, 869 (2003); Lilie/Krüger, Vol. 81 Der Chirurg 787, 788 (2010).

[701] Schutzeichel (2002), p. 108 f.

[702] Cf. Hoyer, Vol. 76 Transplantation 1264, 1265 (2003); German Federal Constitutional Court [Bundesverfassungsgericht], 11 August, 1999, Case No. 1 BvR 218/98, in Vol. 46 Neue Juristische Wochenschrift 3399, 3403 (1999).

[703] Schroth, in Gutmann et al. (ed.) (2003), p. 138 f.

paternalistic,[704] they must ensure that people are protected from significantly harming themselves personally.[705] This has often been criticised and is countered with the arguments that, first, protecting people from harming themselves is not legitimate and, second, paternalism by law is problematic in this case, since a prohibition to donate does not only affect the potential donor, but also another person, namely the organ recipient.[706]

Many argue that people have different motives for donating an organ while being alive. A complete ban of unspecified LOD would lead to the nullification of several good intentions.[707] Opponents are sceptical particularly with regard to donations for unknown recipients.[708] The proponents of an unrestricted donor-recipient relationship, in contrast, claim that empirical evidence shows that people who donate organs to unspecified recipients are "very self-directed and without psychopathology." Accordingly, mistrust only arises because the majority does not aspire to those altruistic principles themselves.[709]

Those opposed to the requirement of a specific relationship between donor and recipient bring out that the original relationship between donor and recipient might change after the transplantation.[710] Even though the donation might have had altruistic motives, the recipient will usually feel obligated to show gratitude towards the donor. This feeling can even be so strong that the intended recipient rejects the LOD offered by an emotional or genetic relative. With respect to the recipient's gratitude, it is possible for the donor to be disappointed if the recipient is not as thankful as expected.[711] With regard to life partners, the partnership could even end, which might lead to the donor regretting his donation. A failed transplantation could also have negative consequences for the relationship. In addition, the organ recipient might feel pressured after the transplantation, particularly if the donor watches how he *treats* the donated organ.[712] The opponents of the restriction of the donor-recipient relationship argue that all these aspects of pressure or any sense of obligation are not possible with regard to LOD if the recipient is unknown.[713]

[704] Gutmann, in Schroth et al. (ed.) (2006), p. 254 f.

[705] German Federal Constitutional Court [Bundesverfassungsgericht], 16 March, 1982, Case No. 1 BvR 938/81, in BVerfGE 60, 123, 132; German Federal Constitutional Court [Bundesverfassungsgericht], 11 August, 1999, Case No. 1 BvR 218/98, in Vol. 46 Neue Juristische Wochenschrift 3399, 3401 (1999).

[706] Gutmann, Vol. 52 Neue Juristische Wochenschrift 3387, 3388 (1999).

[707] Gutmann/Schroth, in Oduncu/Schroth/Vossenkuhl (ed.) (2003), p. 276.

[708] Price (2010), p. 213; Taylor et al., in: Gutmann et al. (ed.) (2004), p. 206.

[709] Price (2010), p. 214.

[710] Cf. Schutzeichel (2002), p. 188 ff.

[711] Ibid.

[712] Ibid.

[713] Spital, in Gutmann et al. (ed.) (2004), p. 136.

This approximate presentation of the debate on a possible donor-recipient relationship makes clear that several arguments exist for each position and explains why the countries approach the aspect in several different ways.

In order to compare the national regulations, the countries considered have to be placed into groups based on whether and to what extent they require a relationship between donor and recipient. None of the considered countries restricts LOD to those who are genetically related anymore. (1) Thus, the classification has to consider the fact that several countries restrict LOD to relatives — genetic and emotional. Within this group of countries, the regulations are still fairly different, making a further categorisation necessary. (a) Several countries list the allowed donor-recipient relationships exhaustively. (b) Others, in contrast, contain a list of allowed relationships, but additionally include an *open clause* that makes an extension of the explicitly stated relationships possible. (2) In contrast, a considerable amount of countries do not restrict the donor-recipient relationship at all. Thus, on principle, those countries allow a LOD for any recipient. However, different procedural safeguards have been established. These countries will be subdivided according to their procedural safeguards.

Following the presentation of the different restrictions regarding the donor-recipient relationship, *special cases* associated with the circle of possible donors are explained. (1) The first case considered is whether a cross-over LOD is allowed. (2) The second possibility taken into consideration is the unspecified LOD. After explaining those types of LOD, the lawfulness of those LODs in the considered countries is analysed.

b) Restrictive Measures

This section compares the extent to which national regulations require a donor-recipient relationship. (1) The countries that require a certain donor-recipient relationship for LOD are presented and afterwards (2) the countries that do not restrict the donor-recipient relationship.

aa) Restriction of the Donor-Recipient Relationship

In the countries that require a specific relationship between donor and recipient, LOD is illegal if there is no relationship between donor and recipient.

The details with regard to the required relationships differ. (1) Some countries define the allowed relationships exactly, (2) while others contain an open clause in addition to the listed relationships.

(1) Exhaustive Enumeration of Donor-Recipient Relationship

Only Bulgaria, Estonia, Greece and Lithuania determine the range of possible relationships between donor and recipient exactly without providing an open clause.

According to the Bulgarian regulation, the living organ donor has to be a spouse or a relative of the recipient on the direct line or by collateral line up to the fourth degree. This also includes kinship occurred on the grounds of adoption, but not earlier than 3 years after the adoption if the recipient is an adoptive parent.[714] Two additional cases are stated by law which allow the donation with a permit from the Ethical Commission: First, if no legal proof of a partnership exists (for example marriage), but the donor has been cohabitating with the recipient for a period of more than 2 years and there is indisputable proof for it, LOD might be allowed as well. Second, LOD may also be possible if the donor is a biological parent of the recipient and has not legitimised the child following the procedure established by law.[715] The rule is nevertheless exhaustive with respect to the allowed relationships, since it only contains the two clearly defined additional cases listed above. It does not contain an *open clause* enabling the interpretation of individual cases.

In Estonia, only LODs for the benefit of the donor's descendant, spouse, cohabite, parent, grandparent or their descendants are lawful.[716]

In Greece, the relatives that are allowed to become a living organ donor are exactly listed. In fact, "a. The patient's spouse b. The patient's natural parents and children c. The patient's sisters or brothers of the whole blood d. The patient's aunts, uncles, cousins and nephews of the whole blood and e. The patient's grandparents or grandchildren on the whole blood" might act as living organ donors.[717]

In Lithuania, a non-regenerative body part can only be removed from a relative or the spouse.[718]

Therefore, Bulgaria, Estonia, Greece and Lithuania are very strict with regard to the donor-recipient relationships. They define the allowed relationships in an exact manner.

[714] Art. 26 (1) Law of the Transplantation of Organs, Tissues and Cells.

[715] Ibid. Art. 26 (2).

[716] Sec. 9 (1) 3) Transplantation of Organs and Tissues Act.

[717] Canellopoulou-Bottis, Vol. 7 European Journal of Health Law 427, 433 (2000).

[718] Art. 10, 5 Law on Donation and Transplantation of Human Tissues, Cells and Organs.

(2) Open Clause in Addition to the Listed Donor-Recipient Relationships

The Czech Republic,[719] Finland,[720] France,[721] Germany,[722] Hungary,[723] Italy,[724] Poland[725] and Sweden[726] also regulate that LOD is only allowed in cases in which a specific relationship between donor and recipient exists. In contrast to Bulgaria, Estonia, Greece and Lithuania, the rules are not as exhaustive. Allowed relationships are listed, but additional donor-recipient relationships are possible due to an open clause.

In Italy, LOD is only permitted for the benefit of parents, children, siblings and half siblings of the recipient. Thus, a genetic relationship is required. If the recipient has no blood relatives as indicated or if none of the mentioned blood relatives is suitable or available, the circle of allowed donors may be extended to other relatives or non-blood related donors.[727]

The regulation of the necessary donor-recipient relationship is phrased in a similar fashion in Finland,[728] France,[729] Germany[730] and Hungary.[731] They require specific donor-recipient relationships, but a LOD might also be legal if there is a close personal relationship between donor and recipient. The details of the rules differ, though.

In France, the law now allows LOD by "emotionally related donors." It therefore does not necessarily require a genetic relationship, but, in case the relationship is only emotional, it needs to be proven that the relationship is strong and stable for at least 2 years.[732]

In Hungary, the living organ donor has to be "a lineal kin of the recipient, a sibling of a lineal kin of the recipient, a sibling of the recipient, a lineal kin of a sibling of the recipient."[733] LOD may also be allowed if a joint request is filed by the donor and recipient. This request is then considered by a hospital ethics committee. It gives consent to the organ removal only if it has established that a close emotional relationship exists between donor and recipient and that the

---

[719] Art. 3 (2) Transplantation Act on Donation, Removal and Transplantation of Organs and Tissue and on the Amendment of some Acts.

[720] Sec. 4 I Act on the Medical Use of Human Organs and Tissues.

[721] Information from B. Sénémaud.

[722] Sec. 8 I 2 Act on the donation, removal and transplantation of organs.

[723] Sec. 206 (2), (3) Act on Health.

[724] Art. 1 Legislation on Donation, Retrieval and Transplantation of Organs and Tissues.

[725] Art. 12, 1., 1) The Cell, Tissue and Organ Recovery, Storage and Transplantation Act.

[726] Sec. 7 Transplant Act.

[727] Art. 1 Legislation on Donation, Retrieval and Transplantation of Organs and Tissues.

[728] Sec. 4 I Act on the Medical Use of Human Organs and Tissues.

[729] Information from B. Sénémaud.

[730] Sec. 8 I 2 Act on the donation, removal and transplantation of organs.

[731] Sec. 206 (2), (3) Act on Health.

[732] Information from B. Sénémaud.

[733] Sec. 206 (2) Act on Health.

donation has taken place without consideration in return, force, duress, coercion or deception.[734] This possibility does not apply to detainees and conscripts.[735]

Following an amendment to the German *Act on the donation, removal and transplantation of organs* in 2007, the removal of one kidney, of part of the liver and of non-regenerative organs is only allowed for the benefit of relatives of the first or second degree, spouses, registered partnerships, fiancés or other persons with whom the donor obviously entertains an especially close personal relationship.[736] The phrase "other persons with whom the donor obviously entertains an especially close personal relationship" allows donor-recipient relationships that are not explicitly named. While this phrase particularly means to include persons living together as cohabitants, it was also emphasised that a regional separation does not exclude the possibility that a personal relationship has increased over a long period of time.[737] However, the types of relationships included by this phrasing are disputed. With regard to the interpretation of the term, the Federal Social Court stated that the question, whether the requirement is fulfilled or not, has to be decided on a case by case basis.[738] Planning for the future together is not necessarily required, but a duration of the existing relationship after the surgery has to be expected.[739]

In Finland, the allowed relationships are not named exactly. However, a certain donor-recipient relationship is required. A competent adult can donate non-regenerative organs only for the benefit of a near relative or other person close to him.[740] That predominantly applies to living kidney donation, but also to living liver donation, even though, so far, this type of LOD has not happened.[741]

The Czech Republic also does not list the allowed donor-recipient relationships exactly. It requires the donor and recipient to have a close personal relationship. Under what circumstances this is the case has to be determined according to Sec. 116, 117 *Civil Code*. Pursuant to Sec. 116 *Civil Code*, relatives of direct degree are considered as being close to the donor and also persons with a familial or similar relationship.[742] An exception of this required relation is provided in case the donor announces the will to donate the organ to a certain recipient. This explicit statement has to be made in writing and the donor's signature has to be witnessed by a notary public. In addition, an Ethical Committee has to approve such a donation.[743]

---

[734] Ibid. Sec. 206 (2), (3).

[735] Ibid., Sec. 206 (4).

[736] Sec. 8 I 2 Act on the donation, removal and transplantation of organs.

[737] Bundestag printed paper 13/4355 (1996), p. 21.

[738] German Federal Social Court [Bundessozialgericht], 10 December, 2003, Case No. B 9 VS 1/01 R, Vol. 59 Juristenzeitung 464, 464 ff. (2004).

[739] Ibid.

[740] Sec. 4 I Act on the Medical Use of Human Organs and Tissues.

[741] Information from K. Salmela.

[742] Norba (2009), p. 294.

[743] Art. 3 (2) Transplantation Act on Donation, Removal and Transplantation of Organs and Tissue and on the Amendment of some Acts.

In Poland, an organ can be removed "for the benefit of a linear relative, brothers and sisters, an adopted person, a wife or husband, and with the reservation of article 13, for the benefit of another person in the event of special personal reasons giving grounds for such conduct."[744] The reservation of article 13 states that cases outside of the stated relationship might be possible after examining the applicant and considering the opinion of the Committee of Ethics of the National Transplantation Council. In those cases, the approval of the district court appropriate for the donor's permanent residence or temporary stay is required.[745] This procedure starts when the person who intends on becoming a donor submits an application to the court. The application must contain a written consent of the intended recipient, an opinion of the Committee of Ethics of the National Transplantation Council and an opinion of the justification and usefulness of the procedure given by the head of the team of physicians performing the transplantation.[746]

In Sweden, non-regenerative material may only be donated if donor and recipient are related or if they are otherwise very close. The regulation does even make further cases possible, since in special cases non-regenerative material may also be donated to a person other than aforesaid.[747] That makes the regulation less strict than the other national rules just presented. Donor and recipient are only considered very close in cases where they are spouses, registered partners or cohabitants, because only persons, who, by virtue of their relationship, have a very strong interest in helping the intended recipient should be considered. If donor and recipient are neither related, nor very close, a non-regenerative organ might still be legally donated, but only if there are special circumstances. Special circumstances include relatives by marriage or very close friends, but also other persons. A special circumstance might exist, for example, if the considered person is a medically suitable donor and no other relative fulfils this requirement.[748] The provision only applies to non-regenerative material. This contains the removal of one kidney, but also the partial transplantation of pancreas or liver.[749]

In Austria, the first sentence in the *Position Paper* states that LOD is only accepted between relatives and other persons that have an emotionally close relationship. The medical practice complies with this.[750] Regarding the binding law, the *Criminal Code* that prohibits violations of good morals has to be considered. Whether a LOD for an unknown recipient is against good morals depends on the interpretation of this term.[751]

---

[744] Art. 12, 1., 1) The Cell, Tissue and Organ Recovery, Storage and Transplantation Act.

[745] Ibid., Art. 13, 1.

[746] Ibid., Art. 13, 3.

[747] Sec. 7 Transplant Act.

[748] The Ministry of Health and Social Affairs Sweden 1997, The Swedish Transplant Act, Guidance on the provisions, Sec. 7.

[749] Ibid.

[750] Information from K. Kozyga.

[751] Information from K. Bruckmueller.

All in all, several national laws restrict the donor-recipient relationship by requiring a certain relationship between donor and recipient. Several of those also contain an open clause to include additional relationships as well.

bb) No Restriction of the Donor-Recipient Relationship

Belgium, Denmark, England, Latvia, the Netherlands, Portugal, Scotland, Spain and Switzerland do not have any restrictions with regard to the donor-recipient relationship required for a legal LOD. These national laws do not explicitly state that organs can be donated to whomever the donor wants. However, some national laws contain evidence that leads us to conclude that there is no legal prerequisite for a specific type of relationship between donor and recipient. Other countries simply do not address the issue at all. As *argumentum e contrario*, no restrictions regarding the donor-recipient relationship exist.

The countries that do not restrict the donor-recipient relationship also hold that the donor and the recipient do not have to be related. This makes LOD for an unknown person possible. However, the regulations also differ. A subdivision is created here to analyse whether procedural safeguards exist.

(1) No Procedural Safeguards

Denmark, Latvia, the Netherlands and Switzerland do not possess any regulation addressing the relationship between donor and recipient. Hence, it can be assumed that the donor-recipient relationship is not restricted, in other words, no particular relationship between donor and recipient is required. Those countries do not contain any specific instruction for the cases in which the donor donates an organ for the benefit of an unknown recipient.

For Denmark, the law and the medical practice are examined. The law does not require a specific donor-recipient relationship. However, there is a common medical practice that used to only accept LOD within the family. Later, it started accepting LODs for friends.[752] It can be concluded that the legal situation and the medical practice differ, since LOD for an unknown recipient is legal, but not actually performed.

In the Netherlands, LOD used to be restricted to close relatives. Currently, donors and recipients do not have to be genetically related.[753]

In Switzerland, donor and recipient also do not have to be related.

The Latvian law does not explicitly state that no special relationship between donor and recipient is required, but addresses the issue by requiring the medical

[752] Information from T. Kjeldsen.

[753] Health Council of the Netherlands (2007).

staff to ensure anonymity between donor and recipient if they are not next of kin.[754] The implication of this requirement is that donor and recipient might not know each other; they can be unrelated. Latvian law does not contain any further specific instructions for cases of unspecified LOD.

Thus, several countries do not demand a specific relationship between donor and recipient and have also not established specific procedural safeguards.

(2) Procedural Safeguards

Belgium, England, Portugal, Scotland and Spain do not require a certain relationship for LOD to be legal, but have established certain procedural safeguards for LOD.

In England, Sec. 33 *Human Tissue Act 2004* restricts all kinds of LOD. It does not differentiate between LOD for relatives and LOD for the benefits of an unknown person.[755] According to Sec. 33 *Human Tissue Act 2004*, LOD is principally illegal; however, the Secretary of State may, according to regulations, permit LOD "in a case where — (a) the Authority is satisfied — (i) that no reward has been or is to be given in contravention of section 32, and (ii) that such other conditions as are specified in the regulations are satisfied, and (b) such other requirements as are specified in the regulations are complied with."[756] Therefore, the approval of the Human Tissue Authority is required for every LOD.[757] In addition, the *Code of Practice 2* about "[d]onation of solid organs for transplantation" mentions the altruistic non-directed and paired or pooled donation. Especially in cases of altruistic, non-directed LOD, no special donor-recipient relationship is necessary.[758]

The Scottish legal regulation is similar. According to Sec. 17 *Human Tissue (Scotland) Act 2006*, LOD is generally illegal, but the Scottish Ministers may by regulations provide that this "does not apply in a case where — (a) the Ministers are satisfied that — (i) no reward has been or is to be given in contravention of section 20; and (ii) such other conditions as may be specified in the regulations are satisfied; and (b) such other requirements as may be specified in the regulations are complied with."

In Portugal, one can donate an organ to a person he is not related to. If a non-renewable organ is concerned, which includes kidney and also the donation of part of the liver,[759] the admissibility for every donation depends upon the positive

[754] Sec. 15 Law on the Protection of the Body of Deceased Human Beings and the Use of Human Tissues and Organs in Medicine.

[755] This used to be different. Before the Human Tissue Act 2004 came into effect, the Human Organ Transplants Act 1989 contained specific restrictions for LODs between persons who were not related (Fateh-Moghadam (2008), p. 282).

[756] Sec. 33 (3) Human Tissue Act 2004.

[757] Cf. Fateh-Moghadam (2008), p. 282 ff.

[758] No. 92 Code of Practice 2.

[759] Information from A. Pereira.

opinion of the Entity of Verification and Admissibility of Collection for Transplantation.[760] Consequently, unspecified LOD is possible. It only has to be proven that the consent is informed, free and not remunerated.[761]

In Belgium, Art. 8 *bis Law on Retrieval and Transplantation of Organs* demands that every organ removal from a living person be the object of a multidisciplinary concertation, which provides a certain safeguard for LOD.

In Spain, the head of the transplant medical unit has to consent to the transplantation.[762] In addition, a comment by the ethics commission of the transplantation hospital is required.[763]

Therefore, Belgium, England, Portugal, Scotland and Spain do not have any restrictions on the donor-recipient relationship, but do have procedural safeguards.

cc) Summary of the Required Donor-Recipient Relationship

None of the countries considered allow LOD between genetically related people only. The countries do have different requirements regarding the relationship between donor and recipient. Several countries only allow LOD for the benefit of a genetic or emotional relative. Most of these countries not only list the possible donor-recipient relationships, but contain an additional open clause. However, there are also countries that do not restrict the donor-recipient relationship in any way. A few of these countries have specific procedural safeguards. Figure 2 presents the proportions of the different regulations of the required donor-recipient relationship.

c) Special Cases

As shown, the relationship between donor and recipient consists of several different forms, for example, a genetic relationship, marriage or friendship. The national laws determine whether the donor and recipient must be related and what forms of relationships between donor and recipient are legally allowed.

The most common type of LOD, the direct LOD, is the subject-matter of this thesis. With regard to the other possible types of LOD that have been listed above, it is examined here whether cross-over LOD and unspecified LOD are legal in the countries considered. Both types of LOD will be considered further in chapter "Analysis of the Normative Arguments That Dominate the Policy Arena About Necessity and Legitimacy of Legal Restrictions in Living Donor Transplantation" along with the other possible types of LOD that have been introduced in section III of this chapter.

---

[760] Art. 6, 3 Law on Harvesting and Transplanting Human Organs and Tissues.

[761] Information from A. Pereira.

[762] Art. 15, 4. Royal Decree 2070/1999.

[763] Ibid., Art. 9, 2.

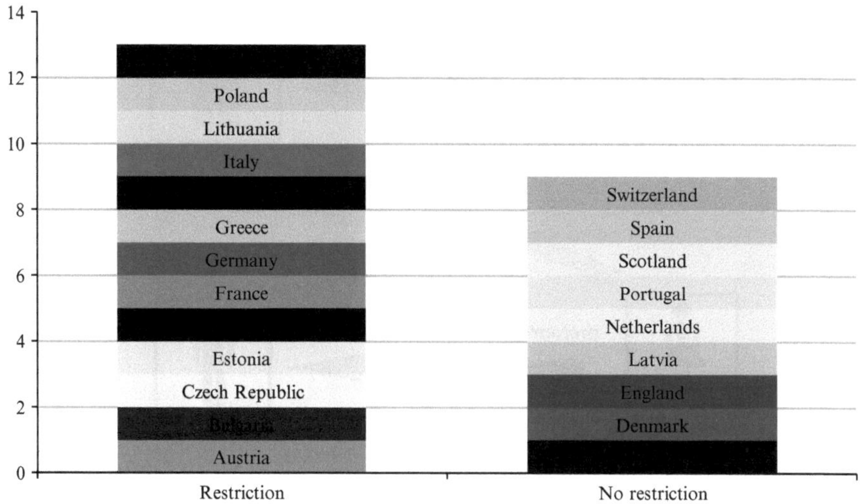

Fig. 2 Donor-recipient relationship

aa) Cross-Over Living Organ Donation

A cross-over donation takes place if a willing person cannot donate an organ to her partner due to immunological reasons and another donor-recipient pair has the same problem. If medically possible, the two recipients can swap places. Recipient A gets the organ from the partner of recipient B, while, in exchange, recipient B receives the organ from donor A. This procedure then allows for two medically incompatible pairs to be converted into two medically compatible pairs.[764]

This is illustrated by Fig. 3.[765]

(1) Debate About Cross-Over Living Organ Donation

Not all countries allow this type of LOD. In fact, pundits remain divided on this issue and the opinions about it are very conflicting. The reasoning brought forth in support of the restriction of the donor-recipient relationship is also valid for this type of LOD, but additional arguments particularly for this type of LOD exist.

The two donors and recipients usually only get to know each other in the context of the LOD. Therefore, opposing scholars criticise the lack of personal relationship

[764] Akkina et al., Vol. 57 American Journal of Kidney Diseases 152, 152 (2011); de Klerk (2010), p. 14; German Federal Social Court [Bundessozialgericht], 10 December, 2003, Case No. B 9 VS 1/01 R, Vol. 59 Juristenzeitung 464, 466 (2004); Ferrari/de Klerk, Vol. 22 Journal of Nephrology 699, 699 (2009); cf. Hilhorst, Vol. 8 Ethical Theory and Moral Practice 197, 205 (2005); cf. Hoyer, Vol. 21 Pediatric Nephrology 1364, 1365 (2006); Menikoff, Vol. 29 The Hastings Centre Report 28, 28 (1999); Ross, Vol. 16 Kennedy Institute of Ethics Journal 151, 152 (2006).

[765] European Urology (2005).

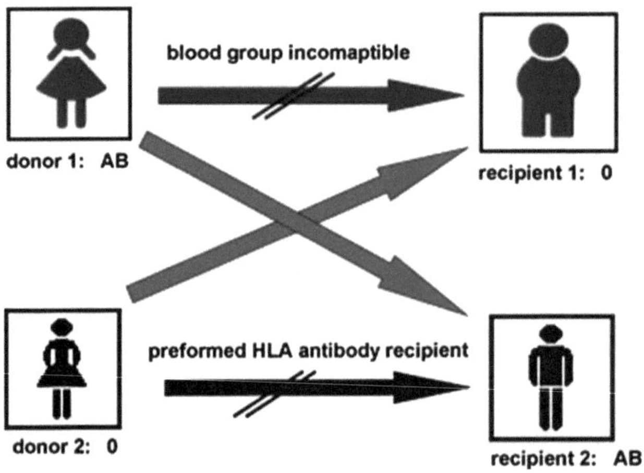

Fig. 3 Cross-over LOD

between the medically fitting pair.[766] They argue that a personal relationship between the theoretical donor and recipient cannot substitute the relationship between the actual donor and recipient.[767]

Psychological aspects are also mentioned in the debate. First, some claim that if the principle of donor exchange was to be implemented universally, potential donors would not have the chance anymore to deny LOD based on medical reasons.[768] Potential donors sometimes refer to medical reasons as an excuse to withdraw their initial consent for LOD without snubbing the intended organ recipient.[769] Therefore, potential donors could be especially burdened if this possibility becomes impossible.[770] Second, others claim that a difficult situation might arise if the donor perceives that his donated kidney functions well, while the kidney his partner received does not.[771]

Another specific problem of cross-over donation mentioned is the fact that even after the first organ has already been removed, the other donor can still withdraw his consent in case the organ removals for the two donors do not happen simultaneously.[772]

[766] Hohmann (2003), p. 86.

[767] Ibid., p. 129.

[768] de Klerk (2010), p. 23; cf. Hoyer, Vol. 21 Pediatric Nephrology 1364, 1365 (2006); Ross, Vol. 16 Kennedy Institute of Ethics Journal 151, 151 (2006); Schutzeichel (2002), p. 221.

[769] Erim, Vol. 81 Der Chirurg 820, 822 (2010).

[770] de Klerk (2010), p. 23; cf. Hoyer, Vol. 21 Pediatric Nephrology 1364, 1365 (2006); Ross, Vol. 16 Kennedy Institute of Ethics Journal 151, 151 (2006); Schutzeichel (2002), p. 221.

[771] de Klerk (2010), p. 23; Gutmann (2004), p. 17.

[772] Price (2010), p. 221; Schroth, Vol. 59 Juristenzeitung 469, 472 (2004).

The legal experts in favour of cross-over donation also address the aspect of a personal relationship between donor and recipient. They argue, in contrast, that a relationship between the assigned donor and recipient in the swapping process exists in some respect, even though this relationship is not based on personal affection. They state that the relationship is based on a strong connection resulting from the same sickness of the partner and the wish of both potential donors to save their partners' lives. One should also note that if the donation for one's own partner fails, it is because of medical reasons[773] — and not because the requested personal relationship does not exist. Therefore, considering all four participants, experts argue that the required close personal relationship exists.[774]

In addition, many claim that the intent of requiring a personal relationship is primarily to prevent organ trafficking.[775] Contrarily, others argue that the danger of organ trafficking is not present in cases of cross-over donation, though, or at least the danger is not higher than in cases of direct LOD.[776] Every donor donates in order to help a person he is closely connected with. Since both pairs are only active because of this reason, the exchange takes place on the basis of equality.[777]

Because arguments in favour of cross-over LOD, as well as arguments against this type of LOD exist, it is not surprising that the countries' regulations differ.

(2) National Regulations on Cross-Over Living Organ Donation

The first cross-over transplantation worldwide took place in South Korea in 1991.[778] Even though South Korea is not generally considered in this thesis, the South Korean situation with regard to organ transplantation is quite interesting. Post-mortal kidney transplantation is, due to cultural and religious reasons, very seldom in South Korea and neither socially nor legally accepted.[779] As a result,

[773] Cf. Akkina et al., Vol. 57 American Journal of Kidney Diseases 152, 152 (2011); Ferrari/de Klerk, Vol. 22 Journal of Nephrology 699, 699 (2009). The intended living kidney donation of willing donor/recipient pairs just fails because of biological incompatibility in approximately 30 % of all cases (Cf. Akkina et al., Vol. 57 American Journal of Kidney Diseases 152, 152 (2011); Ferrari/de Klerk, Vol. 22 Journal of Nephrology 699, 699 (2009)). Incompatibility is either due to blood type mismatch or because of a positive crossmatch, what means that a pre-existing antibody to a donor antigen is present in the recipients body (Akkina et al., Vol. 57 American Journal of Kidney Diseases 152, 153 (2011)).

[774] Hohmann (2003), p. 86.

[775] Bundestag printed paper 13/4355 (2005), p. 20; cf. Hohmann (2003), p. 126 ff.

[776] Hohmann (2003), p. 129.

[777] de Klerk (2010), p. 14.

[778] de Klerk et al., in Gutmann et al. (ed.) (2004), p. 256; Park et al., Vol. 67 Transplantation 336, 336 (1999).

[779] de Klerk (2010), p. 22; cf. Hamza et al., Vol. 45 Der Urologe 60, 62 (2006); cf. Oniscu/Forsythe, Vol. 38 Annals Academy of Medicine 365, 366 f. (2009).

LOD is very important and cross-over transplantation is carried out many times per year.[780]

The first cross-over LOD in Europe took place in Basel in 1999.[781] Cross-over donation is not exercised in Switzerland anymore, though, because transplantations between incompatible people are presently exercised.[782]

With regard to the countries considered in this paper, cross-over donation is legal in the countries that do not require a certain donor-recipient relationship. In Belgium,[783] England,[784] the Netherlands,[785] Scotland[786] and Spain[787] it is also performed on a regular basis.

In Belgium, cross-over donations are carried out, but the situation is quite ambiguous. The law does not explicitly prohibit this type of LOD. However, some legal experts wonder if it would be necessary to invent a particular framework for cross-over donation in order to execute it.[788]

The situation in England changed in 2006 with the implementation of the *Human Tissue Act 2004*. Before this point of time, cross-over LOD was not legally prohibited, but the medical practice did not support it.[789] The legal situation was interpreted rather narrowly by requiring a long standing relationship between donor and recipient. Hence, cross-over LOD was excluded altogether.[790] The *Human Tissue Act 2004* now explicitly mentions cross-over LOD, making an implementation easier.[791] With respect to this change in the law, *Rigg* rememberes "that it may have been discussed in theoretical terms in light of experience elsewhere in the world."[792]

In the Netherlands, the first cross-over donation took place in 2003.[793] Soon afterwards, a cross-over kidney exchange programme was established.[794] This programme has been successful and, thus, is taken here to exemplify a reasonable procedure. Such a programme can only be successful if it contains a large pool of

---

[780] de Klerk et al., in Weimar/Bos/Busschbach (ed.) (2008), p. 236; de Klerk (2010), p. 15.

[781] Bundestag printed paper 15/5050 (2005), p. 39; Swiss Dispatch of 12. September 2001 on a Federal Law on Transplantation of Organs, Tissues and Cells, p. 98.

[782] Information from M. Monnier.

[783] Information from R. Wilmotte.

[784] No. 92 Code of Practice 2.

[785] Hamza et al., Vol. 45 Der Urologe 60, 62 (2006); Information from M. Bos.

[786] Scottish Executive Health Department (2006), p.6

[787] Conde, in: Roxin/Schroth (ed.) (2010), p. 874.

[788] Information from R. Wilmotte.

[789] Information from K. Rigg; information from C. Rudge.

[790] Information from C. Rudge.

[791] Information from K. Rigg.

[792] Ibid.

[793] de Klerk (2010), p. 17 f.; Hilhorst, Vol. 8 Ethical Theory and Moral Practice 197, 205 (2005).

[794] Bos, in Weimar/Bos/Busschbach (ed.) (2008), p. 24; Hilhorst in Weimar/Bos/Busschbach (ed.) (2008), p. 133; de Klerk (2010), p. 17 f.; Weimar et al., Vol. 368 The Lancet 987, 987 (2006).

donors and recipients. This requires it to be organised nationwide. When evaluating this programme, the danger of putting too much pressure on potential donors must be weighed against the positive opportunity the programme offers to motivated donors.[795] The registration procedure of the participants contains four match procedures a year. A computer match programme identifies suitable donor-recipient combinations for an exchange. If one donor-recipient pair can be matched to several pairs, further selection criteria are needed. After a suitable match is found, the local transplant coordinator arranges the exchange.[796] All in all, the living donor kidney exchange programme in the Netherlands has been deliberately designed to allow for a fair matching of pairs and treatment of participants. It can be viewed as an instructive example.

In Portugal, even though the *Law on Harvesting and Transplanting Human Organs and Tissues* does not restrict the donor-recipient relationship, cross-over donations have not yet been performed.[797]

The situation in Denmark is similar, since the law does not require a specific donor-recipient relationship. However, it can be assumed that cross-over LOD is not executed in medical practice. This assumption is due to the common medical practice which used to only accept LOD within the family, but later also started accepting LODs for friends.[798]

The national laws that restrict the donor-recipient relationship do not contain regulations regarding cross-over donation in particular. Thus, they must be interpreted individually.

Bulgaria, Estonia and Lithuania list the allowed donor-recipient relationship exhaustively in their transplant laws. This does not include cross-over donations.

In contrast, the Czech Republic, Finland, Germany, Hungary, Italy, Poland and Sweden list the accepted relationships, but also contain an open clause.

In Finland and Hungary, the open clause does not make cross-over LOD legally possible. In Poland, cross-over LOD does not take place.[799]

The German *Act on the donation, removal and transplantation of organs* does not contain any regulation regarding cross-over donation.[800] In the deliberations of the parliament, this aspect has also not been addressed.[801] Therefore, the law, which requires a close relationship, needs to be interpreted.[802] The debate over cross-over LOD focuses on whether the necessary personal connection is already established due to the special life situation of the couples.[803] In 1997, the Federal Health

[795] de Klerk (2010), p. 17 f.

[796] Ibid., p. 33 ff.

[797] Information from A. Pereira.

[798] Information from T. Kjeldsen.

[799] Information from D. Lewandowska.

[800] Riedel, in Rittner/Paul (ed.) (2005), p. 70.

[801] Nickel/Schmidt-Preisigke/Sengler (2001), p. 104.

[802] Schutzeichel (2002), p. 217.

[803] Sasse, in Barta/Weber (ed.) (2001), p. 119.

Minister, *Horst Seehofer*, stated that cross-over donation is not legal under the German *Act on the donation, removal and transplantation of organs*.[804] In 2001, the Higher Social Court in North Rhine-Westphalia also declared cross-over LOD to be illegal.[805] Pursuant to the Federal Constitutional Court, a close personal relationship must be apparent; this can be observed in the relationship's internal and external characteristics. If the pairs only meet each other for the purpose of the transplantation and the transplantation takes place promptly after the first meeting, a close personal relationship cannot be assumed.[806] The Federal Social Court holds a dissenting view. The court argued that the Higher Social Court in North Rhine-Westphalia interpreted Sec. 8 I 2 *Act on the donation, removal and transplantation of organs* too narrowly, and decided to remand the case. The Federal Social Court argued that the question of whether a close personal relationship exists has to be answered for every potential cross-over LOD individually. It does not necessarily require planning for a shared future, but it is expected that the relationship will last for a duration of time after the surgery is over. Thus, a relationship that is, from the viewpoint of the individuals concerned, meant to remain for an unlimited period of time, is required.[807] The legal situation is obviously unclear in Germany.

Austria requires the donor and recipient to be somehow related to each other. An emotional relationship is considered sufficient, though, and many argue that there is a close relationship between the donor and the indirect recipient in the context of cross-over donation. Consequently, cross-over donation is principally permitted in Austria.[808]

In Italy, cross-over LODs are legal and actually performed, but very seldom.[809]

In Sweden, cross-over donations are not prohibited by law, but only one cross-over LOD has been carried out so far.[810] This could be, because Sweden is a rather small country, making it difficult to match suitable donors and recipients.[811] In addition, Sweden established an extensive programme for ABO-incompatible living kidney transplantations. That, for example, causes that the organ recipient is

---

[804] Rotondo (2004).

[805] Higher Social Court North Rhine-Westphalia [Landessozialgericht Nordrhein-Westfalen], 31 January, 2001, Case No.L 10 VS 28/00.

[806] German Federal Constitutional Court [Bundesverfassungsgericht], 11 August, 1999, Case No. 1 BvR 2181/98, at 60 (insofar not printed in Vol. 52 Neue Juristische Wochenschrift 3399 (1999)).

[807] Cf. Bundestag printed paper 15/5050 (2005), p. 41; German Federal Social Court [Bundessozialgericht], 10 December, 2003, Case No. B 9 VS 1/01 R, Vol. 59 Juristenzeitung 464, 464 ff. (2004); Gutmann (2006), p. 34; cf. Schroth, Vol. 17 Medizinrecht 67, 67 (1999). This is not undisputed, though. Seidenaht, for example, states that no minimal length for the relationship between donor and recipient exists (Seidenaht, Vol. 16 Medizinrecht 253, 256 (1998)).

[808] Information from K. Kozyga.

[809] Information from P. Bruzzone.

[810] Information from A. Lennerling.

[811] Information from C. Möller.

undergoing immunoadsorption treatment prior to the transplantation and afterwards as well.[812]

Thus, some countries allow and perform cross-over LOD, not all though.

(3) Anonymity in Case of Cross-Over Living Organ Donation

There is no noticeable consensus among the countries that permit cross-over donations on whether anonymity between donor and recipient should be maintained.

The first cross-over donation in Switzerland involved one pair from Germany and one pair from Switzerland. In this case, the pairs met prior to the procedure, so their anonymity was not preserved.[813]

In the Netherlands, maintaining anonymity between donor and recipient is not a legal requirement, but in medical practice, transplantations are usually carried out in an anonymous way.[814] A study in the Netherlands showed that all potential cross-over pairs interviewed preferred anonymity.[815]

In contrast, English law even stipulates that the donor must be informed prior to the operation that anonymity has to be maintained, so that anonymity between donor and recipient is preserved at least before the surgeries.[816]

In the general discussion, advantages and disadvantages with regard to maintaining anonymity in cross-over donations have been brought forth. Maintaining anonymity is considered to be favourable for the transplant centres; they do not have to arrange meetings prior to the transplantation.[817] It is also considered to be advantageous in the sense that psychological pressure or conflicts between the pairs cannot arise if the results of the transplantations are not similar.[818] In contrast, the donor may be impacted negatively if she does not receive any information about the functioning of her kidney and is not informed of whether her donation was beneficial for the recipient.[819] It is also partly considered a restriction to autonomy if donor and recipient are forced to maintain anonymity.[820]

Thus, there is no common standard for allowing cross-over donation. The countries that allow this type of LOD also have different rules regarding anonymity.

---

[812] Information from A. Lennerling.

[813] de Klerk (2010), p. 15 f.

[814] Information from M. Bos; Price (2010), p. 221.

[815] de Klerk (2010), p. 32 f.

[816] No. 92 Code of Practice 2.

[817] de Klerk et al., in Gutmann et al. (ed.) (2004), p. 260.

[818] Ibid.

[819] Ibid.

[820] Ibid.

bb) Unspecified Living Organ Donation

Several national transplant laws require a genetic or emotional relationship between donor and recipient (or some form of close connection). An unspecified LOD describes cases where a person decides to donate an organ to an anonymous and unspecified recipient[821]: a living person donates an organ to a stranger.[822]

(1) National Regulations

In the debate about the legitimacy of unspecified LOD, several arguments for and against it are brought forth. Those have already been presented at the beginning of this chapter. Thus, it is not surprising that there is no consensus among the national laws regarding this form of LOD.

This kind of LOD is legal in countries that not do restrict the donor-recipient relationship at all, namely in Belgium, Denmark, England, Latvia, the Netherlands, Portugal, Scotland, Spain and Switzerland. However, unspecified LOD seems to be rather rare in medical practice. Statistics from the United Kingdom show that from April 1, 2008 to March 31, 2009 only 15 of the 927 living kidney donations were unspecified.[823]

In Spain, unspecified LOD is now legal, but it was not practiced medically in the past. The national transplant organisation now supports LOD to an unspecified recipient; it rejected this LOD form in the past.[824]

The situation in Portugal has been similar in the sense that by law no genetic or emotional relationship was required between donor and recipient. LOD to an unspecified recipient is legal, yet no such cases have actually been performed in practice.[825]

This is the same in Belgium[826] and Denmark.[827] The law does not require a certain relationship between donor and recipient, yet unspecified LOD is not exercised in medical practice.

The national laws that restrict the donor-recipient relationship have to be discussed individually again.

In Sweden, solid organs may only be donated within the relationships defined by law or if donor and recipient are otherwise very close. In special cases, a donation to another person than the aforesaid is allowed. As stated above, such a special reason can exist in cases of marriage or a very close friendship, but also with regard to

---

[821] Dor et al., 2011 Transplantation 1 f.

[822] Cf. Henderson et al., Vol. 3 American Journal of Transplantation 203, 203 (2003).

[823] NHS Blood and Transplant (2008–2009).

[824] Information from M. Zardoya Martínez.

[825] Information from A. Pereira.

[826] Information from R. Wilmotte.

[827] Information from T. Kjeldsen.

other persons.[828] In practice, at least fifteen unspecified LODs have taken place since 2004.[829] Strict investigations are carried out to ensure that no money is involved, which does make the process considerably more complicated. In addition, it is not very common among Swedish people to donate an organ to a stranger during their lifetime. However, those who like to do so are taken into account.[830]

In Italy, unspecified LODs are not explicitly prohibited by law and are also actually carried out, but very seldom.[831]

The other laws that restrict the donor-recipient relationship require the donor and recipient to know each other or have any kind of relationship. Consequently, Bulgaria, the Czech Republic, Estonia, Finland, France, Germany, Hungary and Lithuania do not allow unspecified LODs. In Austria, unspecified LODs are also not carried out, since the *Position Paper* requires the donor and recipient to be related or emotionally very close.[832] In Poland, unspecified LODs are not carried out either.[833]

In Germany, it is illegal for living persons to donate an organ to a stranger.[834] However, in 1996, before the German *Act on the donation, removal and transplantation of organs* came into effect, the German transplant surgeon *Hoyer* donated a kidney to a stranger who was selected with computer-based screening. He did so because of the low rates of LOD. His intent was to convince people to consider LOD more often.[835] He thought that it might be a strong statement if a surgeon, who is aware of all risks of LOD, donates an organ while being alive.[836] Since this occurrence took place before the German *Act on the donation, removal and transplantation of organs* came into effect, it did not have any legal consequences.[837]

After a closer examination of the laws that require a specific donor-recipient relationship for LOD, it is clear that only Sweden, in addition to the countries whose laws do not contain any restriction on the donor-recipient relationship (Belgium, Denmark, England, Latvia, the Netherlands, Portugal, Scotland, Spain and Switzerland), allows an unspecified LOD by law.

All in all, unspecified LOD is lawful in a few countries. In medical practice, however, it is only seldom performed. Nevertheless, LOD is the only type of organ

[828] The Ministry of Health and Social Affairs Sweden 1997, The Swedish Transplant Act, Guidance on the provisions, Sec. 7.

[829] Information from A. Lennerling.

[830] Information from C. Möller.

[831] Information from P. Bruzzone.

[832] Information from K. Kozyga.

[833] Information from D. Lewandowska.

[834] Bundestag printed paper 15/5050 (2005), p. 46; Sasse, in Barta/Weber (ed.) (2001), p. 117.

[835] Daar et al., Vol. 11 Transplantation Reviews 95, 98 (1997); Hoyer, Vol. 76 Transplantation 1264, 1264 f. (2003).

[836] Hoyer, Vol. 76 Transplantation 1264, 1265 (2003).

[837] Ibid.

donation that has steadily increased since 1990.[838] Also, a recent investigation of LOD in medical practice shows that unspecified LOD is actually performed. A recent survey with the responses from 113 kidney transplant units from 25 EU Member States and 38 liver transplant units from 19 EU Member States found that unspecified living kidney donation was performed in 33 % of the kidney transplant units and unspecified living liver donation in 27 % of the liver transplant units. Additionally, they ascertained that the content of the medical screening depends on the country being considered, and that all countries agree that there are a few absolute contraindications for LOD.[839]

(2) Anonymity in Case of Unspecified Living Organ Donation

In an unspecified LOD, must anonymity be maintained between the donor and the recipient?

In the Netherlands, anonymity is not a legal requirement for unspecified LOD. Medical practice, however, shows that during the transplantations donor and recipient usually do not know each other.[840]

In England, the donor has to be informed of the requirement to maintain anonymity between him and the recipient before the surgeries; confidentiality must be respected.[841]

In Sweden, maintaining anonymity is a binding rule.[842]

In Switzerland, donor and recipient generally stay anonymous after an unspecified LOD. In case both involved patients wish the anonymity to be lifted after the donation has been carried out, those wishes may be granted after carefully assessing the risks.[843]

Overall, there is no consensus among the countries considered on whether unspecified LOD is legal. Even though this type of LOD is slowly gaining acceptance, the numbers in actual practice are still limited.[844] In the countries that allow this type of LOD, it seems to be medical practice to maintain anonymity between the donor and recipient during the process.

---

[838] Frunza et al., Vol. 9 Journal for the Study of Religions and Ideologies 3, 4 (2010); cf. Hilhorst et al., Vol. 24 Transplant International 1164, 1164 (2011).

[839] EULOD WP 2 (2012), DOW: Deliverable 4, p. 3.

[840] Information from M. Bos.

[841] No. 92 Code of Practice 2.

[842] Information from C. Möller.

[843] The Swiss Acadamies of Arts and Sciences, Live donations of solid organs, Medical-ethical guidelines and recommendations, p. 11.

[844] Roodnat et al., Vol. 10 American Journal of Transplantation 821, 824 (2010).

cc) Interim Result

The legality of cross-over and unspecified LODs varies from country to country. In fact, countries without any restrictions on the donor-recipient relationship permit both types of LOD. The national laws that restrict cross-over and unspecified LODs have to be interpreted individually; both types of LOD are not viewed equally.

d) European Tendencies About Restrictions Regarding the Donor-Recipient Relationship

Several European documents about organ transplantation also address the issue of possible restrictions regarding the donor-recipient relationship.

The *Resolution on Harmonisation of Legislations of Member States relating to Removal, Grafting and Transplantation of Human Substances* from 1978 confines the transplantation of non-regenerative substances to genetically related persons. However, in special cases where there are good chances of success, an exception is possible.[845] Interestingly enough, the resolution is rather old, but nevertheless enables flexibility and openness for further medical development.[846] With regard to the issue of anonymity, the resolution is in favour of maintaining anonymity between donor and recipient if there is no personal or family relation between them.[847]

The *Convention on Human Rights and Biomedicine* from 1997 does not contain any regulation regarding the donor-recipient relationship. However, the *Additional Protocol on Transplantation of Organs* addresses the issue of a required donor-recipient relationship in Art. 10 by stating that "[o]rgan removal from a living donor may be carried out for the benefit of a recipient with whom the donor has a close personal relationship as defined by law, or, in the absence of such relationship, only under the conditions defined by law and with the approval of an appropriate independent body." The normally required close personal relationship, for example, includes the immediate family, especially parents, siblings and spouses, as well as long-standing partners, godparents and close personal friends. The purpose is to protect people without a strong emotional relationship from being pressured into performing a LOD. The regulation also contains an exception, though, since it can legally deviate from the general rule under the conditions defined by law and with the approval of an independent body. Evidence exists that there may be perceptible psychological benefits to the donor who improved the recipient's health or even saved his life. This fact may be taken into account when deciding whether the veil

---

[845] Art. 4 Resolution (78) 29 on Harmonisation of Legislations of Member States relating to Removal, Grafting and Transplantation of Human Substances.

[846] Gutmann/Schroth (2002), p. 73.

[847] Art. 2, 2 Resolution (78) 29 on Harmonisation of Legislations of Member States relating to Removal, Grafting and Transplantation of Human Substances.

of anonymity may be lifted. The national laws have to set out the conditions under which such circumstances may be considered, for example, the establishment of an appropriate body that considers each case.[848]

These European documents are in favour of restricting the donor-recipient relationship to related people. Again, exceptions to this rule are allowed.

6) Subsidiarity

The aspect of subsidiarity is concerned with the question of whether LOD constitutes a primary cure option or merely a subordinate one.

This issue includes two aspects: (1) I discuss whether the option of transplanting an organ donated by a living person is subordinate to transplanting an organ donated by a deceased; this issue is especially crucial when considering whether LOD should only be permitted when no organs donated post-mortem are available for transplantation in reasonable time. (2) In addition, I consider the relation between LOD and alternative therapies. Specifically, I discuss whether LOD should only be allowed if all other possible therapies are exhausted.[849]

a) Background of the Issue of Subsidiarity

Why does the aspect of subsidiarity even arise when regulating LOD?

LOD provokes the situation that a completely healthy person becomes part of a surgery that contains risks.[850] Those risks could be viewed as violating the principle

---

[848] Art. 10 Additional Protocol to the Convention on Human Rights and Biomedicine concerning Transplantation of Organs and Tissues of Human Origin, Explanatory Report.

[849] Gutmann/Schroth (2002), p. 76.

[850] The risks must be assessed for each organ individually since they vary greatly (Childress/Liverman (2006), p. 270; Guillod/Perrenoud, in Gutmann et al. (ed.) (2004), p. 158). Evaluating the mortality risk of the organ donor illustrates that the risk is very low in cases of living kidney donation – below 1 per mill (Muthny et al., 2010 Transplantationsmedizin 106, 107; Siegmund-Schultze (1999), p. 106), but significantly higher in cases of living liver donation, where the stated mortality rates differ. The reason for that is that no uniform classification exists. (Cf. Broering et al., Vol. 45 Current Problems in Surgery 587, 635 (2008); Dreßke et. al., Vol 81 Der Chirurg 813, 817 (2010)). However, the mortality rate of living liver donation is as high as 1 % (Bundestag printed paper 15/5050 (2005), p. 16; Guillod/Perrenoud, in Gutmann et al. (ed.) (2004), p. 158; Neuberger/Price, Vol. 327 British Medical Journal 676, 677 (2003); Schneider (2004)).

This applies to the donation of a right lobe. The mortality rate in cases of the removal of the left lobe is only about 0.1 % (Neuberger/Price, Vol. 327 British Medical Journal 676, 676 (2003)). All in all, a living liver donation of the left lateral contains smaller risks for the donor than the donation of the right lobe (Ben-Haim et al., Vol. 7 Liver Transplantation 948, 951 (2001); Bundestag printed paper 15/5050 (2005), p. 16; Dreßke et. al., Vol 81 Der Chirurg 813, 817 (2010); Ross/Thistlethwaite, Vol. 122 American Academy of Pediatrics 454, 455 (2008)). The left lateral only makes up approximately 20 % of the hepatic mass and is used for living liver transplantation

of nonmaleficence.[851] This view could be opposed by arguing that the potential donor consents to the organ removal, and, due to the right of self-determination, the donor's decision should be respected.[852] If the potential living donor would not be allowed to donate just because a post-mortem donated organ is available, it would be contrary to the right of self-determination. In addition, many claim that the recipient herself should have the opportunity to decide what kind of organ will be implanted into her body. And, there is no objective reason why a recipient should not have the option to choose which organ she will receive.[853]

In light of these two main arguments, namely the principle of nonmaleficence on the one hand and the right of self-determination on the other hand, the countries considered regulate the issue of subsidiarity in different ways. Some countries do not consider LOD as subordinate in any way. Others opine that LOD may only be performed subsidiary to post-mortem donation and also to alternative therapies. Some countries even differentiate between both relationships: living and deceased organ donation and LOD and alternative therapies.

b) Relationship Between Living Organ Donation and Post-Mortem Donation

There is no consensus among the countries considered on whether LOD is only subordinate to post-mortem donation.

aa) Subsidiarity of Living Organ Donation to Post-Mortem Donation

LOD is only legal in several countries when an organ from a deceased person is unavailable.

(1) Interpretation of the Term "Available"

None of the laws of those countries just mentioned defines the meaning of "available." (1) Many question what point of time is decisive when deciding whether an organ donated post-mortem is available. (2) In addition, the term is ambiguous because it is unclear whether an organ is only considered to be "available" when it

---

into a child or small adult. Living liver donation for an adult usually requires the more risky donation of the right lobe, though, which accounts for approximately 60 % of the hepatic mass (Broering et al., Vol. 45 Current Problems in Surgery 587, 601&630 (2008); Brown et al., Vol. 348 The New England Journal of Medicine 818, 819 (2003); cf. Rogier, in Rittner/Paul (ed.) (2005), p. 47).

[851] Gutmann/Land, Vol. 384 Langenbeck's archives of surgery 515, 516 (2000); Schutzeichel (2002), p. 36.

[852] Cf. Beauchamp/Childress (2001), p. 63; cf. Gutmann/Schroth, in Oduncu/Schroth/Vossenkuhl (ed.) (2003), p. 282.

[853] Edelmann, Vol. 50 Versicherungsrecht 1065, 1068 (1999).

is equally suited for the recipient in the same manner the organ potentially donated by a living person is, or if an organ is "available" simply when a compatible transplantable organ exists.

With regard to the first question, namely the essential point of time, the German law states that it is relevant if an organ donated post-mortem is "available at the time when the organ removal takes place."[854] Hence, the point of time of the organ removal is essential, which probably conforms to the medical necessity of the transplantation.[855] Since this moment can be considered as the beginning of the transplant procedure, it seems logical, to assess it as decisive.

However, the question about the exact meaning of the term "available" remains. The national laws do not provide an answer to this. They merely require that no suitable organ of a deceased person is available. This could, on the one hand, mean that any transplantable, suitable organ is sufficient. Since the rule of subsidiarity is connected to the relationship between living and deceased donation, it could, on the other hand, imply that the organ in question must be as suitable as the organ potentially donated by a living person. To answer this question, the whole purpose of the legal regulation needs to be observed. The function of the rule of subsidiarity is to clarify that LOD is only the *ultima ratio*. LOD is considered an exception, while deceased donation is supposed to be the normal case. Therefore, the national laws that consider LOD subordinate to deceased donation have a rather restrictive attitude towards LOD. As a consequence, it can be assumed that LOD is generally already subsidiary if any suitable organ from a deceased person is available.[856]

(2) National Regulations

The laws in the Czech Republic,[857] Estonia,[858] Finland,[859] Germany,[860] Greece,[861] Hungary,[862] Lithuania,[863] Moldova,[864] Portugal[865] and Slovakia[866] contain a provision which prohibits the performance of LOD when an organ from a deceased

[854] Sec. 8, I, 1, No. 3 Act on the donation, removal and transplantation of organs.

[855] Schreiber (2004), p. 100 f.

[856] Cf. Schreiber (2004), p. 101 ff.

[857] Art. 3 (1) b) Transplantation Act on Donation, Removal and Transplantation of Organs and Tissue and on the Amendment of some Acts.

[858] Sec. 9 (1) 4) Transplantation of Organs and Tissues Act.

[859] Sec. 2 II Act on the Medical Use of Human Organs and Tissues.

[860] Sec. 8, I, 1, No. 3 Act on the donation, removal and transplantation of organs.

[861] Canellopoulou-Bottis, Vol. 7 European Journal of Health Law 427, 433 (2000).

[862] Sec. 203 (1) Act on Health.

[863] Article 10, 1.Law on Donation and Transplantation of Human Tissues, Cells and Organs.

[864] Article 15, (1) Law on the Transplant of Human Organs, Tissues and Cells.

[865] Art. 6, 2 Law on Harvesting and Transplanting Human Organs and Tissues.

[866] Sec. 36 (1) (d) Act on Healthcare, Healthcare-Related Services and on the Amendment and Supplementing of Certain Laws.

person is available. In addition to merely defining LOD as subsidiary to post-mortem donation, all of these countries, except for Hungary, clarify that this subsidiarity rule only applies in cases where the available organ of a deceased is suitable/compatible.

The Finnish and the Belgian subsidiarity rules are more detailed than the national laws of the countries listed in the beginning of this section.

The distinctive feature of Finland's regulation is that LOD is also legal in cases in which a suitable organ from a deceased donor is available, as long as the prognosis of a transplantation of an organ donated by a living person is expected to be significantly better than the prognosis of a transplantation of an organ of a deceased.[867]

With regard to this regulation, a comparison of the results of LOD and post-mortem donation seems inevitable.

At least with regard to kidney donation, the short-term and the long-term results of a living kidney donation are better than those of deceased kidney donation.[868] There are several reasons to explain this. First of all, the cold-ischemia time[869] is shorter for LOD.[870] This is because the procedure can be planned[871] and the organ removal and transplantation take place at the same location.[872] In addition, organs from a living donor usually have an ideal quality,[873] while two-thirds of the organs

[867] Sec. 2 II Act on the Medical Use of Human Organs and Tissues.

[868] Esser, in Höfling (ed.) (2003), p. 201; Gutmann, in Schroth et al. (ed.) (2005), Sec. 8 at 22; cf. Land/Gutmann, Vol. 35 Transplantation Proceedings 926, 928 (2003); Lilie/Krüger, Vol. 81 Der Chirurg 787, 787 (2010); Norba (2009), p. 60; Schreiber (2004), p. 102; Teubner (2006), p. 5; Ugowski (1998), p. 22. A statistic from 1995 reveals that the three-year survival rate for kidneys from spouses was 85 %, from living, unrelated donors who were not married to the recipient 81 %, from parents 82 %, but only 70 % for cadaveric kidneys (Terasaki et al., Vol. 333 The New England Journal of Medicine, 333, 333 (1995)). A more recent statistic from 2009 confirms this result by ascertaining that the three-year survival rate from LOD is almost 90 %, while it is below 80 % for post-mortem donated kidneys. The five-year survival rate is also better for living kidney donation, numerically 84.3 %, but only ca. 70 % for cadaveric kidneys (Deutsche Stiftung Organtransplantation (2010)).

[869] This is generally speaking the time between the chilling of an organ after its blood has been reduced or completely cut off and the time it is warmed by having its blood supply restored (Bundestag printed paper 15/5050 (2005), p. 89; National Cancer Institute (2012)). With regard to organ donation, the cold-ischemia time begins when the organ is cooled with a cold perfusion solution after organ procurement surgery, and ends after the tissue reaches psychological temperature during the implantation procedure (Reference.Md (2007)).

[870] Reference.Md (2007).

[871] Esser, in Höfling (ed.) (2003), p. 201; Guillod/Perrenoud, in: Gutmann et al. (ed.) (2004), p. 158; Gutmann/Schroth (2002), p. 1; Hofer (2006), p. 126; Norba (2009), p. 60 f.; Schreiber, in Broelsch (ed.) (2006), p. 23; Schutzeichel (2002), p. 66; Swiss Dispatch of 12. September 2001 on a Federal Law on Transplantation of Organs, Tissues and Cells, p. 97.

[872] Deutsche Stiftung Organtransplantation (2010), p. 66.

[873] Esser, in Höfling (ed.) (2003), p. 201; Siegmund-Schultze (1999), p. 100.

donated post-mortem are damaged.[874] Thus, the organ from a living donor can take up the assigned tasks right away.[875] Furthermore, as mentioned in the introduction, the longer dialysis is carried out, the worse the long-term function of the transplanted kidney is. Therefore, transplantation before even starting the dialysis treatment is favoured.[876]

The argument in favour of living kidney donation, which claims that it leads to better results than post-mortem kidney donation, does not apply to living liver transplantation.[877] However, since the number of livers donated post-mortem is low, transplantation is too late in many cases. The importance of living liver donation should therefore not be underestimated.[878]

To relate the Finnish regulation to the statistics just presented, at least with regard to kidney donation, the provision of subsidiarity seems to be fulfilled rather seldom. The prognosis of living kidney donation is generally better than the prognosis of a transplantation of an organ donated post-mortem. This aspect leads to the assumption that living kidney donation is usually not subsidiary to post-mortem kidney donation.

The Belgian law stipulates a subsidiarity for certain defined cases only: if the removal of an organ may have serious consequences for the donor or if it relates to a non-regenerative organ, LOD can only be conducted if the recipient's life is at risk and the deceased organ donation does not produce an equally satisfactory result.[879] This is held to be justified by the risks LOD imposes on the donor and by the need for protection of living donors against external pressure, such as money offers to sell their organs.[880] Consequently, a living kidney donation is allowed only if the life of the recipient is in danger and if a post-mortem donation is less effective. The comparison of the results of living kidney donation and post-mortem kidney donation showed that post-mortem donation will usually be less effective than LOD. Since renal failure can be treated with dialysis, it is of course questionable in which cases the recipient might be at risk at all. The usual cases are: children; persons who are not easily compatible and waiting for a compatible post-mortem

[874] Bundestag printed paper 15/5050 (2005), p. 15; Neuhaus, Commission printed paper 15/151 (2004), p. 2; Norba (2009), p. 60.

[875] Esser, in Höfling (ed.) (2003), p. 201; Schreiber (2004), p. 102; Koene, in Price/Akveld (ed.) (1997), p. 33.

[876] Swiss Dispatch of 12. September 2001 on a Federal Law on Transplantation of Organs, Tissues and Cells, p. 97.

[877] A statistic from 2009 shows that the three-year survival rate of living liver transplantations is just slightly higher than the survival rate of a liver donated by a deceased. The five-year survival rate is even higher if the liver was donated post-mortem, namely 55.1 %, while the five-year survival rate for a living liver transplantation is 54.8 % (Deutsche Stiftung Organtransplantation (2010)).

[878] Charité (2010).

[879] Art. 6 Sec. 1 Law on Retrieval and Transplantation of Organs.

[880] Nys, The Belgian Law on retrieval and transplantation of organs, p. 10 (Power Point Presentation).

donor is too risky; and persons for whom the dialysis becomes too difficult. Whether the recipient is considered to be at risk is evaluated on a case by case basis by the doctor. He takes all circumstances into account before coming to a decision. In consequence of this regulation, LOD is not very widespread in Belgium. They have very good results for post-mortem organ donation, though.[881]

In the Austrian medical practice, LOD is considered subsidiary to deceased donation when a post-mortem donated organ provides better or at least the same chances for healing.[882] Post-mortem organ donation is the primary focus and LOD is considered as subsidiary.[883] Though Austria established an opting out system for deceased organ donation,[884] a significant number of organs donated by the deceased are available.[885]

Therefore a significant number of countries consider LOD to be subsidiary to post-mortem donation.

(3) German Situation as an Example

As stated at the beginning of this section, one of the countries that considers LOD a subordinate option is Germany. The situation there is noteworthy insofar as legal experts in this country strongly debate the regulation. The dispute in Germany, therefore, demonstrates the criticism passed on the principle of subsidiarity.

The German *Act on the donation, removal and transplantation of organs* states clearly, without ambiguity, that the removal of an organ from a living person is only allowed if a suitable organ from a deceased donor is not available at the time of the organ removal.[886] In favour of this rule, it is argued that LOD impairs the donor's state of health.[887] As a consequence of this aspect, the principle of subsidiarity is held to be justifiable insofar as it indicates that the possibility of LOD should not result in losing sight of endeavouring organs donated post-mortem.[888]

In contrast, many assert that the current German provision is neither necessary nor capable to achieve this aim. Since LOD leads to better results, many claim the consideration of LOD as a subordinated option leads to the implication that the organ recipient primarily dependent on post-mortem donation is forced by law to

[881] Statistically that means that in 2008, 441 post-mortem kidney transplantations took place, but only 45 living kidney transplantations. The amount stayed almost the same in 2009, when 428 post-mortem kidney transplantations were accomplished and only 49 living kidney transplantations (Information from R. Wilmotte).

[882] Hohmann (2003), p. 155.

[883] Information from K. Kozyga.

[884] Gesundheit Österreich GmbH (2009).

[885] Information from K. Bruckmueller.

[886] Sec. 8, I, 3 Act on the donation, removal and transplantation of organs.

[887] Ach/Anderheiten/Quante (2000), p. 91.

[888] Bundestag printed paper 13/4355 (2006), p. 20; Gutmann (2006), p. 69; Schreiber/Wolfslast, Vol. 10 Medizinrecht 189, 193 (1992).

receive the worse therapy.[889] In addition, it is also brought out that a certain amount of patients have concerns about receiving an organ from a deceased person.[890] For those, the notion of living with an organ that was removed from a deceased body seems intolerable.[891] Hence, it is opined that there is no objective reason that the recipient should not have the possibility to choose for herself whether she gets an organ from a deceased or from a living person transplanted.[892] The regulation of subsidiarity is also considered problematic with respect to the fact that an insufficient number of organs are available to cure every patient on the waiting list. Many question why patients who could possibly receive an organ donated by a living person should be forced to get an organ donated post-mortem. This also entails that the next person on the waiting list will not receive an organ.[893]

Independent from the criticism of the principle of subsidiarity with regard to the German regulation of subsidiarity, which has general validity, it is questionable, whether the subsidiary principle has practical relevance at all. First, since LOD leads to better results, one could question if every organ donated post-mortem can be considered as "suitable" in terms of the regulation.[894] The exact meaning of this phrasing is unclear. Second, the decisive point of time in order to decide whether an organ donated by a deceased is available is the moment of the organ removal. The wording of the German provision shows that the patient and the surgeon can reschedule the organ removal or turn the offered organ down if this offer takes place during the preparation of the LOD, but prior to the organ removal.[895] Third, the potential recipient can delete herself from the waiting list.[896] Fourth, a breach of the provision does not result in a criminal action.[897]

Nevertheless, the German Parliamentary Enquete Commission does still argue in favour of the rule of subsidiarity. Even though it admits that the chance to get an organ donated post-mortem is only theoretical due to the perennial waiting time, it does not consider the rule of subsidiarity as unnecessary. This principle aims to encourage efforts to enlarge the number of organs donated post-mortem.[898] However, there is no reason to assume that a liberalisation of LOD would impede the efforts to improve the number of post-mortem donated organs.[899] In addition, the selection of organs that can possibly be donated post-mortem is higher than those

---

[889] Edelmann, Vol. 50 Versicherungsrecht 1065, 1068 (1999); Gutmann, Vol. 15 Medizinrecht 147, 152 (1997).

[890] Gutmann, Vol. 15 Medizinrecht 147, 152 (1997).

[891] Esser (2000), p. 196.

[892] Edelmann, Vol. 50 Versicherungsrecht 1065, 1068 (1999).

[893] Gutmann (2006), p. 74.

[894] Ibid., p. 75.

[895] Gutmann (2006), p. 75.

[896] Bundestag printed paper 15/5050 (2005), p. 50; Gutmann (2006), p. 75.

[897] Gutmann (2006), p. 75.

[898] Bundestag printed paper 15/5050 (2005), p. 75.

[899] Gutmann (2006), p. 81.

that might be donated by living persons, since organs like the heart and cornea can only be donated by a deceased. Experts believe that the effort to enlarge the amount of organs donated post-mortem will not decrease with a liberalisation of LOD, since several organs can only be donated by the deceased and not by living persons.[900] Hence, they desire the use of LOD in addition to post-mortem donation.[901]

Independent of the severe criticism of the rule of subsidiarity, several countries have still established this principle in their national law.

bb) No Subsidiarity of Living Organ Donation as Against Post-Mortem Donation

In contrast to the countries that consider LOD as subordinate to deceased donation, several countries do not have a ranking order for LOD and post-mortem donation.

Bulgaria, England, Italy, Latvia, the Netherlands, Norway, Poland, Scotland, Slovenia, Spain, Sweden and Switzerland do not explicitly state that post-mortem donation and LOD are ranked equally. These countries do not regulate the relationship between LOD and post-mortem donation at all. That leads to the conclusion that these two types of organ donation are equal.

The legal and medical situation regarding LOD has changed over time in both the Netherlands and Switzerland.

In the Netherlands, the general practice did not always accord with the legal situation. In the recent past, several transplant surgeons themselves were not keen on removing an organ from a healthy living person because it offered no direct benefit to this person. This attitude has changed, however, due to the advantages of LOD in comparison to deceased donation. In light of this, LOD as the preferred option is now being actively promoted. In general practice, post-mortem donation is still not completely ignored in cases where a patient is considering LOD. In fact, the particular patient is also put on the waiting list for post-mortem donated organs. If a kidney donated post-mortem becomes available during the work-up, the patient can still decide to implant this kidney and not the kidney donated by a living donor.[902]

The historical development of the Swiss *Federal Act on the Transplantation of Organs, Tissues and Cells* reveals that the primary draft of the law in 1999 contained a provision ruling that LOD may only be applied in cases where the suitable deceased organ is not available and when the recipient cannot be treated with any other therapeutic method with comparable benefit.[903] Many have criticised the subsidiarity of LOD to post-mortem donation. They believe this is unreasonable since LOD leads to better results on average than post-mortem donation. Therefore, according to the recipient's view, it was inapprehensible for him to get a deceased

---

[900] Schutzeichel (2002), p. 298.

[901] Gutmann (2006), 83.

[902] Information from M. Bos.

[903] Draft of the Swiss Transplantation Act of Dec. 1999, Art. 17 I, b, c.

organ which entails a worse prognosis, even though LOD is possible.[904] Since LOD involves risks,[905] experts argued that the medical risks for the donor have to be evaluated prior to the LOD. If LOD is justifiable, it is incomprehensible to think it should not be up to the autonomous decision of the donor and whether he feels up to taking these justifiable risks or not.[906] As a consequence of this critique, the subsidiary principle was deleted in the revised draft.[907]

In Poland, the absence of the subsidiary principle is oftentimes criticised and several legal experts demand a legal rule stating that LOD is only allowed in cases where post-mortem donation is not possible.[908]

It is obviously apparent that the countries considered do not conform in answering the question of whether LOD is subsidiary to post-mortem donation.

c) Relationship Between Living Organ Donation and Alternative Therapies

The second issue regarding subsidiarity, namely whether LOD should be subordinated to alternatives therapies, is also not answered uniformly.

aa) Subsidiarity of Living Organ Donation as Against Alternative Therapies

Several of the examined countries rule that LOD is only lawful when alternative therapies are either not available or less effective. The exact meaning of this is unclear.

---

[904] Swiss Dispatch of 12. September 2001 on a Federal Law on Transplantation of Organs, Tissues and Cells, p. 196.

[905] Ibid., p. 195.

[906] Gutmann (2011), p. 9; Swiss Dispatch of 12. September 2001 on a Federal Law on Transplantation of Organs, Tissues and Cells, p. 71.

[907] Gutmann/Schroth (2002), p. 78 f.

[908] Weigend/Zielinska, Vol. 14 Medizinrecht 445, 451 (1996).

(1) National Regulations

The laws of Bulgaria,[909] the Czech Republic,[910] Finland,[911] Greece,[912] Lithuania,[913] Moldova,[914] Slovakia,[915] Slovenia[916] and Switzerland[917] contain a provision stating that performing LOD is only legitimate when other methods of therapy are less effective or do not provide comparable efficiency. These national laws only regulate LOD insofar as it is illegal in cases where an equally effective alternative therapy exists. No further details are provided.

In contrast, the rule of subsidiarity in the Netherlands is only applicable under certain circumstances. It is not as general as the provisions in the countries mentioned above. The regulation in the Netherlands states that "[i]f it is reasonable to assume that the removal of the organ during the donor's life will have permanent consequences for the donor's health, the organ shall be removed only if the life of the recipient-to-be is at risk and if no other equally suitable treatment option exists."[918] This regulation leads to the assumption that LOD is *ultimum refugium* in the Netherlands. However, in the medical practice that is not the case. While observing the medical practice, one must distinguish between the donation of kidneys and livers. In the Netherlands, LOD almost completely concentrates on the field of kidney donation and rarely considers liver donation.[919]

For a thorough understanding of the Dutch situation, the general background has to be explained. On the one hand, living kidney donation has permanent consequences because the donor has to live with one kidney afterwards. This does not mean that the health and the quality of life of the donor are necessarily reduced. The kidney recipient, on the other hand, can survive with dialysis. He is still at risk, though, because the mortality rates on dialysis are high and the life expectancy on dialysis is greatly reduced. As a result, living kidney donation is actively promoted by patients and donors in the Netherlands, not so much by the

---

909 Art. 4. (2) Law of the Transplantation of Organs, Tissues and Cells states that "[t]ransplantation shall be carried out only when other methods of therapy are less effective or not applicable", thus, refers to transplantation in general, what also includes LOD.

910 Art. 3 (1) b) The Transplantation Act on Donation, Removal and Transplantation of Organs and Tissues and on the Amendment of some Acts.

911 Sec. 2 II Act on the Medical Use of Human Organs and Tissues.

912 Canellopoulou-Bottis, Vol. 7 European Journal of Health Law 427, 433 (2000).

913 Article 10, 1. Law on Donation and Transplantation of Human Tissues, Cells and Organs.

914 Article 3, b) (general) and Article 16, (1) (only for LOD) Law on the Transplant of Human Organs, Tissues and Cells.

915 Sec. 36 (1) (e) Act on Healthcare, Healthcare-Related Services and on the Amendment and Supplementing of Certain Laws.

916 Art. 2 II The Removal and Transplantation of Human Body Parts for the Purpose of Medical Treatment Act.

917 Art. 12, d. Federal Act on the transplantation of organs, tissues and cells.

918 Sec. 3, 3 Organ Donation Act.

919 Information from M. Bos.

government, though. Living liver donation is mostly considered in cases where the patient suffering from liver failure is in imminent danger of dying before a suitable organ donated by a deceased can be found. It is also prominent for paediatric recipients because a child can be at acute risk of dying and the risk for the living donor, who is usually a parent of the sick child, when removing a small left lateral segment, is modest and the liver will regenerate in the long run. Therefore, the health consequences for the living liver donor are temporary (perioperative), not permanent.[920] As a result, it can be observed that LOD, especially living kidney donation, is a rather common procedure and is not used as restrictively as the Dutch legal regulation suggests.

In Austria, practical experiences show that LOD is considered *ultima ratio*; it is only performed when no other alternative therapy is available with the same or a better outcome.[921] In fact, each case is examined individually, taking urgency and other factors into account.[922]

As this section has shown, a considerable number of countries deem LOD as subsidiary to equally efficient therapies.

(2) Exact Meaning of Efficiency of the Alternative Therapy

How can the effectiveness of certain therapies in opposition to living organ transplantation be evaluated? Almost all laws remain silent on this question. The only country that deals with this problem is Switzerland: The Swiss law states that the Swiss Federal Council may specify which other therapeutic methods do not offer a comparable benefit to the recipient.[923] Although Switzerland provides this possibility, the Federal Council has not acted yet, as it did not deem it necessary to complete this regulation.[924] But in the *Swiss Dispatch of 12 September 2001 on a Federal Law on Transplantation of Organs, Tissues and Cells*, it is stipulated that haemodialysis and peritoneal dialysis cannot be considered as therapeutic treatment with comparable benefit, since the quality of life of the patient is impaired due to several hospital stays, a possible job change and with regard to spare time activities.[925]

Deceased organ donation cannot be considered an alternative therapy, because this treatment involves the transplantation of an organ. Therefore, the same kind of therapy is applied and only the source of the organ differs.[926]

---

[920] Ibid.

[921] Hohmann (2003), p. 155.

[922] Information from K. Bruckmueller.

[923] Art. 15^2 Federal Act on the Transplantation of Organs, Tissues and Cells.

[924] Information from M. Monnier

[925] Swiss Dispatch of 12. September 2001 on a Federal Law on Transplantation of Organs, Tissues and Cells, p. 148.

[926] Hofer (2006), p. 125.

Investigating the initial question about the exact meaning of an equally effective alternative therapy makes a comparison of transplantation and alternative therapies inevitable.

The main advantage of dialysis — as the most common alternative therapy for kidney transplantation — is that no third person becomes involved. Dialysis does however have several disadvantages as already mentioned. One important benefit of kidney transplantation is that the patient suffering from renal failure can live independently of dialysis after the transplantation procedure, while a dialysis patient has to undergo dialysis four or five times a week.[927] Transplantation can also make the integration into employment possible.[928] Furthermore, the costs for kidney transplantation are lower than those for dialysis.[929] Therefore, a successful kidney transplantation is the best type of therapy for renal failure.[930] For liver failure, an alternative therapy is only available for a short time, making transplantation the only life-saving option.[931]

As seen, several countries consider LOD as subsidiary to alternative therapies. Unfortunately, these national laws do not provide any further explanation of their rules. It seems unlikely that they would, with regard to renal failure, consider dialysis as an equally efficient therapy.

bb) No Subsidiarity of Living Organ Donation as Against Alternative Therapies

In contrast, Belgium, England, Estonia, Germany, Hungary, Italy, Latvia, Norway, Poland, Portugal, Scotland, Spain and Sweden do not have any regulation regarding the relation of LOD to alternative therapies. Consequently, it is assumed that the two types of treatment are equal.

In conclusion, there is no general standard in the countries considered regarding the relationship between LOD and alternative treatment therapies.

d) International and European Tendencies

Independently of the national laws, a number of international and European organisations and documents have also addressed the matter of subsidiarity.

---

[927] Swiss Dispatch of 12. September 2001 on a Federal Law on Transplantation of Organs, Tissues and Cells, p. 133.

[928] Ibid., p. 34.

[929] Swiss Dispatch of 12.September 2001 on a Federal Law on Transplantation of Organs, Tissues and Cells, p. 58; Weigend/Zielinska, Vol. 14 Medizinrecht 445, 449 (1996).

[930] Swiss Dispatch of 12. September 2001 on a Federal Law on Transplantation of Organs, Tissues and Cells, p. 61.

[931] Bock (1999), p. 53; Norba (2009), p. 33; Pichlmayr, in Toellner (ed.) (1991), p. 22.

In 1991 the WHO stated, "[o]rgans for transplantation should be removed preferably from the bodies of deceased persons."[932] In the WHO's 2010 guidelines, in contrast, the WHO stated, "[d]onation from deceased persons should be developed to its maximum therapeutic potential, but adult living persons may donate organs as permitted by domestic regulations."[933]

Similar to the WHO, the Committee of Ministers of the Council of Europe declares in its *Recommendation on Liver Transplantation from Living Donors* from 30 September 1997 that "LRLD[934] should be considered only when there is a shortage of cadaver organs, i.e., when alternatives that do not carry the risks incurred by a living donor have been exhausted."[935] Furthermore, the Committee states, "[p]otential recipients of a living liver donation should have been previously assessed as suitable for cadaveric transplant and, if considered suitable for LRLT, should still be retained on the cadaveric programme in case a suitable liver becomes available."[936]

The *Convention on Human Rights and Biomedicine* and the *Additional Protocol on Transplantation of Organs* also address the issue of subsidiarity. Pursuant to Art. 9 of the *Additional Protocol on Transplantation of Organs*, organs may only be removed from living persons when neither a suitable organ from a deceased, nor any other alternative therapeutic method of comparable efficiency, is available. The *Convention on Human Rights and Biomedicine* itself contains the same regulation in Art. 19.1. Since LOD always represents risks for the donor, these provisions are held to be justified by those risks.[937] With regard to kidney transplantation, dialysis is not considered to provide similar results for the patient's quality of life in comparison with those obtained by a kidney transplant.[938]

The European Parliament's *Resolution of 22 April 2008 on Organ Donation and Transplantation* approaches the issue of subsidiarity differently by stating that "living donation should be seen as complementary to post-mortem donation."[939] Hence, it stresses that LOD should be considered as complementary to post-mortem donation.[940]

---

[932] WHO (1991), p. 5 ff.

[933] WHO Guiding Principles on Human Cell, Tissue and Organ Transplantation, 2010, Guiding Principle 3.

[934] LRLT = Living Related Liver Transplantation.

[935] Appendix i) Recommendation No.R (97) 16 on Liver Transplantation from Living Donors, Council of Europe, Committee of Ministers, (Sept. 30, 1997).

[936] Ibid., Appendix ii).

[937] Secretary General of the Council of Europe (1996), Art.19, No. 118.

[938] Secretary General of the Council of Europe (1996), Art.19, No. 119; Explanatory Report to the Additional Protocol concerning Transplantation of Organs and Tissues of Human Origin, Art.9, No. 60.

[939] Resolution of 22 April 2008 on organ donation and transplantation: Policy actions at EU level (2007/2210 (INI)), No. 15.

[940] Nys (2009), p.11.

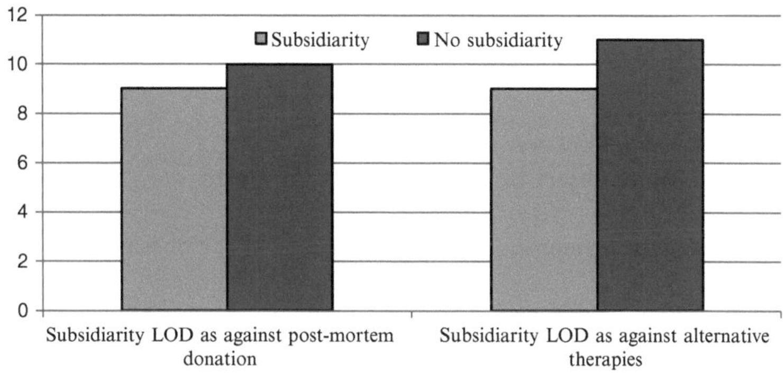

Fig. 4 Summary of the subsidiary principle

In summary, international and European recommendations and contracts tend to support the principle of subsidiarity.[941] However, the European Parliament's *Resolution of 22 April 2008 on Organ Donation and Transplantation*, does not codify a subsidiarity of LOD. Currently, none of these documents are legally binding.

e) Summary of the Principle of Subsidiarity

After evaluating the national laws, it is evident that there is basically no consensus with regard to the issue of subsidiarity. This is also apparent in Fig. 4.

7) Procedural Issues

Several national laws contain rules of procedure to ensure that LOD is carried out in a safe way. Both the degree and the content of those rules in the countries considered differ. Some of the aspects regarded in the following are definitely procedural, while others cannot clearly be classified as either material or procedural.

(1) A significant amount of countries regulates exactly what kind of person is authorised to execute organ removals and transplantations and some countries determine by law (2) where LODs are allowed to take place. (3) Some national laws do not only focus on the donation and transplantation process, but also contain regulations about the period of time after the transplantation by defining the requirement of post-care. (4) Furthermore, almost all countries have some sort of independent commission or committee that is involved in the process of LOD and

[941] Gutmann/Schroth (2002), p. 77.

(5) some national laws provide the possibility of legal protection with regard to the process of LOD.

a) Transplantation Subject to Execution by a Physician

Many of the examined countries determine what kind of person is permitted to carry out LOD.

Belgium,[942] Germany[943] and Moldova[944] explicitly require a physician to carry out the organ removal. In Belgium, this is mentioned for the transplantation as well. In Moldova, a surgeon has to be authorised by the Ministry of Healthcare based on the proposal of the Agency and in accordance with the criteria approved by the Government to carry out organ transplantations.[945]

Sweden does not demand a surgeon by law to perform the LOD. But in the *Guidance on the provisions*, many actions under the *Act* are of the sort that they should be performed only by physicians with special competence and with qualified medico-technical equipment at their disposal.[946]

In Bulgaria, "[a]ll medical establishments, carrying out activities, related to transplantation, shall assign a person from its personnel, who organises, controls and bears responsibility for the expertise, taking, treatment, processing, labelling, storing, provision and implantation of organs, tissues and cells and the announcement of serious adverse reactions and serious incidents."[947] This person must have a higher education with the degree "Master of Medicine" and at least 2 years of professional experience in the particular field.[948] After this person accepts responsibility, he has to attend a training course at least once every 2 years.[949]

Sec. 4 (1) of the Estonian *Handling and Transplantation of Cells, Tissues and Organs Act* states that "[s]pecial medical care providers who hold an activity licence for the handling of cells, tissues and organs which is issued on the basis of this Act (hereinafter handler) have the right to handle cells, tissues and organs." Handling of organs includes the removal for transplantation purposes.[950] Sec. 4 (2) applies to transplantation and determines that the activity licence is issued on the basis of the *Health Services Organisation Act*.

---

942 Art. 3 Law on Retrieval and Transplantation of Organs.

943 Sec. 8 I No. 4 Act on the donation, removal and transplantation of organs.

944 Art. 9 Law on the Transplant of Human Organs, Tissues and Cells.

945 Ibid.

946 The Ministry of Health and Social Affairs Sweden 1997, The Swedish Transplant Act, Guidance on the provisions, Sec. 13.

947 Art. 15 d (1) Law of the Transplantation of Organs, Tissues and Cells.

948 Ibid., Art. 15 d (2).

949 Ibid., Art. 15 (3).

950 Sec. 2 (2) Handling and Transplantation of Cells, Tissues and Organs Act.

Finland only demands that the doctor who treats the recipient to not be involved in decisions about the organ removal.[951] The Austrian *Position Paper* is similar to the Finnish rule.

Thus, several laws determine the person who is allowed to accomplish LOD. It is a general standard for a physician to carry out the organ removal and the transplantation. In addition, the physician who performs the transplantation is oftentimes prohibited from carrying out the organ removal.

b) Requirements Concerning Location of Transplantation

The process of organ transplantation is a medical operation. Consequently, it has to be performed at a location that is suited for such a procedure. A significant amount of the national laws considered regulates permissible locations.

The Portuguese law requires performing the transplantation in a private or public hospital; no additional prerequisites are required.[952]

In Belgium, organ removal and transplantation have to be carried out in a hospital as considered in the *Law about Hospitals*.[953]

In Bulgaria, every medical institution that wants to be involved in organ transplantations needs a specific "permit by the order Art. 48, para 1 of the Law of the medical establishments."[954]

In the Netherlands, a hospital pursuant to the *Organ Donation Act* is "an institution recognized or appointed as a hospital or nursing establishment pursuant to the *National Health Insurance Act* (Bulleting of Acts, Orders and Decrees, 1964, 392) or the *Exceptional Medical Expenses Act* (Bulletin of Acts, Orders and Decrees, 1967, 655) or a department within such an establishment."[955] According to the *Quality Medical Care Act*, the hospitals that are permitted to transplant organs are primarily responsible for the quality of medical care. The medical service provided "has to be effective, efficient, purposeful to the patients and geared to the needs of the patients."[956] In the Netherlands, it is stipulated by law that the amount of transplant centres must not exceed seven. This is to ensure a sufficient quality and to increase efficiency.[957]

Italy also restricts the amount of centres that are allowed to perform organ transplantations. In contrast to the Netherlands, Italy has not appointed a certain limit by law. Rather, it regulates that centres will not receive an extension of their

[951] Sec. 2 III Act on the Medical Use of Human Organs and Tissues.

[952] Art. 3, 1 Law on Harvesting and Transplanting Human Organs and Tissues.

[953] Art. 3 Law on Retrieval and Transplantation of Organs.

[954] Art. 13 (1) Law of the Transplantation of Organs, Tissues and Cells.

[955] Sec. 1, f) Organ Donation Act.

[956] Akveld, Falcone Project, Netherlands, p. 8.

[957] Swiss Dispatch of 12. September 2001 on a Federal Law on Transplantation of Organs, Tissues and Cells, p. 121.

license to perform transplantations if they do not accomplish a particular amount of transplantations. Despite that not all centres reached the prescribed amount, at least until 2001, all authorisations have been renewed.[958]

In Spain, the Ministry of Health explicitly authorises the centres in which human organ recovery can be performed.[959] According to Art. 11 *Royal Decree 2070/ 1999,* the centre must meet several requirements to receive this authorisation, i.e., a sufficient number of medical and custodial employees and proven experience in proper assessment of the donor and with regard to carrying out the removal.[960] If children are involved in the transplantation, a special authorisation is required.[961]

The regulations in Moldova[962] and Poland[963] are similar to the Spanish rule. Organ transplantation is also only allowed in institutions authorised by the Ministry of Healthcare.

The rule in Norway is similar as well, because LOD "may only be carried out at hospitals approved therefore by the King."[964]

The Swedish *Transplant Act* authorises the Government, or the authority appointed by the Government, to prescribe that actions under this *Act* may only be performed in hospitals referred to in Sec. 5 of the *Health and Medical Services Act*, or in another health care unit or the medical sector generally.[965] Permissible institutions include private hospitals and the national departments of forensic medicine.[966]

In Switzerland, transplanting organs in regular hospitals is prohibited. Organs have to be transplanted in transplant centres authorised by the Federal Office. This authorisation can only be achieved if necessary technical and operative requirements are met, a suitable quality assurance system is in place and the quality of the transplantations is ensured.[967]

In Slovakia, the providers responsible for organ donation establish the transplantation centres and removal centres.[968]

Thus, several countries define both the person who is allowed to perform organ removals and transplantations and the location in which those operations may be performed.

---

[958] Ibid.

[959] Art. 9, 6. Royal Decree 2070/1999.

[960] Ibid., Art. 11, 1, b.

[961] Ibid., Annex II, 3.

[962] Art. 9 Law on the Transplant of Human Organs, Tissues and Cells.

[963] Art. 36, 1.The Cell, Tissue and Organ Recovery, Storage and Transplantation Act.

[964] Sec. 5 Act relating to transplantation, hospital autopsis and the donation of bodies etc.

[965] Sec. 13 Transplant Act.

[966] The Ministry of Health and Social Affairs Sweden 1997, The Swedish Transplant Act, Guidance on the provisions, Sec. 13.

[967] Art. 27 Federal Act on the Transplantation of Organs, Tissues and Cells.

[968] Sec. 35 (3) Act on Healthcare, Healthcare-Related Services and on the Amendment and Supplementing of Certain Laws.

c) Post-Care

A specific problem that often arises after the organ donation is the immune system's recognition of the external antigens of the organ transplanted into her body. Therefore, in order to assure the success of an organ donation, immunosuppression[969] is necessary.[970] One must also take into account that the organ removal could be detrimental to the donor's health.

Thus, the post-care requirement is meant to guarantee an ideal medical and psychological care for the donor and recipient, which ensures the success of the transplantation in the long run.[971]

Several national laws have also suggested that the transplanting institution should be held responsible even after the organ transplantation.

aa) International and European Level

Not all international and European LOD documents state the post-care requirement.

For example, both the *Recommendation on Liver Transplantation from Living Donors* and the *Resolution on Harmonisation of Legislation of Member States Relating to Removal, Grafting and Transplantation of Human Substances* do not refer to post-care.

The *Additional Protocol on Transplantation of Organs*, in contrast, states in Art. 7 that "[a]ppropriate medical follow-up shall be offered to living donors and recipients after transplantation." This includes short-term and long-term follow up. A short-term follow up is necessary to assure a complete recovery from the procedure as such. A long-term follow up is essential for the recipient who needs immunosuppressive therapy and for the donor to make the identification of possible long-term effects feasible. Neither donors nor recipients can be forced to accept the recommended long-term follow-up.[972]

The *Directive 2010/53/EU* also attaches importance to post-care.[973]

The WHO *Guiding Principles on Human Cell, Tissue and Organ Transplantation* states in Guiding Principle 3 that the follow-up must be well organised.[974]

---

[969] That means suppressing the defence mechanism of the recipient's body right before or at the moment of the transplantation. This is usually done by the administration of combined pharmaceuticals (Hohmann (2003), p. 43 f.).

[970] Hohmann (2003), p. 43 f.

[971] Esser, in Höfling (ed.) (2003), p. 231.

[972] Art. 7 Additional Protocol to the Convention on Human Rights and Biomedicine concerning Transplantation of Organs and Tissues of Human Origin, Explanatory Report.

[973] Art. 15 (4) Directive 2010/53/EU of the European Parliament and of the Council of 7 July 2010 on standards of quality and safety of human organs intended for transplantation.

[974] WHO Guiding Principles on Human Cell, Tissue and Organ Transplantation, 2010, Guiding Principle 3.

With regard to the considered international and European documents, post-care monitoring of the donor and/or the recipient is not always required, but it is also not completely ignored.

bb) National Laws

Not all national laws make post-care mandatory. The national laws that require post-care differ insofar as some stipulate the requirement for donor and recipient post-care, while others order post-care for one participant only.

In Austria,[975] Germany[976] and the Netherlands,[977] donor and recipient are included in a post-care process.

Only German law stipulates the post-care requirement for donor and recipient. Both participants must declare their willingness to participate in post-operative medical care.[978] However, German law does not sanction the violation of a post-care obligation.[979] In addition, the transplantation centres have to ensure that measures are taken before and after the organ transplantation to provide necessary psychological support to the patients in the hospital.[980] Many consider the imperfect implementation of this regulation in medical practice to be rather problematic. The psychological post-care offered to donors and recipients varies significantly between the transplant centres.[981]

In Austria, according to the *Position Paper*, the transplantation centre has the duty to provide post-care; i.e., the supervision of the postoperative process of recovery until the donor has completely recovered providing long-term medical examinations and medical treatment that is necessary in connection with LOD.[982]

In the Netherlands, the recipient is entitled to appropriate post-care as part of his health insurance benefit package. For the donor, post-care immediately follows his operation. This is mandatory in the sense that the donor's insurance will cover the short-term medical expenses and other expenses incurred by the operation. In the long-term, follow-up care for the donor is not strictly mandatory. However, it becomes more and more common for the transplant centres to encourage the donor to attend yearly outpatient check-ups. The costs are now fully covered by the donor's health insurance.[983]

---

975 Position Paper.

976 Sec. 8 III 1 Act on the donation, removal and transplantation of organs.

977 Information from M. Bos.

978 Sec. 8 III 1 Act on the donation, removal and transplantation of organs.

979 Gutmann, in Schroth et al. (ed.) (2005), Sec. 8 at 51; Nickel/Schmidt-Preisigke/Sengler (2001), p. 112; Walter, in Spickhoff (ed.) (2011), p. 2390.

980 Sec. 10 II Nr. 5 Act on the donation, removal and transplantation of organs.

981 Gutmann, in Schroth et al. (ed.) (2005), Sec. 10 at 17.

982 Position Paper.

983 Information from M. Bos.

In contrast, Bulgaria,[984] the Czech Republic,[985] Portugal[986] and Spain[987] only concentrate on one of the patients.

Bulgaria merely requires the post-care of the recipient. The medical establishment has to observe him continuously, has to control his health status and has to assure the post-treatment.[988]

In contrast, the Czech Republic,[989] Portugal[990] and Spain[991] only focus on the post-care of the donor. In Spain, the donor must receive medical care in order to recover.[992] In Portugal, the donor also has to be informed about the necessary care to observe afterwards.[993]

All in all, only a small amount of countries mention post-care at all and even less countries make it mandatory.

After describing the issues that have both material and procedural aspects, it will be focussed on a commission for LOD — an aspect that is clearly procedural.

d) Commission for Living Organ Donation

Almost every country considered has established some kind of independent authority as a safeguard for LOD.[994] The particular kind of authority and the composition of that authority differ. In addition, the authorities' duties and responsibilities vary. Since the regulations are extremely diverse, groups according to the tasks the authority has to carry out will be established. (1) All the countries that require a permit by an authority for every LOD will be listed and (2) afterwards, the countries that demand such a permit for certain cases will be enumerated. (3) Beyond that, the countries that state that an authority has to supervise the process of organ transplantation, even though the type of supervision differs will be presented. (4) Last, the few countries that do not require the involvement of an authority in the transplant process as such, but stipulate that every transplantation, or certain issues connected thereto, are communicated, will be described.

---

[984] Art. 34 Law of the Transplantation of Organs, Tissues and Cells.

[985] Art. 6 (4) Art. 6 (5) Transplantation Act on Donation, Removal and Transplantation or Organs and Tissues and on the Amendment of some Acts.

[986] Art. 9, 1 Law on Harvesting and Transplanting Human Organs and Tissues.

[987] Art. 9, 7. Royal Decree 2070/1999.

[988] Art. 34 Law of the Transplantation of Organs, Tissues and Cells.

[989] Art. 6 (4) Transplantation Act on Donation, Removal and Transplantation or Organs and Tissues and on the Amendment of some Acts.

[990] Art. 9, 1 Law on Harvesting and Transplanting Human Organs and Tissues.

[991] Art. 9, 7. Royal Decree 2070/1999.

[992] Ibid.

[993] Art. 7 Law on Harvesting and Transplanting Human Organs and Tissues.

[994] Cf. Manyalich et al., Vol. 41 Transplantation Proceedings 2021, 2022 (2009).

aa) Authorisation for Every Living Organ Donation

Many of the examined countries rule that every LOD must first be authorised by a commission. This is the case in Belgium,[995] England,[996] Finland,[997] Moldova,[998] Scotland,[999] Slovakia[1000] and Spain,[1001] where the required permit is only granted if all the legal requirements have been fulfilled.

In Belgium, each removal must be the object of a multidisciplinary consultation.[1002] In addition, the medical officials elected by the King of Belgium must ensure that the relevant legislation is applied.[1003] The establishment of an ethics committee in every Belgian general and psychiatric hospital has been obligatory since 1995. Only small hospitals with less than six physicians are exempted from this obligation.[1004] In 1996, the Ministers of Justice and of Public Health established the Federal Council on Bioethics.[1005] More than one independent authority is thus involved in the process of LOD in Belgium.

In Spain, the head of the transplant medical unit has to give her consent for transplantation.[1006] In addition, the transplanting hospital's ethic commission must comment on the transplantation.[1007]

In contrast, only one authority is involved in LOD in Finland[1008] and Moldova.[1009] In Finland, the National Board of Medicolegal Affairs has to approve every organ removal.[1010] In Moldova, the Approval Commission, an independent authority established by the Ministry of Healthcare, has to grant authorisation.[1011]

In England, the Human Tissue Authority has to approve every LOD through an independent assessment process.[1012] Its members are appointed by the Secretary of State, the National Assembly for Wales and the relevant Northern Ireland

---

995 Art. 8*bis* Law on Retrieval and Transplantation of Organs.

996 Sec. 13 ff. Human Tissue Act; information from D. Price.

997 Sec. 4 II Act on the Medical Use of Human Organs and Tissues.

998 Art. 20 Law on the Transplant of Human Organs, Tissues and Cells.

999 Reg. 2 (3) Human Organ and Tissue Live Transplants (Scotland) Regulations 2006.

1000 Sec. 36 (5) Act on Healthcare, Healthcare-Related Services and on the Amendment and Supplementing of Certain Laws.

1001 Art. 15, 4. Royal Decree 2070/1999.

1002 Art. 8*bis* Law on Retrieval and Transplantation of Organs.

1003 Ibid., Art. 16.

1004 Nys (2009), p. 91.

1005 Ibid., p. 92 f.

1006 Art. 15, 4. Royal Decree 2070/1999.

1007 Ibid., Art. 9, 2.

1008 Sec. 4 II Act on the Medical Use of Human Organs and Tissues.

1009 Art. 20 Law on the Transplant of Human Organs, Tissues and Cells.

1010 Sec. 4 II Act on the Medical Use of Human Organs and Tissues.

1011 Art. 20 Law on the Transplant of Human Organs, Tissues and Cells.

1012 Sec. 13 ff. Human Tissue Act; information from D. Price.

department.[1013] The Human Tissue Authority, hence, is a state commission. Between April 6, 2010 and September 1, 2010, it approved 494 LODs; there is no mention of any rejections or deferred decisions.[1014] The exact procedure differs and depends on whether donor and recipient are genetically or emotionally related or if neither is.[1015]

The Scottish Government asked The Human Tissue Authority to approve transplants from living donors.[1016] In addition, the responsible registered medical practitioner must refer the matter to the Scottish Ministers.[1017]

In Slovakia, every LOD has to be approved. The institution that executes the required approval differs. If the LOD takes place for the benefit of a genetic relative, the board of consultants of the transplantation centre has to approve,[1018] which consists of medical workers. They are assigned by the provider.[1019] In case the organ donation is "for the purposes of transfer into the body of a person remotely genetically related to the donor or a person not related genetically to the donor," the Ministry of Health determines a board of consultants to give its approval.[1020]

Thus, numerous countries consider the approval of an independent authority an inherent part of justifying LODs.

bb) Authorisation Only in Selected Cases

In contrast to the countries that require a permit for every LOD, Bulgaria,[1021] the Czech Republic,[1022] Hungary,[1023] Portugal[1024] and Sweden[1025] demand such an approval merely for defined cases. To turn the argument on its head, it means that LOD only requires a special permit in certain cases.

In Bulgaria, the Ethical Commission has to give its approval only in two cases. Namely if the donor either actually cohabitates with the recipient for more than

[1013] 1 (1) Human Tissue Act 2004, Schedule 2 – The Human Tissue Authority.

[1014] Information from D. Price.

[1015] Fateh-Moghadam (2008), p. 292.

[1016] Information from D. Price.

[1017] Reg. 2 (3) Human Organ and Tissue Live Transplants (Scotland) Regulations 2006.

[1018] Sec. 36 (5) Act on Healthcare, Healthcare-Related Services and on the Amendment and Supplementing of Certain Laws.

[1019] Ibid. Sec. 2 (5).

[1020] Ibid. Sec. 36 (5).

[1021] Art. 12 (4) Bulgarian Law of the Transplantation, of Organs, Tissues and Cells.

[1022] Art. 5 (1) Transplantation Act on Donation, Removal and Transplantation of Organs and Tissues and on the Amendment of some Acts.

[1023] Sec. 206 (3) Act on Health.

[1024] Art. 6, 3 Law on Harvesting and Transplanting Human Organs and Tissues.

[1025] Sec. 8 II Transplant Act.

2 years, or if the donor is the biological parent of the recipient, but has not been legitimised.[1026]

In the Czech Republic, the Ethical Committee is only involved in the organ removal if donor and recipient do not have a close relationship. The Ethical Committee is an independent committee that has at least five members, namely physicians, one clinical psychologist and a lawyer. It should have a certain independence from the responsible health care facility. Therefore, at least two thirds of the committee may not be employed or have a similar working relationship with the acting health care facility.[1027] The law does not contain rules regarding how the Committee has to come to a decision in case a donation between people without a close relationship is intended. It only regulates that the Committee has to find out the donor's reasons for the donation and has to judge those. However, the reasons that would bar an authorisation are not stated.[1028]

In Hungary, a hospital ethics committee only gets involved if the donor is neither a lineal kinsperson of the recipient nor a sibling of a lineal kin of the recipient nor a sibling of the recipient nor a lineal kin of a sibling of the recipient. In this case, donor and recipient have to request a hospital ethics committee's consent to the organ removal after it has established that a close emotional relationship exists between donor and recipient, and that the donation has taken place without consideration in return, force, duress, coercion or deception.[1029]

In Sweden, permission from the National Board of Health and Welfare is necessary if the donor is a minor or a mentally incapacitated adult. This permission is only granted in exceptional cases if the organ is non-regenerative.[1030]

To donate a non-regenerative organ in Portugal, the Entity of Verification and Admissibility of Collection for Transplantation must approve.[1031] This authority is established by the hospitals themselves. It is composed of doctors and nurses with expertise in transplantation. Therefore, the authority does not have expertise in criminal investigation or social work, nor does it have the means to organise a solid investigation of the social conditions behind the donation.[1032]

These countries hold the common view that an established authority is necessary to grant permission for certain defined cases of LOD, but not for every organ donation. The details of the rules are rather diverse. The institutions themselves are composed differently and the situations of LOD that they intervene in also vary.

[1026] Art. 26 (2) Law of the Transplantation of Organs, Tissues and Cells.

[1027] Art. 5 (1), (2) Transplantation Act on Donation, Removal and Transplantation of Organs and Tissues and on the Amendment of some Acts. This Act also contains further details about the formation of the Committee and about recording the activities (Art. 5 Transplantation Act on Donation, Removal and Transplantation of Organs and Tissues and on the Amendment of some Acts).

[1028] Norba (2009), p. 298.

[1029] Sec. 206 (3) Act on Health.

[1030] Sec. 8 II Transplant Act.

[1031] Art. 6, 3 Law on Harvesting and Transplanting Human Organs and Tissues.

[1032] Pereira, Vol. 18 European Journal of Health Law 1, 4 (2011).

The Dutch *Organ Donation Act* neither stipulates any procedure nor establishes a commission for LOD. That, however, does not mean that no procedure takes place. It just means that procedures are not stipulated by the law itself, because they constantly adapt to changing scientific developments. Consequently, procedural rules are either written down in national/uniform protocols or in local/hospital protocols.[1033]

The USA can exceptionally serve as an example as well. The *National Organ Transplant Act* does not contain exact procedures. The Organ Procurement and Transplantation Network (OPTN) that has been created under the *National Organ Transplant Act*[1034] established several guidelines, though, for example *Guidance for the Medical Evaluation of Potential Living Liver Donors* and *Guidance for the Informed Consent of Living Donors.*[1035] In addition, the OPTN/UNOS Living Donor Committee "considers issues relating to the donation and transplantation of organs from living donors to recipients. The committee makes recommendations to improve the process of living donation and transplantation."[1036]

It is hence very different whether the countries require a commissions authorisation for every LOD or for certain LOD's or if they do not require any authorisation. What kind of regulation the countries decided on is independent from the regulation of the donor-recipient relationship.

cc) Supervision

A few countries do not require the permission of an independent authority, but demand surveillance of the donation process.

In Lithuania, one duty of the National Bureau on Transplantation is to supervise the process of donation.[1037] The same task is carried out by the Transplant Agency in Moldova.[1038] In the Netherlands, the Dutch Transplant Foundation intends to increase organ transplantation.[1039] The Foundation functions as an organ centre, which supervises the acquisition, classification and transportation of organs.[1040]

In Germany, the Commission involved does not have to supervise the whole process of LOD. Instead, it only has to give an opinion on whether there are substantiated reasons to assume that the donor's consent is not being given freely or that the organ is the object of prohibited organ trade.[1041] It therefore takes into

---

1033 Information from M. Bos.

1034 Organ Procurement and Transplantation Network (United Network for Organ Sharing (2012).

1035 U.S. Department of Health & Human Services (2009).

1036 Transplant Living (2012).

1037 Art. 2^1 2 Law on Donation and Transplantation of Human Tissues, Cells and Organs.

1038 Art. 4 Law on the Transplant of Human Organs, Tissues and Cells.

1039 Information from R. Coppen.

1040 Sec. 24 Organ Donation Act.

1041 Sec. 8 III 2 Act on the donation, removal and transplantation of organs.

account that living organ donors are especially in need of protection since the LOD is primarily carried out for the benefit of the organ recipient.[1042] A closer look at the Commission's decisions makes clear that they are not binding for the treating surgeons.[1043] He can refrain from the organ removal even though the vote of the Commission was positive, and he can also carry out the organ removal even if the vote was negative.[1044] Nevertheless, the Commission's decision has a considerable impact on the carrying out of the LOD.[1045] That especially applies if the Commission disapproves a certain LOD, since it is very unlikely that a surgeon will perform a LOD in such a case.[1046] The Commission exists in a gray area — namely in between only giving advice and actually deciding.[1047] In addition to examining whether the donor's consent was given freely and ruling out the possibility that the organ could be the object of prohibited organ trade, the Commission often investigates the cases where no formal closeness exists, even though the wording of Sec. 8 III 2 *Act on the donation, removal and transplantation of organs* does not appoint the Commission to do so. Whether the Commission nevertheless has this competence is unclarified. Those in favour of granting the Commissions to do those investigations claimed that the legislator's intent for restricting the donor-recipient relationship was to eliminate involuntariness and organ trade. Consequently, Sec. 8 I 2 *Act on the donation, removal and transplantation of organs* contains a legal presumption that the donor acted involuntary or that organ trade occurred if donor and recipient do not entertain an especially intimate personal relationship.[1048] Many criticise this, arguing that the evaluation of voluntariness and organ trade in light of the required relationship does not provide a suitable basis to ascertain the closeness of donor and recipient.[1049] Independently of this debate, many assert that it would be preferable to clarify the competences of the Commission.[1050] It is also noteworthy that the Commissions are "responsible according to Land legislation."[1051] The German *Act on the donation, removal and transplantation of organs* only regulates the rough composition of the Commission, while "[f]urther details, especially those governing the composition of the Commission, the procedure and

---

[1042] Fateh-Moghadam, Vol. 21 Medizinrecht 245, 245 (2003).

[1043] Fateh-Moghadam, Vol. 21 Medizinrecht 245, 246 (2003); Nickel/Schmidt-Preisigke/Sengler (2001), p. 114.

[1044] Nickel/Schmidt-Preisigke/Sengler (2001), p. 114; Wagner/Fateh-Moghadam, Vol. 56 Soziale Welt 73, 88 (2005).

[1045] Fateh-Moghadam, Vol. 21 Medizinrecht 245, 254 (2003); Fateh-Moghadam et al., Vol. 22 Medizinrecht 82, 90 (2003).

[1046] Fateh-Moghadam, Vol. 21 Medizinrecht 245, 254 (2003).

[1047] Ibid., p. 246.

[1048] Cf. Fateh-Moghadam, in Schroth et al. (ed.) (2006), p. 173 f., who invalidates the argument immediately after presenting it.

[1049] Ibid., p. 174.

[1050] Fateh-Moghadam, Vol. 21 Medizinrecht 245, 254 (2003), p. 91; Fateh-Moghadam, in Schroth et al. (ed.) (2006), p. 176.

[1051] Sec. 8 III 2 Act on the donation, removal and transplantation of organs.

financing shall be regulated by Land law."[1052] The Federal Law thus contains incomplete guidelines and the federal states complied with the order to pass further legislation differently.[1053] Since Germany has 16 federal states, 16 laws of implementation exist[1054] and 23 different commissions are working in Germany.[1055] There are three different models used by the federal states to determine the legal source of the regulation's content. In addition, differences exist with respect to the commission's composition, organisation and responsibility. How the commission is financed and what procedures it must follow before making a decision also differ. Finally each commission uniquely determines how the decision-making works and the level of legal protection.[1056]

dd) Notification

The Bulgarian,[1057] Lithuanian[1058] and Swiss[1059] national laws instruct every LOD to be disclosed.

In Bulgaria, each medical establishment that removes organs from living donors has to inform the Executive Agency for Transplantation thereof at least 7 days in advance.[1060] After the organ removal, the medical establishment has to register the procedure within 7 days with the Executive Agency for Transplantation.[1061] With regard to the practical implementation of this, it was reported that this rule has been ignored in the St Ekaterina hospital in Sofia,[1062] a University National Specializing Hospital for Active Treatment.[1063]

The regulation in Switzerland is similar to the Bulgarian regulation. Any person who removes organs must notify the Federal Office of this.[1064]

In contrast, in Lithuania not every LOD has to be disclosed to the National Bureau on Transplantation, but — in accordance with the procedure established by

[1052] Ibid. Sec. 8 III 3, 4.

[1053] Fateh-Moghadam, Vol. 21 Medizinrecht 245, 246 ff. (2003); Fateh-Moghadam et al., Vol. 22 Medizinrecht 19, 20 (2004); Gutmann (2006), p. 56; Gutmann, in Middel et al. (ed.) (2010), p. 30; Wagner/Fateh-Moghadam, Vol. 56 Soziale Welt 73, 79 (2005).

[1054] Cf. Gutmann (2011), p. 6.

[1055] Fateh-Moghadam et al., Vol. 22 Medizinrecht 19, 21 (2004).

[1056] Fateh-Moghadam, Vol. 21 Medizinrecht 245, 246 ff. (2003).

[1057] Art. 29 Law of the Transplantation of Organs, Tissues and Cells.

[1058] Art. 5^1 Law on Donation and Transplantation of Human Tissues, Cells and Organs.

[1059] Art. 24, 1 Federal Act on the Transplantation of Organs, Tissues and Cells.

[1060] Art. 29 (1) Law of the Transplantation of Organs, Tissues and Cells.

[1061] Ibid., Art. 29 (2).

[1062] Pancevski, Vol. 367 The Lancet 461, 461 (2006).

[1063] University Hospital St.Ekaterina (2004).

[1064] Art. 24, 1 Federal Act on the Transplantation of Organs, Tissues and Cells.

the Ministry of Health — only those connected to an adverse event and/or reaction.[1065]

Thus, stipulating a special supervision by law is rather uncommon in the national laws considered.

e) Legal Protection

In an ideal case, there is no doubt about the existence of all necessary requirements, the transplantation is successful and neither the donor nor the recipient is harmed. In reality, this will not always be the case. Usually the regular legal protection applies to LOD as well. In Switzerland, for example, it is clarified in Art. 68 III *Federal Act on the Transplantation of Organs, Tissues and Cells* that "[t]he right of appeal shall otherwise be governed by the general provisions on the Administration of Federal Justice." In addition to the general legal protection, several laws contain particular legal protection for situations that specifically arise in the field of LOD. (1) On the one hand, a few national laws explicitly state that the process of organ donation and the obedience to the law must be kept under specific surveillance. (2) On the other hand, numerous national laws require a special authorisation to justify LOD. Consequently, some laws offer to challenge the refusal of a necessary authorisation.

aa) Supervision Over Compliance with the Law

In Estonia,[1066] Poland[1067] and Switzerland,[1068] whether the medical practice conforms to the legislation regulating LOD is kept under surveillance.

The supervising person differs. The Health Care Board is responsible in Estonia,[1069] while the minister competent to deal with health matters is in charge in Poland[1070] and the Federal Office in Switzerland.[1071]

bb) Legal Actions

Several countries require a public authority to grant a permit for LOD. If this permit is rejected, this might render the intended LOD impossible. Therefore, it is

[1065] Art. 5[1] Law on Donation and Transplantation of Human Tissues, Cells and Organs.

[1066] Sec. 16 (1) Transplantation of Organs and Tissues Act.

[1067] Art. 42, 2. The Cell, Tissue and Organ Recovery, Storage and Transplantation Act.

[1068] Art. 63 Federal Act on the Transplantation of Organs, Tissues and Cells.

[1069] Sec. 16 (3) Transplantation of Organs and Tissues Act.

[1070] Art. 42, 2. The Cell, Tissue and Organ Recovery, Storage and Transplantation Act.

[1071] Art. 63 Federal Act on the Transplantation of Organs, Tissues and Cells.

important for the persons concerned to have the possibility to take legal action against the negative decision.

In Finland, the National Board of Medicolegal Affairs has to grant a permit for every organ removal.[1072] The board's decision can usually be appealed. However, as an exception, any decision made regarding the permit for an organ removal cannot be appealed.[1073]

In Sweden, an organ removal from a minor or from an adult who is incapable of consenting may only be performed after permission has been granted by the National Board of Health and Welfare.[1074] Those decisions may be contested by appeal to a common administrative court.[1075] Not every person can contest a decision, though, only the person concerned by the decision.[1076]

In Switzerland, appeals may generally be filed with respect to rulings derived from the Swiss *Federal Act on the Transplantation of Organs, Tissues and Cells* and its implementing ordinances.[1077]

In Scotland, reconsideration of the Scottish Ministers' decision is also possible with regard to several decisions made in relation to matters that concern organ donation. The grounds a decision may be reconsidered and the procedure for the reconsideration are also defined in detail in the regulations.[1078]

In England, the Human Tissue Authority may reconsider its decisions if the donor, the recipient or the registered medical practitioner who caused the matter to be referred to the Authority make a request to that effect.[1079] A decision may then actually be reconsidered by the Authority if it finds that any information given for the purpose of the decision was false or misleading, or if there has been any material change of circumstances after the decision.[1080]

In Germany, the situation in the federal states differs. A few federal states have explicitly regulated that no appeals can be made against the expert opinion of the Commission. In contrast, several federal states did not pass any regulation with regard to this issue at all.[1081]

Hence, only a few of the countries that request a specific authority to permit LOD also have specific legal actions to challenge the authority's decision.

[1072] Sec. 4 II Act on the Medical Use of Human Organs and Tissues.

[1073] Ibid., Sec. 26.

[1074] Sec. 8 Transplant Act.

[1075] Ibid., Sec. 17.

[1076] The Ministry of Health and Social Affairs Sweden 1997, The Swedish Transplant Act, Guidance on the provisions, Sec. 17.

[1077] Art. 68 I Federal Act on the Transplantation of Organs, Tissues and Cells.

[1078] Reg. 6 (1) Human Organ and Tissue Live Transplants (Scotland) Regulations 2006.

[1079] Sec. 13 (2) & (3) Human Tissue Act 2004 Regulations 2006.

[1080] Ibid., Sec. 13 (1).

[1081] Fateh-Moghadam, Vol. 21 Medizinrecht 245, 252 (2003).

Table 4 Summary of the procedural issues (and those in the middle of being material and procedural)

Considered aspect	Result after comparing the countries
Person who performs the transplantation	• Physicians commonly carry out the organ removal and the transplantation, but further details differ
Location of the transplantation	• LOD is typically performed at a hospital or transplant centre, but the detailed requirements differ
Post-care	• Several national transplant laws require post-care for donor and/or recipient
Commission for LOD	• Several countries only allow LOD after a commission has authorised it, but there are also countries that only require an approval by a commission in certain defined cases • A few countries explicitly demand surveillance of the donation process • A small amount of countries demands that every LOD be disclosed
Legal protection	• A few national transplant laws stipulate special surveillance of the implementation of the relevant law • A few countries contain the possibility to challenge the authority's decision involved in LOD

f) Summary of the Procedural Issues (and Those in the Middle of Being Material and Procedural)

As seen, in addition to the requirements regarding the content of living organ transplantation, several national laws also contain requirements that are both material or procedural or that are clearly only procedural. The implementation and the scope differ greatly among the examined countries. An overview of the results will be given in Table 4.

8) Social Security Regulations for Donors

LOD causes considerable expenses.[1082] It not only involves the organ removal and the organ transplantation, but also medical preparation and medical post-care. In the majority of cases it would be impossible for the patients to cover these expenses all by themselves.

Usually, health insurance pays the expenses of the insured. Since LOD also affects the donor — a healthy person — a regulation must also clarify who pays the donor's expenses.

In the majority of countries, (1) either a state institution or (2) an insurance company pays the expenses, but the details differ. (3) Furthermore, there are countries with a completely different regulation.[1083]

[1082] Cf. Neft, Vol. 20 Neue Zeitschrift für Sozialrecht 566, 566 (2011).

[1083] The following presentation is only fragmentary, as it does not include all countries considered.

a) Institution on the Part of the State

In Belgium,[1084] Bulgaria,[1085] Hungary[1086] and Latvia,[1087] a state-owned or state-controlled institution pays the incurred expenses. However, the acting state institution and the expenses it covers vary.

In Belgium, the donor is compensated by the government or the social security institutions. The King of Belgium establishes the concrete rules. With regard to the amount of compensation, both the incurred expenses and the loss of income are covered as long as they are a direct result of the renunciation of organs.[1088]

In Bulgaria, the Ministry of Health primarily compensates the expenses incurred by the medical institutions.[1089] If the medical institution has a contract with the National Health Insurance Fund and the companies for voluntary health insurance, then they can finance the activities related to transplantation.[1090] With regard to the amount compensated, expenses of extraction and/or examination, treatment, processing, labelling, storage, provision and implantation of organs and all related expenses of both patients, as well as the expenses for diagnostics and treatment of the recipient in the post-transplantation period, are covered.[1091]

In Hungary, the state is ultimately responsible for expenses. It has to compensate for the donor's loss of income related to the donation and for the justified costs incurred in connection with donating and travelling, which are not reimbursed under her social insurance coverage. If the organ donation impairs the donor's health, results in bodily harm, disablement or the donor dies, and if this consequence cannot be imputed to the healthcare worker carrying out the procedure, she or her dependant relatives will be compensated by the state for the damages that are not reimbursed under her social insurance coverage.[1092]

The Latvian regulation is very generous for the donor. She is entitled to the state-guaranteed medical aid free of charge until the end of her life.[1093]

Thus, several countries consider the state as being responsible for the donor's costs.

[1084] Art. 4, Sec. 2 Law on Retrieval and Transplantation of Organs.

[1085] Art. 16 Law of the Transplantation of Organs, Tissues and Cells.

[1086] Sec. 207 (2) Act on Health.

[1087] Sec. 19 Law on the Protection of the Body of Deceased Human Beings and the Use of Human Tissues and Organs in Medicine.

[1088] Art. 4, Sec. 2 Law on Retrieval and Transplantation of Organs.

[1089] Art. 16 (1) Law of the Transplantation of Organs, Tissues and Cells.

[1090] Ibid., Art. 16 (3).

[1091] Ibid., Art. 16 (1).

[1092] Sec. 210 Act on Health.

[1093] Sec. 19 Law on the Protection of the Body of Deceased Human Beings and the Use of Human Tissues and Organs in Medicine.

b) Recipient's Health Insurance

Since LOD is for the benefit of the organ recipient, in the Czech Republic, Estonia, Germany and Switzerland, the recipient's insurance company pays the donor's costs.[1094]

In the Czech Republic, the recipient's health insurance has to cover all costs connected to the LOD.[1095] In addition, the medical institution has the duty to take out insurance for the donor that covers all injuries that might result due to the organ removal.[1096]

The donation is normally financed by the recipient in Estonia. This no longer applies if the obligation to compensate the medical expenses is assumed by the Estonian Health Insurance Fund.[1097] This obligation exists for organ transplantation to recipients covered by health insurance.[1098] Regarding the amount of compensation, donors and recipients covered by health insurance also receive worker's compensation by the Estonian Health Insurance Fund.[1099]

The German regulation is changing right now. So far, the health insurance company of the recipient had to compensate the expenses as long as no complications occured.[1100] This covered all preparation procedures, the organ removal and post-care, including life-long check-ups of the living donor. Travel expenses, accommodation costs and loss of income of the donor had been accounted for as well.[1101] The situation was different if complications occured. The donor did not get recompensed by the recipient's health insurance company in those cases; rather the social accident insurance had to pay.[1102] This regulation was doubtful. The person concerned is part of the insured group of people, but according to Sec. 7 in connection with Sec. 8 *Code of Social Law VII*, the insured event is only present in case of an *accident*. Consequences of LOD, however, are in most cases

[1094] As an argument in favour of this, it is stated that LOD saves money for the recipient's health insurance. First, LOD saves about 180,000 € in ten years in comparison to dialysis (Teubner (2006), p. 4. The calculation is based on Landtag of Bavaria printed papers 14/1450, p. 3). Second, the recipient can resume working after living organ transplantation and therefore pays a higher amount of premium to her health insurance (Teubner (2006), p. 4; cf. Viehbahn, in Höfling (ed.) (2008), p. 10).

[1095] Sec. 35 a Law on the Public Health Insurance.

[1096] Sec. 22 II, Sec. 21 I h Transplantation Act on Donation, Removal and Transplantation of Organs and Tissues and on the Amendment of some Acts.

[1097] Sec. 14 Transplantation of Organs and Tissues Act.

[1098] Ibid., Sec. 15 (1).

[1099] Ibid., Sec. 15 (2).

[1100] Bundestag printed paper 15/5050 (2005), p. 60; German Federal Social Court [Bundessozialgericht], 12 December, 1972, Case No. 3 RK 47/70; cf. Neumann/Neuhaus/Schmeding, Vol. 81 Der Chirurg 804, 811 (2010).

[1101] Bundestag printed paper 15/5050 (2005), p. 60 f.; Hohmann (2003), p. 132; cf. Neft, Vol. 20 Neue Zeitschrift für Sozialrecht 566, 567 (2011).

[1102] Sec. 23 Transplantation Act in conjunction with Sec. 2 I Nr. 13 b Code of Social Law VII.

not included by the definition of the term accident in Sec. 8 *Code of Social Law VII*. Hence, several of the post-surgical consequences have not been part of the insurance coverage of social accident insurance.[1103] The German Federal Social Court strengthened the rights of living organ donors, though.[1104] In the case from May 2012, the donor's left abdominal wall was partly paralysed.[1105] He asked for a pension for injured people from his social accident insurance company. The donor had a 20 % reduction in earning capacity.[1106] The social accident insurance company refused any payment by stating that he voluntarily became part of the surgery. In addition, the insurance company stated that a person is only protected if a further event that externally acts on the body takes place.[1107] The German Federal Social Court did not agree. The LOD was an insured action. The cut by the scalpel that caused the donor's left abdominal wall to become partially paralized has been qualified as an external act on the body. Consequently, the requirements for an industrial accident are fulfilled.[1108] The Court clarified, however, that the social accident insurance does not have to pay for all damages that are connected to the organ removal.[1109] In addition to the German Federal Social Court ruling in favour of living organ donors, the regulation is currently being modified. The German Federal Health Minister *Bahr* announced that the situation of living organ donors must be improved.[1110] Every living organ donor is supposed to be entitled to the recipient's health insurance company with respect to health treatment, pre- and post-care, rehabilitation and sick pay.[1111] The Federal Council demands social accident insurance companies to cover all complications that might result from the LOD.[1112] The *Law on Continued Remuneration* (*Entgeltfortzahlungsgesetz*) has also been changed in favour of living organ donors. It now rules that living organ donors who are unable to work as a consequence of the LOD receive continued remuneration from employers for 6 weeks.[1113] The recipient's health insurance has to compensate the costs accrued by the donor's employer.[1114]

In Switzerland, the insurer responsible for the costs of treating the recipient suffering from organ failure has to pay all insurance costs and must also provide appropriate compensation for loss of income or other expenses incurred by the donor in connection with the removal. This also applies to removals or

[1103] Gutmann (2006), p. 90 f.; Gutmann, in Middel et al. (ed.) (2010), p. 31; Gutmann (2011), p. 7.

[1104] Ärztezeitung (2012).

[1105] Ärztezeitung (2012); Jus@Publicum (2012).

[1106] Ärztezeitung (2012).

[1107] Ärztezeitung (2012); Jus@Publicum (2012).

[1108] Ärztezeitung (2012).

[1109] Ibid.

[1110] Bundestag (2012).

[1111] Ibid.

[1112] Ibid.

[1113] Bundesrat printed paper 292/12 (2012), p. 11.

[1114] Ibid.

transplantations that cannot be carried out. In case the recipient's insurer is unknown, the Confederation has to recompense the expenses.[1115] In addition to this regulation, details have been specified by the Federal Council in the *Transplantation Ordinance*. According to Sec. 11 *Transplantation Ordinance*, the person who removes an organ from a living person has to ensure that this person is insured for the disability or death that could occur as a consequence of the removal at least 12 months after the organ removal. In cases where the donor dies, the benefits payable to surviving dependants amount to 250 000 Swiss francs. In cases where the donor develops a disability, a sum of 250 000 Swiss francs has to be insured as well. In addition, according to Sec. 12 *Transplantation Ordinance*, the donor is compensated for all expenses he had in connection to the donation. Thus, in Switzerland the insurance for an organ donor is very extensive.

All in all, it is common for the state to cover the donor's costs or for the recipient's health insurance to pay the donor's expenses.

c) Miscellaneous

A few countries do not hold a state institution or the recipient's insurance company liable to pay the expenses.

In Austria, all costs that are connected to the organ removal and the preparation are covered by the donor's health insurance, despite the fact that the donor is healthy and his life does not depend on LOD.[1116] In the occurrence of an event that is insured, the organ donation can be equated with an illness pursuant to Sec. 120 a *General Social Insurance Law*.[1117]

In Portugal, the donor has the right to medical assistance until he is fully recovered. He is also fully compensated for the damage suffered.[1118] Insurance is mandatory for the donor to be supported by the transplanting establishments.[1119]

In Moldova, not one single institution is responsible for paying the costs of the transplantation. Actually, those costs may be covered by first the National Health Insurance Company, second, the state budget, third, payments/fees for medical services that pertain to the patients, and fourth, the donations of charity organisations or other persons, and individuals, not involved directly in a specific transplant process.[1120]

The actual Finnish regulation is unknown, but some claim that the living donor should be free of charge so that financial circumstances will not prevent donation. They suggest that the employer pays the donor's salary during the time the donor is

[1115] Art. 14 Federal Act on the Transplantation of Organs, Tissues and Cells.

[1116] Sec. 120 II General Social Insurance Law.

[1117] Information from S. Füszl.

[1118] Art. 9, 1 Law on Harvesting and Transplanting Human Organs and Tissues.

[1119] Ibid., Art. 9, 2.

[1120] Art. 29 Law on the Transplant of Human Organs, Tissues and Cells.

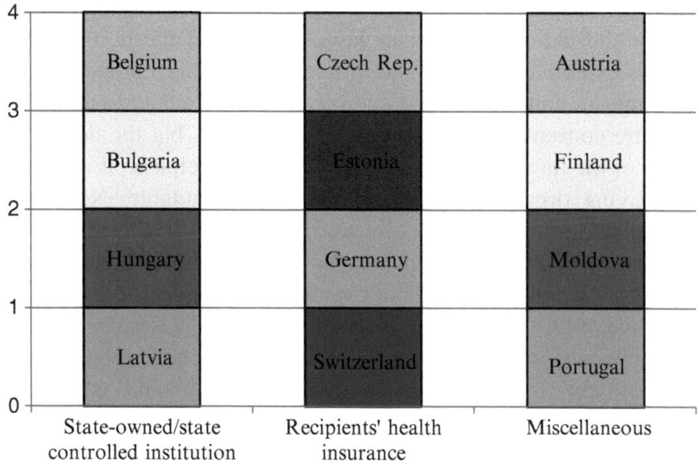

Fig. 5 Summary of the social security regulations

unable to work; this means the daily allowance under the *Health Insurance Act* can be paid to the employer, instead of the donor. The overall aim is to reduce the costs incurred by the living donor.[1121]

d) Summary of the Social Security Regulations

In conclusion, it is common for countries to regulate the requirements to justify LOD and to regulate how the procedure will be financed. Some countries do this in the transplant laws directly. Others, in contrast, have not established a specific regulation in the transplant law, but simply apply the normal legal rules for the occurrence of an insured event. However, with regard to the institution that has to recompense the expenses and the amount of costs that are covered, the national laws are different. Figure 5 provides an overview of the results.

III. Summary

Organ shortage is a problem apparent in all of the countries considered. A few suggestions were made to solve this problem. One idea was to deregulate LOD, hence increasing the amount of LODs. The greatest challenge for deregulation is the — sometimes severe — differences in the legal framework of the countries considered. As shown, adjusting the national transplant laws might help to achieve

[1121] Finnish Ministry of Health and Social Affairs (2010).

this goal. The first step taken to achieve this goal was providing the comprehensive comparison of the national transplant laws. The initial thesis, that the national transplant laws exhibit similarities and differences, has been proven true.

(1) The countries commonly hold that no citizen has a duty to donate an organ. They all require consent and disclosure to justify LOD, but the details differ. (2) Since the aim of the intended LOD is to heal the sick recipient, it is necessary that the planned living organ transplantation has to be suitable. Not all countries explicitly require this, though. (3) Minors and mentally incapacitated adults are mostly completely prohibited from becoming living organ donors, or are only allowed to act as a living organ donor under several specific requirements. (4) One central issue of LOD by mentally capable adults is the restriction of the donor-recipient relationship. None of the countries demand a genetic relation between the donor and recipient. It differs, however, whether any kind of relationship is necessary or whether donor and recipient do not have to be related at all. (5) Some countries do neither consider LOD as subordinate towards deceaded donation nor towards alternative therapies. Other countries, in contrast, opine that LOD may only be performed subsidiary to post-mortem donation and also to alternative therapies. Some countries, however, even differentiate between both relationships: living and deceased organ donation and LOD and alternative therapies. (6) In addition to the content of LOD, several national laws also regulate aspects that are material and procedural or clearly procedural. It is especially very common among the countries considered to involve an independent commission in the process of LOD. However, differences exist with regard to its composition and the assigned tasks. (7) With respect to the social security regulations for donors, it can be summarised that, in the majority of countries, either a state institution or an insurance company pays the expenses. Nevertheless, countries with a completely different regulation exist as well.

LOD is thus restricted in very different ways and to a different extent in the countries. Several arguments are brought forth in favour of restricting LOD, but also against restricting LOD. All arguments will be analysed in the next chapter.

Analysis of the Normative Arguments That Dominate the Policy Arena About Necessity and Legitimacy of Legal Restrictions in Living Donor Transplantation

Chapter "Comparative Analysis of European Transplant Laws Regarding Living Organ Donation" has just shown that almost all of the countries considered established specific legal rules for LOD. However, only some regulations are similar, substantial differences exist as well. The countries especially differ with respect to restricting LOD. Several arguments are brought forth in favour of restricting LOD, but also against restricting LOD. Those will be analysed in this chapter.

I. Definition of the Problem

As already emphasised, there are not enough donor organs available to help all patients suffering from organ failure. Especially concerning LOD, there are, however, also legal reasons for this shortage. LOD regimes create, in part, an *artificial scarcity*.[1] As the German Constitutional Court declared in its 1999 ruling on LOD, "[w]henever legal regulations entail that a citizen in need of treatment is denied a therapy which, according to the state of medical research, could be provided to him or her, and which would entail a prolonging of life or at least a substantial mitigation of the ailment, the constitutionally guaranteed basic negative right to life and health is infringed upon."[2] The *CFREU* also protects the right to life[3] and the person's right to integrity.[4] It does not mention LOD in particular, but the

[1] Cf. Evans, Vol. 15 Journal of Medical Ethics 17, 19 (1989); Fateh-Moghadam, (2011); cf. Radcliffe Richards, in Weimar/Bos/Busschbach (ed.) (2011), p. 41.

[2] German Federal Constitutional Court [Bundesverfassungsgericht], 11 August, 1999, Case No. 1 BvR 218/98, in Vol. 46 Neue Juristische Wochenschrift 3399, 3400 (1999).

[3] Art. 2 (1) Charter of Fundamental Rights of the European Union.

[4] Ibid., Art. 3 (1).

DOI 10.1007/978-3-642-33799-4_4, © Springer-Verlag Berlin Heidelberg 2013

inclusion of these two rights in the *CFREU* is nonetheless important for LOD. These two rights should therefore not be ignored in LOD.

Living donor kidney transplantation has become an established medical procedure worldwide.[5] As seen in Table 1 in the "Introduction," the rates of LOD differ in the countries considered, meaning that LOD can still be considered as an *underutilised resource* in many countries.[6] It is therefore crucial to establish an appropriate legal framework for LOD (to guarantee the voluntariness of donors), separate from unjustified legal barriers and misguided rationales. This chapter systematically presents the arguments for the necessity and the legitimacy of legal restrictions on LOD, and subjects them to closer scrutiny.

Fifteen years ago, it was said that Europe suffered from a disease called HIL ("Highly Inappropriate Legislation") in the field of LOD.[7] This may no longer be completely true. Finding the best regulatory model for LOD, however, requires a critical examination of the principles underlying different policy options and legal models.

II. Arguments Serving to Restrict Living Organ Donation

The analysis of the arguments in favour of legal restrictions on LOD begins with the arguments which focus on substantive (not procedural) boundaries. The arguments reviewed in this chapter mainly address LOD in general, but there are also specific arguments regarding special types of LOD. The following aspects will be considered: (1) Does LOD violate the rule to do no harm? (2) Can LOD be voluntary? (3) How does potential social pressure influence LOD? (4) The problem of organ trade is considered, alongside (5) the aspects concerning justice and equality. (6) After presenting the arguments that apply to LOD in general, special groups of donors are examined individually and (7) arguments against specific methods and programmes to increase LOD are presented. (8) After describing the relevant material aspects, procedural issues are addressed.

There is no single or unified moral theory to date. Moral reasoning and approaches to medical ethics are as pluralistic as our societies. Ethical theories are composed of varied, partly incompatible, sets of premises and intuitions, each containing diverse strengths and weaknesses. To create a common ground for intertheoretical moral discussion and a practical approach for ethical (and policy) decision-making, mainstream medical ethics often rely on principles to inform its reasoning process. The most relevant of these principles are (a) *respect for persons*, including their autonomous choices and actions; (b) *beneficence*, including both the obligation to benefit others (positive beneficence) and to maximise good

[5] Price (2010), p. 196.

[6] EULOD WP 2 (2012), DOW: Deliverable 4, p. 4.

[7] Nielsen, in Price/Akveld (ed.) (1997), p. 73.

consequences — i.e., to do the greatest good for the greatest number (utility); (c) *justice*, the principle of fair and equitable distribution of benefits and burdens and finally (d) *nonmaleficence* (*non nocere*), the obligation not to inflict harm. In conflicting cases, these principles have to be applied to specific circumstances and balanced against each other.[8] Although *principalism* is severely criticised in moral and legal theory, it serves as a means to structure the debates for the purposes of this thesis.

1) The Harm-Argument

That LOD is sometimes considered a violation of the principle of nonmaleficence,[9] or the rule not to do harm to anyone else (*primum non nocere*),[10] has already been stated.[11] Many disagree that LOD is a violation of this principle, citing the principle of *volenti non fit iniuria* for support.[12] They argue that the donor herself voluntarily decides to take part in the surgery, thus exercising her right of autonomy and therefore cannot be considered harmed.[13] Hence, an ethical conflict between respecting autonomy and the rule to do no physical harm to healthy persons exists.[14]

(1) The principle of autonomy is introduced in the following section, and (2) whether LOD contravenes the principle of nonmaleficence is discussed.

a) Autonomy

In this analysis of the principle of autonomy, (1) the principle is introduced in general, before (2) discussing its role in LOD. (3) The concept of informed consent

---

[8] Cf. Bakker, in Price/Akveld (ed.) (1997), p. 27; Beauchamp/Childress (2001), p. 12 ff.; cf. Donnelly/Price (ed.) (1997), p. 51 f.; cf. Lamb, in Price/Akveld (ed.) (1997), p. 43 f.; cf. Schroth, in Schroth et al. (ed.) (2006), p. 79 ff.

[9] Beauchamp/Childress (2001), p. 113; Bechstein/Moench, Vol. 24 Transplant International 1162, 1163 (2011); Potts/Evans, in Weimar/Bos/Busschbach (ed.) (2008), p. 377.

[10] Bakker, in Price/Akveld (ed.) (1997), p. 26; Bechstein/Moench, Vol. 24 Transplant International 1162, 1163 (2011); Fateh-Moghadam, in Rittner/Paul (ed.) (2005), p. 132; cf. Hessing, in Price/Akveld (ed.) (1997), p. 99; cf. Land, Vol. 2 Transplant International 168, 169 (1989); Wagner/Fateh-Moghadam, Vol. 56 Soziale Welt 73, 78 (2005).

[11] P. 8 f.

[12] This principle says that, in case the person concerned consents, no injury is done (Benke/Meissel/Luggauer (1997), p. 406; Daar et al., Vol. 11 Transplantation Review 95, 101 (1997); Lieberwirth (2007), p. 302).

[13] Cf. Fateh-Moghadam, in Rittner/Paul (ed.) (2005), p. 134; Schroth, in Schroth et al. (ed.) (2006), p. 86.

[14] Cf. Cronin, Vol. 343 British Medical Journal (2011); cf. Land, Vol. 2 Transplant International 168, 169 f. (1989); Pöltner (2002), p. 97.

that has been introduced in chapter "Comparative Analysis of European Transplant Laws Regarding Living Organ Donation" will be reviewed and (4) the scope and the restrictions of autonomy will be examined.

aa) The Concept

The origin of the word autonomy is Greek. It is a composition of *autos*, which means self, and *nomos*, which means rule.[15] It originally referred to the self-rule of independent city states.[16] Later, *Kant* gave a philosophical meaning to the term.[17] According to *Kant*, *moral* autonomy refers to the capacity to impose the (objective) moral law on oneself, while *personal* autonomy refers to a trait that individuals can exhibit relative to any aspects of their lives; it is not limited to questions of moral obligation.[18] The term autonomy describes the ability of persons to decide for themselves.[19] In this sense, autonomy implies that people are able to make individual choices.[20] Its most important meaning in moral, legal and political philosophy, however, is that it denotes a set of rights expressive of one's sovereignty over oneself.[21] This notion is based on a concept of respect for persons.[22] It consequently expresses that every person has a right to self-determination,[23] especially in critical life situations.[24] In the Western liberal tradition of ethical and political thought, respect for a person's autonomy is the respect for her voluntary choice as the sole rightful determinant of her actions, except where the interests of others need protection from her.[25] Autonomous human beings are concerned with more than their "experiential" interests (including interests in health and integrity of the body). They also have important "critical" interests that reflect their personalities; critical interests play a role in one's choices and thus determine how one gives meaning to her own life.[26] This even applies to self-harming behaviour.[27] The right

[15] Beauchamp/Childress (2001), p. 57 f.; Hügli/Lübcke (ed.) (1997), p. 71; Pieper, in Korff (ed.) (2000), p. 289.

[16] Beauchamp/Childress (2001), p. 57 f.; Pieper, in Korff (ed.) (2000), p. 289; cf. Prechtl/Burkhard (2008), p. 56; Ritter (ed.) (1971), p. 701.

[17] Mittelstraß (ed.) (2005), p. 319; Ritter (ed.) (1971), p. 708.

[18] Christman (2011); Dworkin (1988).

[19] Ritter (ed.) (1971), p. 707.

[20] Beauchamp/Childress (2001), p. 57 f.

[21] Feinberg, in Christman (ed.) (1989).

[22] Benn (1988); cf. Lamb, in Price/Akveld (ed.) (1997), p. 48.

[23] Dreier, in Dreier (ed.) (2004) , before Art. 1 GG at 75, Art. 2 II at 20; Gessmann (ed.) (2009), p. 72; Hügli/Lübcke (ed.) (1997), p. 71; cf. Joint Working Party of The British Transplantation Society and The Renal Association (2011), p. 25; Pöltner (2002), p. 94.

[24] Feinberg (1986), p. 54.

[25] Feinberg (1986); cf. Gutmann, in Schroth et al. (ed.) (2006), p. 219 f.; cf. Wagner/Fateh-Moghadam, Vol. 56 Soziale Welt 73, 83 (2005).

[26] Dworkin (1993); cf. Glannon, Vol. 343 British Medical Journal (2011).

[27] Gutmann, in Schroth et al. (ed.) (2006), p. 220.

to self-determination in principle includes the right to decide about her own body.[28] Given the weight of the concept of autonomy, one could argue that the donor's autonomous decision is not only necessary, but also a sufficient condition to justify LOD, including unspecified LOD.[29] As will be explained in the following, the role of autonomy in LOD has also been heavily criticised.

bb) Radical Critique of the Role of Autonomy in Living Organ Donation

Many ethical positions claim that personal autonomy cannot lead to the moral justification of LOD. In the past, Christian theologians held that we ourselves are not the rightful owners of our bodies, rather God is. Traditionally, many believed that a person was not allowed to commit suicide to save another person[30]; he was also not allowed to deprive himself of an organ for the benefit of another person. Such an act was considered to be self-mutilation.[31] Catholic moral theology brought forward a *totality theory* with regard to LOD, stating, "[A]nybody severability is an impermissible interference with nature and God's order, affecting the physical completeness of a human being. [. . .] Only a non-healthy body part may be removed for the benefit of the organism as a whole."[32] From a traditional theological standpoint, LOD should be completely prohibited. *Kant* transformed this Christian notion into a philosophical one at the end of his life. Although he obviously did not directly address LOD, he actually did discuss tooth transplantation in 1797. He stated: "To deprive oneself of an integral part or organ (to maim oneself)—for example, to give away or sell a tooth to be transplanted into another's mouth (. . .) and so forth—are ways of partially murdering oneself."[33] Applying this statement to LOD leads one to assume that *Kant* would have criticised the procedure as immoral.[34]

Both positions seem irrelevant to current policy-making practices. First, a deeper examination of *Kant's* concept of autonomy reveals that it could be compatible with most types of LOD. However, it could also be used to provide strong moral and legal-philosophical arguments against allowing the type of behaviour (including

[28] Bundestag printed paper 15/5050 (2005), p. 34; Forkel, Vol. 23 JURA 73, 73 (2001); Gutmann/Schroth (2002), p. 110; cf. Joint Working Party of The British Transplantation Society and The Renal Association (2011), p. 25.

[29] Fateh-Moghadam et al., Vol. 22 Medizinrecht 19, 20 (2004); Gutmann, Vol. 15 Medizinrecht 147, 147 (1997); Ross, Vol. 30 The Journal of Law, Medicine and Ethics 440, 441 (2002).

[30] Ermecke (1961), p. 250.

[31] Bundestag printed paper 15/5050 (2005), p. 34; cf. Elsässer, Vol. 12 Transplantationsmedizin 184, 185 (2000); Ermecke (1961), p. 250 f.

[32] Cf. Donnelly/Price (ed.) (1997), p. 51.

[33] Kant, in Gregor (ed.) (1998), p. 547.

[34] Price (2010), p. 207; cf. Donnelly/Price (ed.) (1997), p. 50

LOD) that may render the donor incapable of moral agency.[35] Second, *Kant* explicitly made clear that the (theoretically contested[36]) conception of 'duties to oneself' he draws upon in his example is confined to morality and not applicable in policy making and the law.[37] Christian (especially Catholic) doctrine has given up its *totality theory* as far as LOD is concerned. Organ donation in general is now seen as an act of love within Christianity.[38] LOD is approved by Christian theologies, it is even referred to as "a supremely generous and [...] positive act."[39] Classifying LOD as self-mutilation does not respect the character of LOD. Due to the exceptional circumstances and the benevolent intention, LOD should be distinguished from simply harming oneself.[40] Most importantly, even if Christian theologies had not revised their doctrine, it would still be illegitimate for the liberal, religiously neutral constitutional state to give (exclusively) religious reasons for policies which infringe on the freedom and, especially, on the right to autonomy.[41]

Hence, traditional Christian and philosophical viewpoints used to regard the principle of autonomy as insufficient to justify LOD. These viewpoints are now obsolete, though.

cc) Informed Consent

The principle of autonomy is very general. The concept of "informed consent" is more specific to medical interventions.[42] Chapter "Comparative Analysis of European Transplant Laws Regarding Living Organ Donation" showed that informed consent for LOD is a standard requirement in the countries considered. It also showed that in England, France, Italy and Spain, the patient's consent is required to justify LOD, but is not held to be a sufficient condition. Rather, they require an additional legitimising element.[43]

---

[35] Cf. Kant (1968 [1797]), p. 555; Seelmann, in Amelung/Beulke/Lilie (ed.) (2003), p. 853–867. I am indebted to A. Pascalev for this notice.

[36] See e.g. Singer, Vol. 69 Ethics, 202, 202 ff. (1959).

[37] Information from T. Gutmann.

[38] Cf. Gründel, Vol. 5 Zeitschrift für Transplantationsmedizin 70, 72 (1993); Roff, Vol. 33 Journal of Medical Ethics 437, 440 (2007); Sharp/Randhawa/Kaur-Bola, in Weimar/Bos/Busschbach (ed.) (2011), p. 114.

[39] Dor et al., 2011 Transplantation 1, 3; cf. Elsässer, Vol. 12 Transplantationsmedizin 184, 185 f. (2000).

[40] Bundestag printed paper 15/5050 (2005), p. 34; Weber (1999), p. 164.

[41] Rawls (1993).

[42] Eyal, in Zalta (ed.) (2011).

[43] The exact meaning of this and what these countries require in detail can be looked up in chapter "Comparative Analysis of European Transplant Laws Regarding Living Organ Donation".

dd) Scope and Restriction of Autonomy

In contrast, Austria, Germany, the Netherlands, Portugal and Switzerland consider consent as adequate to justify LOD.[44] Competent adults are seen as having the right to expose themselves to risks and are also seen as being capable of accepting any negative consequences that may accompany their risks, so long as their behaviour does not harm any third person or the general public.[45] An autonomous person makes his own judgment concerning his bodily integrity.[46]

Nevertheless, it has been described that even in these countries, the scope of the patient's autonomy is still limited. The most important, generally shared limitation is that LOD is illegal if it would end the donor's life or would be directly life-threatening.[47] In addition, several countries declare consent that goes against 'good morals' void.[48]

Thus, LOD is not at variance with the principle of autonomy. Since this principle is in an ethical conflict with the principle of nonmaleficence, I examine whether LOD contravenes the principle of nonmaleficence is examined in the following.

b) Living Organ Donation Contravenes the Principle of Nonmaleficence

The principle of nonmaleficence, prominent in the Hippocratic tradition of medical ethics,[49] must be differentiated from beneficence, which describes the obligation to actively help others.[50]

Before one can usefully apply the principle of nonmaleficence to LOD, the term *harm* must be defined. A person is harmed if his interests are adversely affected. According to this definition, it is irrelevant whether one's rights are violated, or an injustice is inflicted upon the person concerned.[51] The principle of nonmaleficence is thus significant when surgery is performed on a healthy person, as in the case of LOD.

---

[44] P. 28 ff.

[45] Gutmann, Vol. 52 Neue Juristische Wochenschrift 3387, 3388 (1999); Gutmann/Schroth, in Oduncu/Schroth/Vossenkuhl (ed.) (2003), p. 277; Parzeller/Henze/Bratzke, 87 Kritische Vierteljahresschrift 371, 383 (2004).

[46] Hilhorst et al., Vol. 24 Transplant International 1164, 1165 (2011).

[47] Dickens/Fluss/King, in Chapman/Deierhoi/Wright (ed.) (1997), p. 98; cf. Gutmann/Schroth (2002), p. 49; Pfeiffer (2004), p. 45; Weber (1999), p. 164.

[48] P. 48 ff.

[49] The Hippocratic Oath states, inter alia:" I will apply dietetic measures for the benefit of the sick according to my ability and judgement; I will keep them from harm and injustice." See e.g. Edelstein (1943).

[50] Beauchamp/Childress (2001), p. 113 ff.; cf. Joint Working Party of The British Transplantation Society and The Renal Association (2011), p. 25.

[51] Beauchamp/Childress (2001), p. 116.

LOD is unique insofar as it involves two patients,[52] and, in most cases, does not have any benefit for the physical health of the donor.[53] Although, especially with regard to living kidney donation, the risks are fairly modest, they are in no case negligible.[54] Therefore, a contravention of the principle of nonmaleficence may be assumed.

This argument may be rebutted if one uses a more complex notion of benefits and harms. Even according to the principle of nonmaleficence, it is morally acceptable for a person to become a living organ donor if the benefits are expected to outweigh the expected physical (and possibly also psychological) risks *to him*. All things considered, the overall well-being of a person may be improved by becoming a living organ donor. Well-being is more than physical health. Many argue that well-being has "to be taken as part of a whole package which includes the psychic, perhaps moral, perhaps spiritual betterment and achieving an objective that he himself has which is to make the donation."[55] Psychological and emotional advantages and an increase of self-esteem may counterbalance the donor's pain, discomfort, anxiety and risk.[56] This is also true in the case of unspecified LOD. When evaluating the harm to the organ donor, *harm* must not be interpreted too narrowly. Moreover, potential donors might also suffer "psychosocial and moral harms if they are prevented from serving as a donor."[57] Finally, if harm means adversely affecting one's interests,[58] then the donor's *critical* interests in restoring someone else's health, and in *being the person* who acts in this way, must also be taken into account.[59]

On the other hand, when using complex notions of benefits and harms to assess the interests of persons in a vulnerable position, those who are heavily dependent on the donor (e.g., underage children) may try to discourage the donor from donating; it might even serve as a *moral contraindication* in special cases.[60]

Nevertheless, the principle of nonmaleficence is not necessarily against LOD. This is evident even if one does not take competing principles into account.

---

[52] Bakker, in Price/Akveld (ed.) (1997), p. 30; Daar et al., Vol. 11 Transplantation Review 95, 100 (1997); Swiss Dispatch of 12. September 2001 on a Federal Law on Transplantation of Organs, Tissues and Cells, p. 30.

[53] Bundestag printed paper 15/5050 (2005), p. 35; Esser, in Höfling (ed.) (2003), p. 247; Norba (2009), p. 56; Papachristou et al., Vol. 78 Transplantation 1506, 1506 (2004); Schreiber (2004), p. 18; Schutzeichel (2002), p. 100; Ugowski (1998), p. 24.

[54] Bakker, in Price/Akveld (ed.) (1997), p. 28; Glannon, Vol. 343 British Medical Journal (2011); cf. Kirste, Vol. 81 Der Chirurg 778, 786 (2010); Schroder et al., Vol. 18 Progress in Transplantation 41, 41 (2008).

[55] George (2006).

[56] Daar et al., Vol. 11 Transplantation Review 95, 102 (1997).

[57] Ross, Vol. 30 The Journal of Law, Medicine and Ethics 440, 440 f. (2002).

[58] Beauchamp/Childress (2001), p. 113 ff.

[59] Cf. Gutmann, in Schroth et al. (ed.) (2006), p. 246 f.

[60] Cf. Hilhorst et al., Vol. 24 Transplant International 1164, 1165 (2011).

In the following, (1) I describe the consequences for LOD that are drawn from the application of the principle of nonmaleficence. (2) Afterwards, I present the position of the surgeon and the transplant team in the process of LOD. (3) An acceptable risk-benefit ratio in LOD is examined (4) and the role of common welfare in LOD is explained. (5) I ask if a special relationship is necessary for LOD. (6) Last, the principle of subsidiarity is seen from the angle of restricting LOD.

aa) Consequences

Notwithstanding the view taken here, conclusions drawn from the application of the principle of nonmaleficence differ. (1) Some argue against LOD in general, (2) while others only conclude that LOD has to be restricted to direct LOD.

(1) Stopping Living Organ Donation in General

Those interested in putting an end to LOD in general pursue an absolutist approach. They believe every LOD is a violation of medical ethics and do not see any reason to allow this violation in particular cases.[61] It is considered "morally wrong for a physician to be involved in such a practice."[62]

This *Hippocratic fundamentalism* in the transplant setting is inadequate.[63] It completely fails to take into account the normative weight of both the principle of respect for autonomous choices and the principle of beneficence.[64] Concerning the latter, the overall benefit of the donor and the recipient in the harm-benefit equation is increased in LOD. LOD is often the only possibility for the recipient to receive a donor organ. With regard to the donor, one cannot solely consider the (physical) harm inflicted on him. Psychological and moral beneficence may override the (physical) harm involved.[65]

(2) Restricting Living Organ Donation to Direct Living Organ Donation

Some argue that there should be a complete prohibition of LOD. Others suggest that LOD, because it may contravene the principle of nonmaleficence, should be restricted to direct LOD. They believe a restriction would minimise the risks to the donor's health.[66]

---

[61] George (2006); Potts/Evans, in Weimar/Bos/Busschbach (ed.) (2008), p. 377 ff.

[62] Potts/Evans, in Weimar/Bos/Busschbach (ed.) (2008), p. 378.

[63] Gutmann/Land, Vol. 384 Langenbeck's archives of surgery 515, 516 (1999).

[64] Cf. Gutmann/Schroth (2002), p. 109; cf. Price (2000), p. 227.

[65] Patel/Chadha/Papalois, Vol. 3 Experimental and Clinical Transplantation 181, 183 (2011).

[66] Ross, Vol. 30 The Journal of Law, Medicine and Ethics 440, 440 (2002).

Opponents rightfully point out that the donor's risks are equally high in direct, indirect and unspecified LOD.[67] Consequently, a stringent application of the rule not to cause (physical) harm would prohibit all LODs, including those within families.[68]

Those in favour of restricting LOD to direct LOD admit that the risks may be the same. However, they claim surgeons, or the community at large, might *perceive* risks differently for direct LOD than they do for indirect and unspecified LOD. The death of a relative as a consequence of a LOD might be considered "as a valiant, but tragic attempt to save the life of a loved one."[69] In contrast, if a living organ donor dies while donating or shortly after the donation to a stranger, the involved institution might be perceived as performing unethical practices,[70] especially in the media.

One may consider how others will perceive the risks. But, ultimately, restricting LOD to direct LOD based, on the principle of nonmaleficence does not withstand close examination. One cannot deny that equal risks may be felt more strongly and may have more normative weight in cases involving unspecified LOD. This reflects our tendency to link special privileges and duties to close relationships. This, however, is not a valid reason for prohibiting unspecified LOD altogether. One must also question how far media reactions should be considered.[71]

bb) A Reminder on the Difference Between Liberties and Duties

LOD directly concerns not only the organ recipient and the donor, but also the transplant team. They all have different interests that have to be regarded and balanced.[72]

The (arguments in favour of) legal and policy restrictions of LOD, which curtail the general legal possibility of these forms of organ transplantation will be analysed. It is important to note, however, that removing legal barriers to LOD does not oblige physicians to actually perform everything they are allowed to perform.

A donor does not have the moral or legal right to donate at a transplant centre if the centre's assessment shows that the risk is too high or the risk-benefit ratio is not

---

[67] Health Council of the Netherlands (2003); Hilhorst et al., in Weimar/Bos/Busschbach (ed.) (2011), p. 380; Matas et al., in Gutmann et al. (ed.) (2004), p. 195; Patel/Chadha/Papalois, Vol. 3 Experimental and Clinical Transplantation 181, 183 (2011); Veatch (2000), p. 189.

[68] Patel/Chadha/Papalois, Vol. 3 Experimental and Clinical Transplantation 181, 183 (2011); Ross, Vol. 30 The Journal of Law, Medicine and Ethics 440, 440 (2002).

[69] Ross, Vol. 30 The Journal of Law, Medicine and Ethics 440, 442 (2002).

[70] Ibid.

[71] Ibid.

[72] Cf. Hohmann (2003), p. 25.

good enough.[73] LOD requires the participation of the transplant centre. In addition to the patient's autonomy, the autonomy of the transplant team and surgeon is also respected.[74] The transplant team and surgeon have to decide if it is morally justifiable to help the person concerned to accomplish her decision.[75] They thus decide whether the surgery will be performed or not.[76] Many suggest that the operation should be refused if "the proposed procedure is not consistent with ethical principles or the estimation of risk exceeds what is acceptable by a physician's experience and by reported data."[77] Experiences in the UK show that patients accept a surgeon's refusal to perform the operation.[78] Similarly, the United Network for Organ Sharing (USA) rightly concluded that the autonomy of a transplant professional or group to decide the appropriateness of a given donor–recipient transplant, based on members' analysis of the potential risks and benefits, has equivalent or greater weight than that of the donor.[79]

Transplant teams have to make their own professional judgements in any case,[80] and removing legal barriers to LOD does not entail that physicians must perform an operation, simply because it is allowed.[81] Moreover, the transplant centre is not obligated to carry out a donor's life project.[82] The principle of nonmaleficence, however, does not distinguish between different types of LOD in this regard. Rather, the transplant team has to decide for each case individually — for direct LOD as well as for indirect and unspecified LOD[83] — because the same harms and risk exist in both types of LOD.[84]

It is important to note, however, that the risk-benefit ratio of any proposed LOD is determined not only by medical (or psychological) facts, but ultimately by personal value judgements. These judgements should generally be made by the

[73] Cf. American Medical Association, Vol. 284 The Journal of the American Medical Association 2919, 2925 (2000); cf. Gutmann/Schroth (2002), p. 111; cf. Hippen, Vol. 30 Journal of Medicine and Philosophy 593, 612 (2005); Roff, Vol. 33 Journal of Medical Ethics 437, 440 (2007).

[74] American Medical Association, Vol. 284 The Journal of the American Medical Association 2919, 2925 (2000); Joint Working Party of The British Transplantation Society and The Renal Association (2011), p. 28; Pellegrino (2006); cf. Price (2000), p. 227.

[75] Donnelly/Price (ed.) (1997), p. 54.

[76] Elliot, Vol. 21 Journal of Medical Ethics 91, 95 (1995); Gutmann/Schroth (2002), p. 111; cf. Pellegrino (2006).

[77] Cf. Bechstein/Moench, Vol. 24 Transplant International 1162, 1162 (2011); Delmonico/Surman, Vol. 76 Transplantation 1257, 1259 (2003); cf. Hilhorst et al., Vol. 24 Transplant International 1164, 1165 (2011).

[78] Information from C. Rudge.

[79] United Network for Organ Sharing, Vol. 24 Transplantation Proceedings 2236, 2236 ff. (1992).

[80] Hilhorst et al., in Weimar/Bos/Busschbach (ed.) (2011), p. 380.

[81] Cf. Hilhorst et al., Vol. 24 Transplant International 1164, 1169 (2011); cf. Ross, Vol. 30 The Journal of Law, Medicine and Ethics 440, 442 (2002).

[82] Cf. Hilhorst et al., Vol. 24 Transplant International 1164, 1169 (2011).

[83] Cf. Ross, Vol. 30 The Journal of Law, Medicine and Ethics 440, 442 (2002).

[84] Cf. Hilhorst et al., Vol. 24 Transplant International 1164, 1169 (2011).

one most affected by the outcome — the prospective donor. She alone must decide what is worth the risk in order to attain a certain good.[85] Therefore, a professional refusal to fulfil a potential donor's wishes must be based on sound moral reasons.[86] If the physician refuses to perform the surgery because she disagrees with the patient's values, she is not showing respect for the patient's dignity and autonomy. Also, it is hard to see how a paternalistic argument ("We know better than the donor.") could be strong enough to justify a refusal.[87] Although it may be morally problematic for the transplant team's interests to overrule the informed wish of the potential donor and deny significant benefits for the organ recipient,[88] the transplant team as moral actor is perfectly entitled to take into account its responsibilities for all patients and the transplant system as a whole. Complications and bad outcomes for the donor may in fact have a negative effect on the living donation programme as a whole. It might lead to "suspended operations, wide public debate, doubts about the integrity of the surgeons."[89] This also must be taken into account in the decision-making process of every LOD.[90]

To what extent do these arguments distinguish the different types of LOD? Those in favour of only allowing direct LOD argue "that doctors, who seek and expect societal support for their practices, are uncertain about societal expectations in unrelated cases."[91] Unspecified LOD is still rather seldom[92] and cannot be considered as an everyday practice or as a customary procedure. As a consequence, surgeons and transplant centres might fear that the social support is rather dissatisfying for unspecified LOD.[93] However, surgeons — from a medical and professional perspective — always proceed in the same way, regardless of the type

---

[85] This thesis has special importance for split-liver living donation and other types of LOD with higher peri- and postoperative risks for the donor (Information from T. Gutmann).

[86] Cf. Hilhorst et al., Vol. 24 Transplant International 1164, 1169 (2011), considering the case if a living kidney donor who, after having donated a kidney anonymously, also wanted to donate part of his liver; Neuberger, Vol. 24 Transplant International 1159 (2011).

[87] Cf. Wagner/Fateh-Moghadam, Vol. 56 Soziale Welt 73, 83 (2005); cf. Gutmann, Vol. 52 Neue Juristische Wochenschrift 3387, 3388 (1999); Hilhorst et al., Vol. 24 Transplant International 1164, 1165 (2011).

[88] Cf. Hilhorst et al., in Weimar/Bos/Busschbach (ed.) (2011), p. 380.

[89] Cf. Hilhorst et al., in Weimar/Bos/Busschbach (ed.) (2011), p. 380; cf. Hilhorst et al., Vol. 24 Transplant International 1164, 1166 (2011).

[90] Cf. Hilhorst et al., in Weimar/Bos/Busschbach (ed.) (2011), p. 380.

[91] Cf. Hilhorst et al., in Weimar/Bos/Busschbach (ed.) (2011), p. 381; cf. Hilhorst et al., Vol. 24 Transplant International 1164, 1164 (2011).

[92] Cf. Hilhorst et al., in Weimar/Bos/Busschbach (ed.) (2011), p. 381. In the United Kingdom, e.g., in 2009–2010, 920 directed LODs were performed and only 16 unspecified LODs (NHS Blood and Transplant (2010)). A survey was conducted on 113 kidney transplant units and 39 liver transplant by EULOD WP 2. Their results show that unspecified living kidney donation has occurred in 41 centres, and that only 8 unspecified living liver donations have been performed. (EULOD WP 2 (2012), DOW: Deliverable 4, p. 10 f.).

[93] Cf. Hilhorst et al., in Weimar/Bos/Busschbach (ed.) (2011), p. 381.

of LOD.[94] This clarification, along with the argument for *felt risks* mentioned above, leads to assume, at least when it comes to policy decisions, that public misperceptions of ethically justified practices in LOD should not lead to a legal restriction of these practices, but to educative efforts and attempts to change misled public attitudes.[95]

Hence, the surgeon and the transplant team must also be viewed as autonomous agents. Now that the rights of the surgeon and transplant team have been explained, it will be focussed on the acceptable risk-benefit ratio with respect to LOD.

cc) Limits to the Acceptable Risk-Benefit Ratio

Those who regard the principle of nonmaleficence as relative instead of absolute rightfully point to the fact that organ removal contains certain risks. Consequently, a risk-benefit ratio is required. The benefits to the donor and recipient resulting from the LOD must necessarily outweigh the risks accompanied by it.[96] As I have already clarified, a risk-benefit ratio of any proposed LOD cannot be determined by medical (or psychological) facts alone without taking into account the donor's (and the recipient's) personal value judgements. The physician (or the transplant centre) is a moral actor in his own right. As stated by *Delmonico* and *Surman*: "If equipoise is not affirmatively achieved in the risk-benefit calculation for the donor and the recipient, then sound medical judgment should override all other concerns."[97]

This proves that, apart from the clearest criterion generally applied to LOD, namely that it cannot be justified if it kills the donor, the legal limits placed on LOD vary. In conclusion, those who regard the principle of nonmaleficence as relative require a risk-benefit ratio for LOD. Next, it will be considered whether concerns of common welfare affect LOD as well.

dd) Common Welfare

It is difficult to enunciate a general definition for common welfare.[98] Nevertheless, it is not just an empty phrase.[99] When trying to define common welfare, its opposite, self-interest, is utilised.[100] One then attempts to use a negative delimitation to form

[94] Cf. Hilhorst et al., Vol. 24 Transplant International 1164, 1166 (2011).

[95] Cf. ibid., p. 4.

[96] American Medical Association, Vol 284 Journal of the American Medical Association 2919, 2920 (2000).

[97] Delmonico/Surman, Vol. 76 Transplantation 1257, 1259 (2003).

[98] Böckenförde, in Münkler/Fischer (ed.) (2002), p. 44 ff.; Folke Schuppert/Neidhardt (2002), p. 21; Hasenöhrl (2005); Münkler/Fischer, in Münkler/Fischer (ed.) (2002), p. 9.

[99] Böckenförde, in Münkler/Fischer (ed.) (2002), p. 62; Hasenöhrl (2005).

[100] Hasenöhrl (2005); cf. Münkler/Fischer, in Münkler/Fischer (ed.) (2002), p. 11.

a positive definition. As a result, when aiming at common welfare, the intent is to promote the interests of everyone (in a country) and to promote the people's common interests, instead of only one person or a few.[101]

The German Federal Constitutional Court uses common welfare as one justification for the restrictive orientation of the German transplant law. The Court states that it "is part of the concern of common welfare to protect people from significantly harming themselves."[102] Common welfare is not only regarded in the formulation of laws.[103] As shown above, in certain cases the physician will also consider the patient's common welfare before making a decision. The physician's decision could even be detrimental to the individual welfare of her patient.[104]

How the notion of *common welfare* might serve as an argument to restrict LOD and to distinguish between different types of LOD is unclear. LOD increases the risk-benefit ratio of all persons directly concerned, and it (at least in the case of kidneys) massively reduces costs in the health system. LOD also acknowledges that persons care for others in need and shows solidarity. Moreover, to imply that living donors are "significantly harming themselves" is inadequate. This approach can be classified as hard paternalism, which is in conflict with the donor's right of self-determination — a right guaranteed by constitutional law in many countries.[105]

The common welfare argument for the restriction of LOD is simply not convincing. Next, it will be analysed whether it is convincing to require a special relationship for LOD.

ee) Special Relationship Is Required for Living Organ Donation

The WHO and the Council of Europe Committee of Ministers were in favour of requiring a special relationship for LOD.[106] (1) I consider whether unrelated donors do not benefit as much from LOD as related ones. (2) Afterwards, I introduce an even more radical thesis, according to which *only* the special relationship between donor and recipient justifies LOD.

---

[101] University of Alaska (2003).

[102] German Federal Constitutional Court [Bundesverfassungsgericht], 11 August, 1999, Case No. 1 BvR 218/98, at 3b; cf. Walter, in Spickhoff (ed.) (2011), p. 2388, who also refers to the judgement of the German Federal Constitutional Court.

[103] Cf. German Federal Constitutional Court [Bundesverfassungsgericht], 22 January, 2011, Case No. 1 BvR 699/06, at 47.

[104] Schneider (2006).

[105] Gutmann, in Middel et al. (ed.) (2010), p. 28.

[106] Council of Europe Committee of Ministers, Resolution CM/Res(2008)4 on adult-to-adult living donor liver transplantation, 4b; World Health Organization (2010), WHO Guiding Principles on Human Organ Transplantation, Guiding Principle 3. Interestingly, in a version of the WHO Guding Principles on Human Organ Transplantation from 1991, it was written that "adult living person may donate organs, but in general such donors should be genetically related to the recipients."

(1) Unrelated Donors Do Not Benefit as Much from Living Organ Donation as Related Ones

The expected benefit is essential to justify LOD. Many argue that unspecified donors do not benefit as much from LOD as direct donors.[107] They conclude that unspecified donors should not be permitted to take the same risks as family members.[108] They give the following reasons.

Many state that the donor, in an unspecified LOD, does not have the opportunity to witness the benefits that arise from the act of donation for the organ recipient.[109] Direct LOD, as opposed to unspecified LOD, might be a win-win situation, if, for example, the increase of independence from the dialysis is directly beneficial for the organ donor and the organ recipient.[110]

Cognitive reasons are also given. Many claim that the real benefit for the organ donor might only become apparent on the psychological level.[111] This benefit is the easiest to estimate and to comprehend the more intense the relationship between recipient and donor is. Thus, affection between donor and recipient is thought to be decisive. For example, if a parent donates an organ to her child, potential physical harm due to the organ removal may clearly be compensated by a socio-psychosocial gain, namely by making it possible for the child to survive or to improve her quality of life.[112]

The possible effects of prohibiting a particular LOD are also mentioned. If a potential donor is barred from donating an organ to a stranger, this prevents her from exercising her right to choose. This applies if a person is not allowed to donate for a relative as well. In addition, relatives often feel a very strong motivation to help each other in times of need; this motivation is oftentimes perceived as a duty or obligation. This typically does not apply in the same way to unspecified LOD. Thus, not being allowed to donate is a greater burden for direct donors.[113]

Furthermore, related people share goals. The prohibition of direct LOD interferes directly with the goals of the potential organ donor.[114] This argument

---

[107] Cf. Patel/Chadha/Papalois, Vol. 3 Experimental and Clinical Transplantation 181, 181 (2011); cf. Persson et al., in Weimar/Bos/Busschbach (ed.) (2008), p. 271; cf. Roff, Vol. 33 Journal of Medical Ethics 437, 439 (2007).

[108] Silva/Wright, in Weimar/Bos/Busschbach (ed.) (2008), p. 291 ff.

[109] Cf. Adams et al., Vol. 74 Transplantation 582, 587 (2002); Dew et al., in Weimar/Bos/Busschbach (ed.) (2008), p. 213; Matas et al., in Gutmann et al. (ed.) (2004), p. 195; cf. Virzì et al., Vol. 39 Transplantation Proceedings 1791, 1793 (2007).

[110] Cf. den Hartogh (2008), p. 85; Schneider (2004).

[111] Donnelly/Price (ed.) (1997), p. 51, 65; cf. Glannon, Vol. 343 British Medical Journal (2011); Kasiske et al., Vol. 7 Journal of the American Society of Nephrology 2288, 2295 f. (1996); cf. Kranenburg et al., Vol. 38 Psychological Medicine 177, 178 (2008); Lamb, in Price/Akveld (ed.) (1997), p. 46; cf. Price, in Price/Akveld (ed.) (1997), p. 121.

[112] Bundestag printed paper 15/5050 (2005), p. 35.

[113] Ross et al., Vol. 74 Transplantation 418, 420 (2002).

[114] Ross, Vol. 30 The Journal of Law, Medicine and Ethics 440, 442 (2002).

can be countered by referencing the manifold motives of organ donors.[115] Many motives are actually independent of the recipient's experience of benefits.[116]

Several arguments are brought forth to explain why unrelated donors do not benefit as much from LOD as related ones. After presenting the even more radical view, that only the special relationship between donor and recipient justifies LOD, I will evaluate whether the arguments are persuasive.

(2) Only the Special Relationship Between Donor and Recipient Justifies Living Organ Donation

The main argument for restricting LOD to special donor-recipient relationship asserts that the special relationship between donor and recipient leads to a positive obligation between donor and recipient.[117] This leads one to feel a sense of responsibility to help their sick relative. Consequently, certain responsibilities to help a sick relative exist.[118]

Second, love between donor and recipient is mentioned.[119] Love might lead to the wish to help a suffering relative. Even if the LOD involves risks, this gesture is considered as comprehensible. Families are oftentimes even expected to help if a relative is seriously ill. For example, if a sister offers to donate an organ to her brother, she only needs to explain that she is offering for the sake of her brother. In contrast, if a person offers to take the risks connected to LOD to help a stranger, the offer is not so easily understood. Rather, a special explanation for this decision is expected.[120]

Third, some claim that the suffering of another person is felt more intensively if the person concerned is a relative or a close friend.[121]

---

[115] A good overview about the several motives to donate can be seen at Lennerling et al., Vol. 19 Nephrology Dialysis Transplantation 1600, 1600 ff. (2004); Price, in Price/Akveld (ed.) (1997), p. 115 ff. also lists several possible motives to donate an organ.

[116] Gutmann/Schroth, in Oduncu/Schroth/Vossenkuhl (ed.) (2003), p. 276.

[117] Glannon/Ross, Vol. 11 Cambridge Quarterly of Healthcare Ethics 153, 155 f. (2002); den Hartogh (2008), p. 81; Forsberg et al., Vol. 8 Pediatric Transplantation 372, 378 (2004); cf. Kluge, Vol. 19 Hastings Center Report 10, 12 (1989); Ross, in Weimar/Bos/Busschbach (ed.) (2008), p.185; cf. Spital, Vol. 12 Cambridge Quarterly of Healthcare Ethics 116, 116 (2003).

[118] To emphasise this, the following statement can serve as an example: "If we've got five people here, all of whom could efficiently use one of my kidneys, and one of them is my daughter, I not only don't think I have to randomise the choice among those five, I think I ought not randomise the choice among the five. That is to say I think there are certain kinds of special obligations that weigh in a way that some sort of general altruism doesn't" (Meilaender (2006)).

[119] Elliot, Vol. 21 Journal of Medical Ethics 91, 94 (1995); Land, Vol. 2 Transplant International 168, 170, 172 (1989); Papachristou et al., Vol. 78 Transplantation 1506, 1509 (2004); cf. Spital, Vol. 12 Cambridge Quarterly of Healthcare Ethics 116, 116 (2003).

[120] Elliot, Vol. 21 Journal of Medical Ethics 91, 94 (1995).

[121] German Federal Constitutional Court [Bundesverfassungsgericht], 11 August, 1999, Case No. 1 BvR 218/98, at 76; Kruse (1986), p. 257.

Fourth, others state that the donor's motivation is generally influenced by the quality of the donor-recipient relationship.[122] If the donor and recipient are close, the donor often wants to maintain the existing relationship.[123]

Whether the two opinions just introduced, (1) that unrelated donors do not benefit as much from LOD as related donors and (2) that only the special relationship between donor and recipient justifies LOD are convincing will be examined in the following analysis.

(3) Analysis

There seems to be a flaw in both arguments. Both give reasons to assume that prohibiting direct LOD creates a heavier burden for both donors and recipients, intensely infringes on their right to autonomy (which includes protecting the interest not to be kept from fulfilling duties and responsibilities) and, in the case of the potential recipient, infringes on his legally protected interest in his life and health.[124] For this reason, in preparing the German *Act on the donation, removal and transplantation of organs*, it was generally agreed upon that forbidding direct LOD between persons in close personal relationships would, under all circumstances, be an unconstitutional infringement of basic rights, both of the donor and the recipient; the infringement of basic rights in cases involving indirect and unspecified LOD is less clear.[125]

None of these arguments for the prohibition of unspecified LOD are convincing. There are special responsibilities in close personal relationships. The deduction, however, that these responsibilities are the *only* normative foundation of LOD, is a *non sequitur*.[126] The special relationship in direct LOD is not the "morally relevant key feature that provides a justification for living transplantations."[127] The special responsibility argument does not respect the person's ability to make moral choices. It disregards the fact that the moral value of beneficence, of saving lives and restoring health, does not depend on there being a close relationship between the parties involved.[128] Following the logic of the special responsibility argument, the biblical *good Samaritan* was acting without a moral foundation.[129] It is only relevant that the donor cares for the needs of other, which applies to (non-

[122] Papachristou et al., Vol. 78 Transplantation 1506, 1509 (2004); Walter et al., Vol. 11 Medical Science Monitor 503, 504 (2005).

[123] Papachristou et al., Vol. 78 Transplantation 1506, 1508 (2004).

[124] These rights are, for example, included in the Charter of Fundamental Rights of the European Union.

[125] Cf. Bundestag printed paper 15/5050 (2005), p. 34 f.

[126] Information from T. Gutmann.

[127] Hilhorst et al., Vol. 24 Transplant International 1164, 1167 (2011).

[128] Information from T. Gutmann.

[129] Ibid.

commercial) LOD in all cases, regardless of the relatedness of donor and recipient.[130] *Hilhorst's* evaluation of the motivations and outcomes of direct, indirect and unspecified LOD has proven that no significant differences exist.[131] According to the medical evaluations of unspecified LOD, donors are described as truly generous and selfless.[132] If one compares the act of donating to benefit a stranger to the act of donating to a relative, the act of donating to a stranger could be seen as the highest expression of altruism. *Evans* states, "[I]t might be thought that altruism receives its highest expression in the absence of personal relationships."[133] If LOD was restricted to relatives only, the restriction "would rule out altruism of this supreme kind."[134] The psychological outcomes recorded after organ donation to a stranger are clearly encouraging and are testimony to unspecified LOD.[135] All this leads to assume that both types of LOD should be treated equally.[136] Each LOD should be judged individually. Does the application of the principle of subsidiarity lead to a different result?

ff) The Principle of Subsidiarity

Considering LOD as subsidiary option entails that LOD is only viewed as a last resort; LOD is thus restricted by the principle of subsidiarity. This is mentioned for LOD in general, but it especially applies to unspecified LOD.[137] Proponents of the subsidiary principle are typically in favour of restricting LOD to specified LOD, or even to direct LOD.[138]

The principle of subsidiarity is often justified by the potential living donors' need for protection.[139]

Chapter "Comparative Analysis of European Transplant Laws Regarding Living Organ Donation" already showed that the principle of subsidiarity is also subject to

---

[130] Hilhorst et al., Vol. 24 Transplant International 1164, 1167 (2011).

[131] Hilhorst et al., Vol. 24 Transplant International 1164, 1167 (2011); Hilhorst et al., in Weimar/Bos/Busschbach (ed.) (2011), p. 381; cf. Neuberger, Vol. 24 Transplant International 1159, 1160 (2011).

[132] Neuberger, Vol. 24 Transplant International 1159, 1160 (2011).

[133] Evans, Vol. 15 Journal of Medical Ethics 17, 19 (1989).

[134] Ibid.

[135] Massey et al., Vol. 10 American Journal of Transplantation 1445, 1445 ff. (2010).

[136] Cf. Hilhorst et al., in Weimar/Bos/Busschbach (ed.) (2011), p. 383.

[137] Ross, Vol. 16 Kennedy Institute of Ethics Journal 151, 167 (2006).

[138] Cf. German Federal Constitutional Court [Bundesverfassungsgericht], 11 August, 1999, Case No. 1 BvR 218/98, at 77.

[139] German Federal Constitutional Court [Bundesverfassungsgericht], 11 August, 1999, Case No. 1 BvR 218/98, at 77; Gubernatis (2002). Several other arguments in favour of the principle of subsidiarity of LOD have already been listed in chapter "Comparative Analysis of European Transplant Laws Regarding Living Organ Donation".

severe critique.[140] In addition to the criticism that has already been laid down, it is criticised that the principle of subsidiarity jumbles the macro- and the micro-level.[141] On the macro-level there is good reason to believe that increasing organ procurement from deceased donors is vital. The availability of a post mortem organ is important because it could be another option for the potential donors. This, however, is no reason to interfere with a living donor's decision on the micro-level inter alia.[142] The recipient himself should — based on his right of self-determination — have the opportunity to decide what kind of organ will be implanted into his body.[143] The donor has a right of self-determination as well. If he is capable of giving valid consent, and does so after being sufficiently informed, the donor's right of self-determination is infringed upon if he is kept from helping a suffering person, even though the intended LOD does not involve any major risks.[144]

To use the principle of subsidiarity as an argument to restrict LOD and to prohibit unspecified LOD is just as incomprehensible as all other approaches connected to the principle of nonmaleficence.

2) *Lack of Voluntariness*

If a normative concept of LOD is primarily based upon the principle of respect for autonomous choices, then a lack of voluntariness is, so to speak, the worst case.[145] *Price* writes: "The voluntariness of a donor's decision to donate is central to LDT as a matter of law and ethics."[146] The German *Act on the donation, removal and transplantation of organs* also considers the voluntary decision of the donor to be the central factor of authorisation for LOD.[147] The most recent European document on LOD, the *Directive 2010/43/EU*, also demands the Member States to ensure that LOD is voluntary.[148] Therefore, this aspect will be analysed next by (1) introducing the notion of (in)voluntariness (2) and addressing afterwards whether there is a connection between the donor-recipient relationship and (in)voluntariness. (3)

[140] Edelmann, Vol. 50 Versicherungsrecht 1065, 1068 (1999); Gutmann, Vol. 15 Medizinrecht 147, 152 (1997); Swiss Dispatch of 12. September 2001 on a Federal Law on Transplantation of Organs, Tissues and Cells, p. 195.

[141] Several further points of criticism have already been presented in chapter "Comparative Analysis of European Transplant Laws Regarding Living Organ Donation".

[142] Cf. Gutmann (2006), p. 81 f.; Gutmann, in Middel et al. (ed.) (2010), p. 32 f.

[143] Edelmann, Vol. 50 Versicherungsrecht 1065, 1068 (1999).

[144] Gutmann/Schroth (2002), p. 80.

[145] Eyal, in Zalta (ed.) (2011); cf. Faden/Beauchamp (1986), p. 152 ff.; Information from T. Gutmann.

[146] Donnelly/Price (ed.) (1997), p. 58; Price, in Price/Akveld (ed.) (1997), p. 112.

[147] Fateh-Moghadam et al. in Schroth et al. (ed.) (2006), p. 120; cf. Schneewind/Sedlmayer, in Schroth et al. (ed.) (2006), p. 16; Wagner/Fateh-Moghadam, Vol. 56 Soziale Welt 73, 79 (2005).

[148] Article 13 1. Directive 2010/53/EU of the European Parliament and of the Council of 7 July 2010 on standards of quality and safety of human organs intended for transplantation.

Subsequently, the possible gender related issues related to (in)voluntariness will be described.

a) Notions of (In)Voluntariness

Chapter "Comparative Analysis of European Transplant Laws Regarding Living Organ Donation" showed that all countries require the donor to act voluntarily. None of the national laws contain a definition of the term voluntariness, though.[149] That also applies to the *Directive 2010/53/EU*. According to Art. 13, "[m]ember States shall ensure that donations of organs from deceased and living donors are voluntary and unpaid." It does not contain any clarification of the term voluntariness. It only (rightfully) states that an involuntary donation might jeopardise the whole process of donation. If LOD is not done with the sole aim of improving a sick person's quality of life, then the process of donation could be compromised.

A review of the relevant literature on voluntariness reveals that no consensus on a definition of the term exists.[150] According to *Beauchamp* and *Childress*, "a person acts voluntarily to the degree that he or she wills the action without being under the control of another's influence."[151] This definition will provide a basis for analysing the aspect of voluntariness in connection with LOD.

According to the definition by *Beauchamp* and *Childress*, the organ donor acts voluntarily if she is not exposed to any force or coercion. In addition, a person does not act voluntarily if she has been deceived.[152] The person must be able to comprehend what she is getting into.[153] She must come to an autonomous decision,[154] she must have the opportunity to comprehend that she can choose between two alternatives and she must be able to act according to her will.[155] In how far (in) voluntariness of the donor is connected to the donor-recipient relationship will be analysed next.

b) Connection Between Donor-Recipient Relationship and Involuntariness

Opinions on how voluntariness is affected by the donor-recipient relationship differ. (1) Some argue that *related* LOD is specifically haunted by the danger of involuntary decisions, while (2) others claim *unrelated* LOD is connected to involuntariness. (3) After presenting these two approaches, a conclusion will be drawn.

---

[149] Cf. Lilie/Krüger, Vol. 81 Der Chirurg 787, 788 (2010).

[150] den Hartogh, in Weimar/Bos/Busschbach (ed.) (2008), p. 221.

[151] Beauchamp/Childress (2001), p. 93.

[152] Cf. Fateh-Moghadam et al., Vol. 22 Medizinrecht 19, 28 (2004).

[153] Gutmann/Schroth (2002), p. 37.

[154] Schutzeichel (2002), p. 107.

[155] Land, in Oduncu/Schroth/Vossenkuhl (ed.) (2003), p. 225.

aa) Connection Between Related Living Organ Donation and Involuntariness

Some believe that there is a factual connection between *related* LOD and involuntariness.[156] Several academics state that the donor might be pressured within the family.[157] This should be taken seriously. Coercion and undue influence may occur within the family institution.

Experts claim that relatives generally "do not have the same degree of free choice as anonymous donors." Existing social bonds could lead to certain expectations, which might be internalised by the person concerned.[158] Family members might openly express a certain expectation towards the potential donor.[159] A refusal to donate might seriously affect the potential donor's status in the family. It is therefore considered very difficult for the potential donor to disengage herself from the family's pressure to donate.[160]

Independently of the fact that related donors might be subject to external pressure, they might also feel a certain degree of moral obligation,[161] or what could be considered as 'internal pressure'.[162] A survey conducted on living parental liver donation concluded that the parent's decision in favour of donating part of their liver to their child could not be called a free choice in the sense of a choice made willingly and easily. One donor, for example, stated: "I felt I was obliged to do it. I wanted to do it. There was really no choice. It was all about trying to save my daughter. They could have whatever they needed."[163]

If one applies these arguments to LOD, it could quickly lead to its complete prohibition. But, if family members are thought of as non-autonomous donors, the notion of voluntariness is distorted, as it conflates hard choices with involuntary ones.[164] There is no reason to assume that some critical decisions lack voluntariness

---

[156] Bakker, in Price/Akveld (ed.) (1997), p. 28; Forsberg et al., Vol. 8 Pediatric Transplantation 372, 377 (2004); Ross et al., Vol. 74 Transplantation 418, 419 (2002).

[157] Bock (1999), p. 118; cf. Dworkin, Vol. 33 The Modern Law Review 353, 359 (1979); Erim et al., Vol. 35 Transplantation Proceedings 909, 909 (2003); Evans, Vol. 15 Journal of Medical Ethics 17, 19 (1989); Gutmann et al., Terasaki (ed.) (1995), p. 356; Gutmann/Schroth (2002), p. 62; Heuer/Conrads, Vol. 15 Medizinrecht 195, 201 (1997); Keller, Vol. XXXII Stetson Law Review 855, 869 (2003); Lilie/Krüger, Vol. 81 Der Chirurg 787, 788 (2010); Sanner/Dew/Busschbach, in Weimar/Bos/Busschbach (ed.) (2008), p.195; Schreiber, in Rittner/Paul (ed.) (2005), p. 63.

[158] Cf. Choudhry, Vol. 29 Journal of Medical Ethics 169, 170 (2003); Ross et al., Vol. 74 Transplantation 418, 419 (2002).

[159] Schutzeichel (2002), p. 108 f.

[160] Ibid.

[161] Cf. Lennerling/Forsberg/Nyberg, Vol. 76 Transplantation 1243, 1244 (2003); Ross et al., Vol. 74 Transplantation 418, 419 (2002).

[162] Gutmann (2006), p. 26; cf. Price (2000), p. 295.

[163] Forsberg et al., Vol. 8 Pediatric Transplantation 372, 374 (2004).

[164] Fateh-Moghadam, in Schroth et al. (ed.) (2006), p. 170; Gutmann/Schroth (2002), p. 114; Gutmann (2004), p. 8; Schroth, in Schroth et al. (ed.) (2006), p. 84; cf. Spital, Vol. 12 Cambridge Quarterly of Healthcare Ethics 116, 116 (2003). See also Eyal, in Zalta (ed.) (2011).

if there is no perceived alternative, such as LOD to save the life of one's child. It is still possible to act voluntarily, even if one feels one *has* to act a certain way in a critical situation. There is no contradiction in the statements, "There was really no choice" and "I wanted to do it." Concerns about the moral 'pressure' on donors that are related to the recipient are understandable. Nevertheless, that does not imply that related donors cannot exercise their will. Calling donors' decisions in cases like those mentioned "involuntary" is incorrect; it would prohibit actions which are experienced by the potential donors as morally self-evident and disregard choices which are of utmost importance to the person involved.[165] Moreover, it is not in a potential donor's best interest to be denied the possibility to help a person she is close to.[166] It has even been observed that the relatives that feel the strongest motivation to donate an organ feel the least choice at the same time. This means "that parents who are donating to their children typically display the least ambivalence of all donors."[167] The Commentary on the *WHO Guiding Principles on Human Cell, Tissue and Organ Transplantation* supports related LOD by stating that a genetic donor-recipient relationship as well as well as a so-called legal relationship (e.g., spouses) can confirm that "the donor is motivated by genuine concern for the recipient."[168]

That pressure in families might exist cannot be denied, though. To ensure that relatives really act voluntarily, careful screening procedures are necessary. As *ultima ratio*, the involved medical professionals can provide a *medical excuse* in case the potential donor does not want to donate, but does not have the heart to tell the potential recipient about her unwillingness.[169]

---

[165] Gutmann/Schroth (2002), p. 114 f.; cf. Gutmann, in Schroth et al. (ed.) (2006), p. 246 f.; cf. Price (2000), p. 298.

[166] Gutmann, in Schroth et al. (ed.) (2006), p. 247.

[167] Price (2000), p. 298.

[168] WHO Guiding Principles on Human Cell, Tissue and Organ Transplantation, 2008, Commentary on Guiding Principle 3.

[169] American Medical Association, Vol. 284 The Journal of the American Medical Association 2919, 2921 (2000); cf. Cotler et al., Vol. 9 Liver Transplantation 637, 640 (2003); Dworkin, Vol. 33 The Modern Law Review 353, 359 (1979). Cotler et al. performed a study in which they asked surgeons what, according to their view, should be told to the potential recipients and the families if a potential donor withdraws his consent. The result was that 87 % stated that they would give a medical or technical excuse, while only 9 % suggested to tell the truth and only 4 % would give no explanation (Cotler et al., Vol. 9 Liver Transplantation 637, 640 (2003)). However, the suggestion to provide a medical excuse to relief the donor is also criticised. It is stated that "[p]eople all the time have to make decisions which, [...], will evoke predictable unwelcome reactions from others, and this fact may or may not be relevant information for them. Such unwelcome reactions are simply part of the normal context of decision-making, [...]" (den Hartogh, in Weimar/Bos/Busschbach (ed.) (2008), p.226).)

bb) Connection Between Unrelated Living Organ Donation and Involuntariness

In contradiction to the previous opinion, some argue that there is a factual connection between *unrelated* LOD and involuntariness. The German Bundestag justified the restriction of the donor-recipient relationship by stating that, in general, only a family relationship or another close personal relationship guarantees voluntariness in LOD.[170]

Some (strangely) claim that LOD to a stranger can be considered a supererogatory action, its exercise being an indication for involuntariness or a lack of competence.[171] In addition, it is sometimes feared that the donor who wants to be part of an unspecified LOD might be mentally unstable.[172] If these claims are true, they could possibly lead to the prohibition of unrelated LOD.

Opponents assert that these doubts do not reflect the actual situation. Social research demonstrates that organ donors with a competent, informed and voluntary decision to act altruistically towards strangers can be found in great numbers.[173] It has also been analysed that an unspecified LOD in which the donor has been carefully screened does not have any negative psychological consequences for him.[174] Furthermore, in comparison to related LOD, pressure or coercion from a family member does not have to be feared in case of unrelated LOD. *Hoyer*, a German transplant surgeon who participated in an unspecified LOD, for example, states that "[v]oluntariness is beyond doubt."[175] In any case, there is a growing consensus about the standards for the psychological evaluation of unspecified LOD.[176]

cc) Conclusion

The fact that some experts believe that related LOD is especially connected to involuntariness, while others believe that unrelated LOD is connected to

---

[170] Bundestag printed paper 13/4355 (1996), p. 20; see also German Federal Constitutional Court [Bundesverfassungsgericht], 11 August, 1999, Case No. 1 BvR 218/98, at 3b.

[171] Cf. Roff, Vol. 33 Journal of Medical Ethics 437, 439 (2007).

[172] Cf. Henderson et al., Vol. 3 American Journal of Transplantation 203, 203 (2003); cf. Kranenburg et al., Vol. 38 Psychological Medicine 177, 178 (2008); cf. Massey et al., in Weimar/Bos/Busschbach (ed.) (2011), p. 370 f.

[173] Cf. German Federal Constitutional Court [Bundesverfassungsgericht], 11 August, 1999, Case No. 1 BvR 218/98, at 25; Hilhorst, Vol. 8 Ethical Theory and Moral Practice 197, 200 (2005); Landolt et al., Vol. 76 Transplantation 1437, 1441 (2003); cf. Massey et al., Vol. 10 American Journal of Transplantation 1445, 1445 (2010); cf. Spital, Vol. 71 Transplantation 1061, 1063 (2001).

[174] Cf. Jendrisak et al., Vol. 6 American Journal of Transplantation 115, 119 (2006); Massey et al., Vol. 10 American Journal of Transplantation 1445, 1450 (2010).

[175] Hoyer, in Weimar/Bos/Busschbach (ed.) (2008), p. 234.

[176] Kranenburg et al., Vol. 38 Psychological Medicine 177, 177 ff. (2008).

involuntariness, shows that there might be something wrong with both arguments. In both cases, involuntary decisions — due to incompetence, coercion, or misinformation — are possible.[177] Both groups of LOD might contain cases that should not be performed, while both types of LOD also contain cases that are without doubt the result of a voluntary, well informed decision. So, the distinction between genetically related and unrelated donors is manifestly unjustifiable, being "a totally inadequate proxy for the underlying and ethically justifiable distinction between free and coerced donors."[178] An evaluation of each individual case is preferable.[179] Apart from voluntariness being relevant for LOD in general, voluntariness has to be considered with respect to the gender of the organ donor.

c) Gender Related Issues of Voluntariness

In Europe, there is a gender bias with regard to the amount of LODs. Statistics indicate that 2 out of 3 kidneys are donated by women. This is the opposite with regard to the recipients, where 2 out of 3 are male.[180]

Table 1 gives a more precise overview of the gender bias in the United Kingdom[181] and Table 2 of Switzerland.[182]

Several different reasons are mentioned for this imbalance, although no clear explanation exists.

First, societies that are characterised by a high degree of male dominance are referred to. There, a tendency "to choose the least valuable, least productive family member" as donor, might exist.[183] Men are still, for the most part, the breadwinners in the family. Hence, a downtime, as a result of the LOD, is expected of women more often than of men.[184] However, given the social facts concerning the labour

[177] Gutmann (2006), p. 28 ff.

[178] Choudhry, Vol. 29 Journal of Medical Ethics 169, 169 (2003).

[179] Cf. Choudhry, Vol. 29 Journal of Medical Ethics 169, 169 (2003); cf. Gutmann, Vol. 27 Zeitschrift für Rechtspolitik 111, 113 (1994) who primarily refers to the distinction between genetic and emotional relatives; Gutmann/Schroth (2002), p. 63; Price (2000), p. 328; cf. Sanner/Dew/Busschbach, in Weimar/Bos/Busschbach (ed.) (2008), p. 194 f.

[180] Cf. Gutmann/Schroth (2002), p. 94; Schneider (2004). In the United States, a gender disparity has been observed as well (Bloembergen et al., Vol. 7 Journal of the American Society of Nephrology 1139, 1139 f. (1996); Kayler et al., Vol. 73 Transplantation 248, 248 ff. (2002); US Department of Health & Human Services (2011); cf. Thiel/Nolte/Tsinalis, Vol. 37 Transplantation Proceedings 592, 592 (2005). It can be assumed that this does not apply to the illegal market, though, since all pictures known show men (Information from C. Rudge).

[181] Fuggle et al., Vol. 89 Transplantation 694, 696 (2010).

[182] Thiel/Nolte/Tsinalis, Vol. 37 Transplantation Proceedings 592, 592 (2005).

[183] Schneider (2004); Sheper-Hughes, Vol. 7 American Journal of Transplantation 507, 508 (2007).

[184] Biller-Andorno/Kling, in Gutmann et al. (ed.) (2004), p. 227; cf. Bloembergen et al., Vol. 7 Journal of the American Society of Nephrology 1139, 1143 (1996); Bundestag printed paper 15/5050, p. 36; Kirste (2004), p. 39; cf. Thiel/Nolte/Tsinalis, Vol. 37 Transplantation Proceedings 592, 593 (2005).

Table 1 Living kidney donation in the United Kingdom[a]

	Women	Men
Unrelated living kidney donation from 2000 to 2007	565 (61 %)	355 (39 %)
Related living kidney donation from 2000 to 2007	1,172 (52 %)	1,072 (48 %)
Recipients of unrelated living kidney donation	324 (35 %)	596 (65 %)
Recipients of related living kidney donation	929 (41 %)	1,315 (59 %)

[a]Fuggle et al., Vol. 89 Transplantation 694, 696 (2010). More recent statistics for unspecified LOD are different, though, since from July 2007 until July 2011, 59 % of the donors of unspecified LOD were male (Cronin, Vol. 343 British Medical Journal (2011); NHS Blood and Transplant (2010))

Table 2 Living kidney donation in Switzerland[a]

	Women	Men
Living kidney donation in 1999	56 %	44 %
Living kidney donation in 2001	72 %	28 %
Living kidney donation in 2003	65 %	35 %

[a]Thiel/Nolte/Tsinalis, Vol. 37 Transplantation Proceedings 592, 592 (2005)

market, it may also be a perfectly rational choice for a wife to donate for a family member instead of letting her husband do it. In most European cases, female organ donors are not *chosen*, instead they actively choose the role themselves.[185]

Second, women could feel more pressure than men.[186] Or, in other words, perhaps it is easier for men to resist pressure.[187] This is connected to the fact that women are still expected to perform unpaid duties of care,[188] and also consider themselves as having a certain responsibility as care givers.[189] That makes it more difficult for them to evade the pressure to become a living organ donor.[190]

Third, some assume that women have a more altruistic attitude.[191] When men and women were confronted with emergency scenarios, women expressed greater empathy.[192] However, a prospective German study showed that (in contrast to mothers who gave an organ to their child) wives who acted as organ donors for

[185] Schneewind et al., Vol. 12 Transplantationsmedizin 164 ff. (2000).

[186] Biller-Andorno, Vol. 5 Medicine, Health Care and Philosophy 199, 202 (2002); cf. Thiel/Nolte/Tsinalis, Vol. 37 Transplantation Proceedings 592, 594 (2005).

[187] Biller-Andorno/Kling, in Gutmann et al. (ed.) (2004), p. 227.

[188] Schneider (2004).

[189] Biller-Andorno/Kling, in Gutmann et al. (ed.) (2004), p. 226; Gilligan (1982), e.g. p. 155 & p. 162; cf. Gutmann/Schroth (2002), p. 94; cf. Lennerling et al., Vol. 19 Nephrology Dialysis Transplantation 1600, 1603 (2004); cf. Thiel/Nolte/Tsinalis, Vol. 37 Transplantation Proceedings 592, 594 (2005).

[190] Schneider (2004).

[191] AOK Bundesverband (2004), p. 3; Bundestag printed paper 15/5050 (2005), p. 17; Robert Koch-Institut (ed.) (2003), p. 17.

[192] Biller-Andorno/Kling, in Gutmann et al. (ed.) (2004), p. 226; Russel/Mentzel, Vol. 130 The Journal of Social Psychology 309, 309 (1990).

their husbands primarily followed reasonable self-serving motives: Their aim was to get their 'functioning' partner back.[193]

Fourth, when deciding whether to become a living organ donor or not, some claim that men are more rational.[194] Potential male donors might also be more uncertain and this might be accepted to a greater extent by those involved.[195] As a consequence, female relatives might decide to donate in the meantime.[196]

Fifth, medical reasons are used to explain the gender disparity. Men suffer more often from hypertension and coronary artery disease, leading to their exclusion as donors.[197] In addition, more men suffer from end-stage renal disease[198] and diabetes than women,[199] and consequently, are in need of a donor organ. It is very common for the spouse to act as living organ donor. It therefore seems consistent, from a medical point of view, for more women to become living organ donors and for more men to receive an organ donated by a living person.[200]

Many suggest that attention must be paid to the social pressures exerted on women to act as living organ donors.[201] However, it does not seem helpful to initiate a reduction of female donors. Rather, enhancing the willingness of men to donate seems to be the more desirable solution.[202]

In the end, there are many possible autonomous motives a woman could have which drive her to become an organ donor. Ultimately, it must be devoted attention to each case in order to ensure that every living organ donor acts voluntarily. This is the only way to determine that each donor makes an authentic, well-reasoned decision, regardless of their gender.[203] It must be secured effective protection, particularly for the vulnerable potential donors,[204] keeping in mind that there may be gender specific vulnerabilities in certain circumstances. However, as long as every single act of LOD is justified in itself, the donor's gender seems morally irrelevant. So, gender imbalance in LOD is no *malum in se*, and neither is an

---

[193] Schneewind et al., Vol. 12 Transplantationsmedizin 164 ff. (2000); cf. Schroth, in Schroth et al. (ed.) (2006), p. 81 f.

[194] Achille/Vaillancourt/Beaulieu-Pelletier, in Weimar/Bos/Busschbach (ed.) (2008), p. 321; Biller-Andorno/Kling, in Gutmann et al. (ed.) (2004), p. 227.

[195] Biller-Andorno/Kling, in Gutmann et al. (ed.) (2004), p. 227.

[196] Ibid.

[197] Bloembergen et al., Vol. 7 Journal of the American Society of Nephrology 1139, 1143 (1996); cf. Kayler et al., Vol. 73 Transplantation 248, 252 (2002).

[198] Bloembergen et al., Vol. 7 Journal of the American Society of Nephrology 1139, 1143 (1996); Thiel/Nolte/Tsinalis, Vol. 37 Transplantation Proceedings 592, 593 (2005).

[199] Thiel/Nolte/Tsinalis, Vol. 37 Transplantation Proceedings 592, 592 f. (2005).

[200] Information from C. Rudge.

[201] Sheper-Hughes, Vol. 7 American Journal of Transplantation 507, 508 (2007).

[202] Robert Koch-Institut (ed.) (2003), p. 17; Thiel/Nolte/Tsinalis, Vol. 37 Transplantation Proceedings 592, 594 (2005).

[203] Cf. Gutmann (2006), p. 52.

[204] Biller-Andorno, Vol. 5 Medicine, Health Care and Philosophy 199 (2002).

"equitable donor-recipient-ratio in living organ donation"[205] an important goal *in itself*.

All in all, voluntariness of the donor should be evaluated for each case individually, independently of the donor-recipient relationship and the donor's gender.

3) The Social Pressure Argument

The focus has been on the coercion and undue influence that could affect the donor. (1) The fact that LOD is the only option for several sick patients because of the severe organ shortage already puts pressure on all potential living donors. (2) In addition, complex interactions between donor and recipient exist that go beyond the problem of pressure on the donor in case of a related LOD. (3) The donor and recipient's social environment should not be ignored.

a) Organ Shortage Increases the Pressure on All Potential Living Organ Donors

LOD is the only possibility for several sick patients to receive an organ. Hence, the pure availability of it increases potential 'pressure' on living individuals to consider it. As a consequence, the more people that get the legal chance to become a living donor means more people will face the burden of *deciding* whether to donate or not.[206] Concerns are raised that a general legalisation of unspecified LOD would increase the 'mental and social pressure' on potential living donors in the society as a whole.[207] Furthermore, many worry that this would abolish the easiest option to say "no" ("Sorry, I am not allowed by law to be a donor"). The German Bundestag's Parliamentary Commission decided against a general legalisation of unspecified LOD.[208] The commission drew the conclusion that LOD should be restricted to a minimum and that paired donations and unspecified LOD should remain prohibited.[209]

This argument is startling. Its rationale is that it is more important for the state to protect its citizens from having to face hard choices than to give them the chance to choose whether to help or even save a person in need (which, in the case of cross-over-donation, is often the potential donor's husband, wife, or partner — an issue that was specifically addressed by the commission). This argument presents a

[205] Ibid.

[206] Gutmann/Schroth (2002), p. 115; Gutmann (2006), p. 26.

[207] Bundestag printed paper 15/5050 (2005), p. 40, 42 f., 48, 73 f.; cf. Schroth, in Schroth et al. (ed.) (2006), p. 82, who immediately after presenting this argument invalidates it.

[208] Bundestag printed paper 15/5050 (2005), p. 40, 42 f., 48, 73 f.

[209] Information from T. Gutmann.

degrading image of citizens who are incapable of deciding for themselves in critical situations. It cannot even be justified paternalistically, because it is clearly not in the best interest of competent adults to be generally withheld the chance to save a person (and maybe a close person) in need.[210] In addition, allowing unspecified LOD in particular does not inevitably raise the pressure on all citizens, because the legal allowance to act as unspecified living organ donor cannot be equated with the duty to actually become an unspecified living organ donor.

Hence, it cannot be agreed upon the view that LOD should be restricted because its pure availability increases the pressure on potential donor. Next, the extent LOD influences a particular relationship between donor and recipient will be analysed.

b) Interactions Between Donor and Recipient

LOD has an essential impact on the donor-recipient relationship.[211] The interactions between donor and recipient caused by a potential LOD are very diverse and can lead to several problems. Issues might arise if the potential living organ donor refuses to donate, but also if the LOD takes place.

If the donor can decide autonomously, his decision might be not to perform LOD. The recipient does not want the donor to feel pressured to donate. In case of a refusal, however, this might cause nasty effects on the relationship and might lead to a painful sense of guilt and failure.[212] Therefore, the possibility of LOD as such might already have a negative influence on the donor-recipient relationship.

The donor's consent to donate might also lead to problems. Stress and strain commonly occur before and after the transplantation. The decision-making process itself is especially stressful. If a person decides to donate, it is normal for the potential recipient to be ambivalent towards this offer. He often worries about the donor's health or wonders why the potential donor would offer such a noble gift.[213] In a few cases, the potential recipient felt urged to actually accept the organ.[214]

The situation after the transplantation can be difficult as well, since the outcome can be a moral and psychological burden for both. In cases of failure, the donor might become demoralised,[215] and the recipient might be afraid of disappointing the donor.[216] Even a successful transplantation can lead to difficulties, especially

[210] Gutmann/Schroth (2002), p. 115.

[211] Ummel/Achille, in Weimar/Bos/Busschbach (ed.) (2011), p. 304.

[212] Schneider (2004).

[213] Ummel/Achille, in Weimar/Bos/Busschbach (ed.) (2011), p. 304.

[214] Donnelly/Price (ed.) (1997), p. 78; Fox/Swazey (1999), p. 35.

[215] Virzì et al., Vol. 39 Transplantation Proceedings 1791, 1793 (2007). Other studies, in contrast, showed that the donor even in case the donation failed stated that he would donate again (Otto, in Rittner/Paul (ed.) (2005), p. 60).

[216] Ummel/Achille, in Weimar/Bos/Busschbach (ed.) (2011), p. 304; Waterman et al., Vol. 16 Progress in Transplantation 17, 17 (2006).

for the organ recipient. The LOD can be considered as an extraordinary gift[217]: It is not reciprocal and has no physical or symbolic equivalent. This might lead to a complicated situation in which "the giver, the receiver, and their families may find themselves locked in a creditor-debtor vise that binds them one to another in a mutually fettering way."[218] The recipient also often feels guilty that the donor had to undergo an operation,[219] and might have a persistent feeling of indebtedness to the donor.[220]

All in all, it is common for a potential LOD, as well as the actual performance of it, to influence the donor-recipient relationship. This, however, should not only be considered as a problematic aspect of LOD, since positive effects occur as well. The LOD does not only affect the donor and the recipient, but the social environment as well. That will therefore be described next.

c) Social Environment

LOD mainly concerns donor and recipient, but the social environment of donor and recipient should also not be completely ignored in the process.[221]

The donor and recipient typically share the same social environment if they are related. While resistance from the family in cases of direct LOD is not entirely impossible, it happens very seldom.[222] A conflict is imaginable, though, if a married person considers donating to a genetic relative and his spouse, or own immediate family, has concerns.[223]

In contrast, unspecified LOD often leads to a conflict within the social environment of the donor. Decisions to donate an organ to a stranger are usually disliked by family and friends of the donor. They usually do not understand why their relative intends to risk his life for a stranger.[224] Even though family and friends do not directly take part in the LOD, their help and support is an important factor in the whole process and should not be ignored.

As seen, social pressure may be manifold. It arises due to the shortage of donor organs. It also arises in the interactions between donor and recipient and with

[217] Fox/Swazey (1999), p. 40; Sanner, in Weimar/Bos/Busschbach (ed.) (2008), p. 283.

[218] Fox/Swazey (1999), p. 40.

[219] Forsberg et al., Vol. 8 Pediatric Transplantation 372, 372 (2004).

[220] Fox/Swazey (1999), p. 40; cf. Joint Working Party of The British Transplantation Society and The Renal Association (2011), p. 39; cf. Sharp/Randhawa/Kaur-Bola, in Weimar/Bos/Busschbach (ed.) (2011), p. 113.

[221] Cf. Hilhorst et al., Vol. 24 Transplant International 1164, 1165 (2011); cf. Ismail et al., Vol. 14 Medicine, Health Care and Philosophy (2011); cf. Fox/Swazey (1999), p. 31; cf. Massey et al., Vol. 10 American Journal of Transplantation 1445, 1450 (2010).

[222] Matas, in Gutmann et al. (ed.) (2004), p. 199.

[223] Price (2000), p. 300.

[224] Massey et al., Vol. 10 American Journal of Transplantation 1445, 1450 (2010); Matas, in Gutmann et al. (ed.) (2004), p. 199.

respect to the social environment of donor and recipient. LOD can be problematic; every LOD should be carefully evaluated and discussed with the persons involved. Despite its problems, it is not convincing that the option of LOD should be generally withheld from any subgroup among competent adults.

4) Organ Trade

Preventing organ trade and avoiding the commercialisation of organ donation is an understandable aim. Beyond the prohibition of organ trade by law, many claim that LOD also has to be restricted to prevent organ trade. (1) Some assume that unspecified LOD leads to organ trade, and also (2) that cross-over LOD is a hidden form of organ trade.

a) Unspecified Living Organ Donation Leads to Organ Trade

The German Federal Constitutional Court opines that unspecified LOD leads to organ trade. It stated that the German legislator is right to assume that banning unspecified LOD is necessary to prevent organ trade.[225] Another reason for legal systems to restrict unspecified LOD is to avoid commercialism.[226] For the majority of people, so it is argued, the decision to donate an organ to a stranger is not understandable. Unspecified LOD is, therefore, denominated "a Trojan horse of the market — a *legal fiction* capable of passing commercial transactions for altruistic donations."[227]

The main concern of this view is not to prohibit unspecified LOD as such, but to point out that allowing this possibility could lead to commercialism. It can thus be classified as a "slippery-slope" argument. As *Daar* states, "[t]hese 'slippery-slope' arguments are encountered in popular discourse, but are generally frowned upon by public policy-makers and professional ethicists and philosophers because they are fundamentally flawed."[228] It would therefore be too rash to infer commercialism from unspecified LOD. In addition, the possibility that unspecified LOD might cause commercialisation and organ trade can be nipped in the bud if the donation occurs within a pool of organs that is handled by a neutral agency, which coordinates the organ transplantation, organ allocation and maintains the

---

[225] Cf. Evans, Vol. 15 Journal of Medical Ethics 17, 18, (1989); German Federal Constitutional Court [Bundesverfassungsgericht], 11 August, 1999, Case No. 1 BvR 218/98, at 3b.

[226] Cf. First, Vol. 29 Transplantation Proceedings 67, 69 (1997); cf. Gohh et al., Vol. 16 Nephrology Dialysis Transplantation 619, 621 (2001).

[227] Epstein/Danovitch, Vol. 24 Nephrology Dialysis Transplantation 357, 358 (2009).

[228] Daar et al., Vol. 11 Transplantation Review 95, 100 (1997).

anonymity of the donor.[229] The proper response is to attack the exploitation of the act of donation, not the act itself.[230] In any case, the accumulated experience with different models of LOD between persons who are not closely related, including LOD between strangers (especially in Europe and in North America), clearly shows that that there is no uncontrollable danger of commercialisation attached to them.[231]

b) Cross-Over Living Organ Donation Is a Hidden Form of Organ Trade

Cross-over LOD[232] is sometimes associated with the danger of organ trade. A German parliamentary commission, for example, stated that "[a]llowing cross-over-donations leads to the danger of [covert commercial] brokers and of trading in organs."[233] Cross-over LOD is even considered as involving "a 'hidden' type of kidney sale."[234] Some argue that neither of the two participating donors would donate if the other donor would not do the same; this emphasises the fact that cross-over LOD does not consist of "two separate and unrelated gifts."[235]

The view that cross-over LOD comprises the danger of organ trade is heavily contested.[236] Opponents argue that the danger of organ trade is not present in cases of cross-over LOD, or that the danger is at least not higher than in cases of direct LOD.[237] The motives of the donors are often referred to, since every donor only donates to help a person he is related to.[238] The exchange, thus, takes place on the basis of equality.[239] It is also difficult to understand who would actually benefit financially if two pairs with the same wish for a cross-over LOD are introduced.[240]

Thus, neither unspecified LOD nor cross-over LOD can be considered as causing organ trade.

---

[229] Gutmann/Schroth (2002), p. 16, p. 121 f.; cf. Schreiber, in Rittner/Paul (ed.) (2005), p. 64.

[230] Evans, Vol. 15 Journal of Medical Ethics 15, 19 (1989).

[231] Danovitch, in Weimar/Bos/Busschbach (ed.) (2011), p. 392; cf. Gutmann, Vol. 15 Medizinrecht 147, 150 (1997); Gutmann (2011), p. 2 f.; cf. Swiss Dispatch of 12. September 2001 on a Federal Law on Transplantation of Organs, Tissues and Cells, p. 98.

[232] This type of LOD has already been described in chapter "Comparative Analysis of European Transplant Laws Regarding Living Organ Donation".

[233] Bundestag printed paper 15/5050 (2005), p. 42, 73 f.

[234] Menikoff, Vol. 29 The Hastings Centre Report 28, 28 (1999).

[235] Menikoff, Vol. 29 The Hastings Centre Report 28, 28 (1999); Price (2010), p. 227 f.

[236] German Federal Social Court [Bundessozialgericht], 10 December, 2003, Case No. B 9 VS 1/01 R, in Vol. 9 Juristenzeitung 464, 466 (2004); Gutmann/Schroth (2002), p. 15; Gutmann (2006), p. 33; Neft, Vol. 13 Neue Zeitschrift für Sozialrecht 519, 521 (2004); Nickel/Preisigke, Vol. 22 Medizinrecht 307, 308 (2004); Schroth, Vol. 9 Juristenzeitung 469, 470 (2004).

[237] Hohmann (2003), p. 86.

[238] de Klerk (2010), p. 14; Gutmann/Schroth (2002), p. 118; Veatch (2000), p. 187.

[239] de Klerk (2010), p. 14.

[240] Gutmann (2006), p. 33.

5) *Justice and Equality*

This dissertation does not give a thorough examination of the state of philosophical debate about the notions of *justice* and *equality*. These terms, however, serve as a mere headline to address whether benefits and burdens are and/or can be shared between the persons involved in a fair and equitable manner in LOD. (1) I examine whether LOD in general can be judged to be just and equal and (2) I analyse whether justice and equality can be answered in the affirmative with regard to certain groups of potential living donors and models of donation.

a) Justice and Equality of Living Organ Donation in General

Some question whether LOD as such is just and equal.[241] It is sometimes claimed that LOD is an unfair procedure in general, because not all patients in need have a potential living organ donor.[242]

The consequences they infer from this view are questionable. The most compelling conclusion would be to prohibit LOD in general. This means that some patients would no longer have the opportunity to receive a healing organ from a living organ donor, since others do not have this possibility. This solution would not improve the situation for any of the waiting patients. It merely impacts the situation for patients who would have the option negatively, while the situation of patients without a potential living donor remains the same. This 'levelling-down argument' does not make much sense. The current situation might be unequal, but it is not unfair in itself.[243] Also, the argument that LOD is an unfair procedure in itself, because not all patients in need have a potential living organ donor, comes close to the thesis that we should ban marriage as long as not everyone has found a suitable partner. In other words: The luck people have in finding caring partners — for life, or at least for a LOD — is no subject matter for social justice discussions.[244] If this situation is considered unfair, then all living donors that want to donate for a relative could donate the organ to an impartial pool. This idea is immediately dismissed by stating that "[i]t misunderstands the unique partiality involved in living donation and is therefore neither adequate nor effective."[245] Even if not every person suffering

[241] With regard to LOD, the aspect of whether the allocation is just and equal would only become relevant if the amount of unspecified LODs increases (Cf. Segev/Montgomery, Vol. 8 American Journal of Transplantation 1051, 1051 (2008)).

[242] Cf. Bakker, in Price/Akveld (ed.) (1997), p. 29, who also states right away that it seems very unlikely that a surgeon would condemn a patient on the waiting list when he has the option for a LOD; German Association of Dialysis Patients' motion before the German Federal Constitutional Court, 1999.; cf. Hilhorst, Vol. 8 Ethical Theory and Moral Practice 197, 201 (2005).

[243] Hilhorst, Vol. 8 Ethical Theory and Moral Practice 197, 211 (2005).

[244] Information from T. Gutmann.

[245] Hilhorst, Vol. 8 Ethical Theory and Moral Practice 197, 210 (2005).

knows a person willing to act as a donor, each LOD still benefits all people waiting for an organ. The waiting time for all other patients on the waiting list is reduced if there is one less person in need of an organ.[246] LOD in general cannot be considered as unjust or unequal. Whether this is also the case with certain groups of potential donors is analysed next.

b) Justice and Equality with Regard to Certain Groups of Potential Donors

Certain groups of potential living donors might be more at risk for unjust and unequal treatment. (1) As already mentioned, more women than men act as living organ donors. Whether this gender inequality can be considered as being unjust and unequal is analysed. (2) LOD with regard to minorities is also addressed and (3) the effect of age on LOD.

aa) Gender Inequality

More women become living donors than men. If the inequality exists because women are being exploited, this has to be abolished. If that is not the case, the difference does not seem problematic.[247] Because of the organ shortage, people who are willing to donate an organ should not be kept from doing so if they are acting voluntarily and if all other requirements are fulfilled. In fact, it could be considered as unjust if a person is prohibited from acting as living organ donor just because she is a woman. In contrast, it could be considered unjust to pressure a man to act as a living organ donor, without him voluntarily deciding to do so. Again, an "equitable donor-recipient-ratio in living organ donation"[248] is no goal *in itself*. The legal situation, in contrast, is equal, since the laws as such treat men and women equally.

bb) Minorities

Not all groups of people are treated equally with regard to LOD.

In the United Kingdom, almost 1 in 4 patients who are waiting for a kidney transplant are from a minority ethnic group.[249] This is because South Asians and

---

[246] Donnelly/Price (ed.) (1997), p. 127; Hilhorst in Weimar/Bos/Busschbach (ed.) (2008), p. 133.

[247] This has already been clarified above on p. 170 ff.

[248] Biller-Andorno, Vol. 5 Medicine, Health Care and Philosophy 199 (2002).

[249] More precisely, 14% of the people who are waiting for a kidney transplant are South Asian, even though they comprise only 4% of the general population. The situation is similar with regard to African-Caribbean. Over 7% of the people waiting for a kidney transplant have this ethnicity, while they are only 2% of the general population (Cronin et al., in Weimar/Bos/Busschbach (ed.) (2011), p. 62 f.; cf. Morgan/Deedat/Kenten, in Weimar/Bos/Busschbach (ed.), p. 100).

African-Caribbeans are more prone to diabetes and more likely to develop end stage renal failure than Caucasians.[250] They consequently are especially in need of organ donations — deceased donation as well as LOD. The situation for them is even more aggravated due to a lack of organ donors from the South Asian and African-Caribbean populations. One could argue that there is a lack of awareness concerning organ donation in these groups of minorities, and that there are also potentially low rates of referrals to the Intensive Care Unit.[251] Also, these groups of patients suffering from organ failure have to wait about twice as long as Caucasians for a donor organ,[252] even though the majority of those patients receive donor organs from Caucasian donors.[253] The aim should be (finding ways) to increase the pool of suitable organs for South Asians and African-Caribbeans by increasing the amount of donor organs that come from these ethnic communities. In addition, the problem should also be approached at an earlier stage. The number of South Asians and African-Caribbeans who eventually get sick from end stage renal failure must be reduced.[254]

The situation in the United States[255] is similar with regard to the relation between African Americans and Caucasians.[256] African American patients claim that they are less likely to learn about transplantation; they are rarely told by a physician about the possibility of LOD.[257] It has also been reported that African Americans are more concerned about having a living person as organ donor.[258] Whether this dismissive attitude is based on the lack of education by the physicians,

---

[250] Cronin et al., in Weimar/Bos/Busschbach (ed.) (2011), p. 62 f.; cf. Morgan/Deedat/Kenten, in Weimar/Bos/Busschbach (ed.), p. 100; Randhawa, in Weimar/Bos/Busschbach (ed.) (2008), p. 393.

[251] Cronin et al., in Weimar/Bos/Busschbach (ed.) (2011), p. 64; Randhawa, in Weimar/Bos/Busschbach (ed.) (2008), p. 399 ff.

[252] Cronin, in Weimar/Bos/Busschbach (ed.) (2011), p. 403.

[253] Randhawa, in Weimar/Bos/Busschbach (ed.) (2008), p. 392. With regard to this it has to be considered that non-Caucasians and Caucasians have a different genetic background. This often causes different tissue types, so that HLA matching is more difficult (Randhawa, in Weimar/Bos/Busschbach (ed.) (2008), p. 392).

[254] Cronin et al., in Weimar/Bos/Busschbach (ed.) (2011), p. 64; Randhawa, in Weimar/Bos/Busschbach (ed.) (2008), p. 394.

[255] The United States is usually not considered here, but will be regarded at this point of time exceptionally to show the inequality of LODs with regard to minorities.

[256] Arnason, Vol. 21 Hastings Center Report 13, 13 (1991); Cronin, in Weimar/Bos/Busschbach (ed.) (2011), p. 403; First, Vol. 29 Transplantation Proceedings 67, 67 (1997); Myaskovsky et al., in Weimar/Bos/Busschbach (ed.) (2008), p. 366; cf. Segev/Montgomery, Vol. 8 American Journal of Transplantation 1051, 1051 (2008). From 1988 until now, 79,530 Caucasians became living organ donors and 14,017 African Americans (US Department of Health & Human Services (2011)).

[257] Ayanian et al., Vol. 341 The New England Journal of Medicine 1661, 1663 (2011); Myaskovsky et al., in Weimar/Bos/Busschbach (ed.) (2008), p. 368.

[258] Cf. Ayanian, Vol 341 New England Journal of Medicine 1661, 1663 (2011); Ross, Vol. 16 Kennedy Institute of Ethics Journal 151, 161 f. (2006).

or if there is another reason, has not been surveyed. However, other studies show that a significant amount of African Americans believe that they are discriminated by the government and within the health care system.[259] African-Americans are also under-represented among recipients for a donor organ of unspecified LOD.[260] The aim to improve the situation for African American patients should include "providing more systematic education about transplantation, offering greater encouragement to undergo evaluation for transplantation and to consider potential living donors, and monitoring and informing physicians and medical groups about racial differences in referral rates among their own patients."[261]

A survey in Rotterdam came to the conclusion that non-European patients are less likely to donate an organ while alive,[262] and are also less likely to receive a living donor kidney transplant than European patients. Between 2000 and 2009, 84 % of all living kidney donations were performed for the benefit of European patients, even though there were proportionally more non-European patients on the Eurotransplant waiting list for a deceased donor kidney transplant.[263] The attempts to explain this inequality are rather vague. Some mention that religion, e.g., an Islamic background, could be a reason for the inequality.[264] A possible lower social economic status is also mentioned as a potential explanation.[265] Others criticise that a significant amount of people from minority groups seem to be ignorant about LOD.[266]

Overall, the situation of LOD seems to be rather unjust and unequal for minorities. This problem appears to be universal and is not limited to individual countries or to certain minority groups.[267] Claiming that a specific religious attitude leads to the inequality is untenable since not all members of minorities are religious, e.g., Islamic.[268] In addition, experts find that religion usually promotes helping and saving lives. Most religious patients report that their religion has a positive attitude towards LOD.[269] A less-advanced education and the lower income of minority

[259] Cf. First, Vol. 29 Transplantation Proceedings 67, 67 (1997); Myaskovsky et al., in Weimar/Bos/Busschbach (ed.) (2008), p. 367.

[260] Segev/Montgomery, Vol. 8 American Journal of Transplantation 1051, 1052 f. (2008).

[261] Ayanian et al., Vol. 341 The New England Journal of Medicine 1661, 1668 (2011).

[262] Roodnat et al., Vol. 89 Transplantation 1263–1269 (2010).

[263] Maasdam et al., in Weimar/Bos/Busschbach (ed.) (2011), p. 109.

[264] Ibid., p. 110.

[265] Cf. Ismail et al., Vol. 14 Medicine, Health Care and Philosophy (2011); cf. Roodnat et al., Vol. 89 Transplantation 1263, 1268 (2010).

[266] Cf. Ismail et al., Vol. 14 Medicine, Health Care and Philosophy (2011).

[267] Cf. Ismail et al., Vol. 14 Medicine, Health Care and Philosophy (2011); Maasdam et al., in Weimar/Bos/Busschbach (ed.) (2011), p. 110; Roodnat et al., Vol. 89 Transplantation 1263, 1267 (2010).

[268] Cf. Maasdam et al., in Weimar/Bos/Busschbach (ed.) (2011), p. 110.

[269] Abouna, e.g., states that "[c]urrently most Islamic countries approve organ donation from living related donors" (Abouna, in Weimar/Bos/Busschbach (ed.) (2008), p. 320); cf. Bruzzone, in Weimar/Bos/Busschbach (ed.) (2008), p. 327; Ismail et al., Vol. 14 Medicine, Health Care and Philosophy (2011).

ethnic communities seem to be the most plausible culprits of the inequality.[270] Therefore, raising awareness in the relevant communities seems helpful.[271] In general, effort should be made to increase the amount of organ donations from people that belong to racial and ethical minorities.[272]

cc) Age

In direct LOD, the person who acts as donor and the intended recipient have already been determined. There is no opportunity for unjust or unequal treatment based on age to occur.

Age might become an issue in cases of cross-over LOD, where the person who donates her organ directly to the recipient has not yet been determined. A potential organ recipient might insist that the person, whom she directly receives the organ from, is not significantly older.[273] If a couple considers cross-over LOD, it depends on another couple. If recipients refuse the cross-over LOD because the potential donor is significantly older, this might lead to a discrimination against older patients.

Hence, LOD as such is neither unjust nor unequal. Neither can the existing gender inequality be generally considered as problematic. The low rates of LOD among minorities should be raised, though. Issues of discrimination based on age might occur in cross-over LOD.

6) Special Groups of Potential Donors

The national transplant laws contain general rules for competent adults; specific regulations are added for (1) minors and (2) mentally incapacitated adults.

a) Minors

It has been ascertained in chapter "Comparative Analysis of European Transplant Laws Regarding Living Organ Donation" that most national transplant laws, not all, though, completely prohibit minors from acting as living organ donors.

In medical practice, LODs by minors for the benefit of family members have been executed successfully. This procedure, nevertheless, remains controversial. LOD for family members might increase the pressure placed on the potential

---

[270] Cf. Danovitch, in Weimar/Bos/Busschbach (ed.) (2011), p. 394.

[271] Ismail et al., Vol. 14 Medicine, Health Care and Philosophy (2011).

[272] Cf. Arnason, Vol. 21 Hastings Center Report 13, 15 (1991).

[273] Price (2010), p. 221.

donors.[274] Most assume that a competent adult can handle the pressure better than a more immature person. Furthermore, a competent adult does not usually depend on the family to the same degree a minor does.[275] The particular case involving a minor who donates for the benefit of his sibling is also important. Since minors are unable to provide valid consent for the intended LOD, their parents have to do so. That might lead to a conflict of interests for the parents.[276]

The opposing view argues that there might be cases in which the organ removal of a minor is ethically justifiable. Cases are imaginable in which a prohibition to donate affects the minor more negatively than being allowed to donate.[277] That is possible, for example, if a minor wants to donate for the benefit of his sibling, because he, understandably, wants to help a sick brother or sister. Prohibiting the minor from donating would probably be very hard for him to understand and accept.

b) Mentally Incapacitated Adults

Mentally incapacitated adults also lack the ability to give valid consent and are, furthermore, like minors, in danger of exploitation. Because the argumentation is very similar to that which is presented for minors, it will not be repeated. This equality of treatment is also in accordance with the WHO *Guiding Principles on Human Cell, Tissue and Organ Transplantation*, which states, "What is applicable to minors also applies to any legally incompetent person."[278]

The most arguments in favour of a complete prohibition of minors and mentally incapacitated adults to act as living organ donors address direct LOD, while unspecified LOD is not even considered.

7) Arguments Against Specific Methods and Programmes to Increase Living Organ Donation

There are different types of LOD. Their unique features make the individual consideration of each type necessary. (1) Some types of LOD involve more than one donor-recipient pair, while (2) others involve the donation to an anonymous

---

[274] Cf. Gutmann, in Schroth et al. (ed.) (2005), Sec. 8, at 7; Swiss Dispatch of 12. September 2001 on a Federal Law on Transplantation of Organs, Tissues and Cells, p. 145.

[275] Swiss Dispatch of 12. September 2001 on a Federal Law on Transplantation of Organs, Tissues and Cells, p. 145.

[276] American Medical Association, Vol. 284 The Journal of the American Medical Association 2919, 2924 (2000); Lamb, in Price/Akveld (ed.) (1997), p. 49.

[277] Cf. Price (2000), p. 336.

[278] WHO Guiding Principles on Human Cell, Tissue and Organ Transplantation, 2008, Guiding Principle 4.

recipient or a stranger (as opposed to direct donation). (3) In addition, the type of LOD that connects LOD with post-mortem organ donation will be presented.

a) Living Organ Donations That Involve More Than One Donor-Recipient Pair

Several programmes are specifically intended to increase LOD by connecting several pairs, namely (1) cross-over LOD, (2) unbalanced living paired exchange and (3) living paired cascade exchange.

aa) Cross-Over Living Organ Donation

In addition to the fear that cross-over LOD is a hidden form of organ trade, several other critical aspects raised by opponents have already been presented in chapter "Comparative Analysis of European Transplant Laws Regarding Living Organ Donation." Even more critical aspects are raised, though.

One main concern is that the pressure on the potential donor might be higher than it is in case of a direct LOD, because the medical borders that possibly eliminate LOD for the particular patient are inapplicable.[279] In addition, since the organ is not directly transplanted into the sick relative, but rather into a stranger, the imposition on the donor may be perceived as greater.[280]

After the cross-over LOD has been executed, several further problems connected to the donor-recipient relationship may arise. First, the main benefit for the donor is psychological. This benefit arises especially in cases of successful LOD. If the LOD fails, however, the donor at least knows that he did everything in his power to help the sick person. In cross-over LOD, the donor's organ is not directly transplanted into the recipient, so the psychological benefit might be more diffused in these cases than it is in a direct LOD.[281] Second, the fact that the donor does not have a personal relationship with the direct recipient might increase the possibility of the donor regretting his decision.[282] Third, if the success of the two executed transplantations is different, this might cause a feeling of inequality.[283]

The situation for the organ recipient has to be considered as well. On the one hand, he depends on the LOD. On the other hand, it is a high psychological burden

[279] Bundestag printed paper 15/5050 (2005), p. 42; Price (2010), p. 221; Ross, Vol. 16 Kennedy Institute of Ethics Journal 151, 152 (2006); Veatch (2000), p. 192.

[280] Ross et al., Vol. 336 The New England Journal of Medicine 1752, 1755 (1997).

[281] Ibid., 1754.

[282] Bundestag printed paper 15/5050 (2005), p. 42.

[283] Siegmund-Schultze (1999), p. 215.

for him to ask a relative for a donation, and it may be even more demanding to ask the relative to take part in a cross-over LOD.[284]

Chapter "Comparative Analysis of European Transplant Laws Regarding Living Organ Donation" also showed, however, that good counter-arguments to the criticism exist. When comparing direct LOD and cross-over LOD, the expected benefit that causes the donor's willingness to donate is the same.[285] However, the arguments against cross-over LOD, especially the fear of increased pressure towards the living organ donor, have to be taken seriously. Cross-over LOD rules out the donor's potential medical excuse, namely that he is medically incompatible. One can assume, though, that potential donors want to be suitable, not that they are hoping for a *medical excuse* in such a case.[286] Most potential donors consider the result of incompatibility as bad news; they welcome the possibility of cross-over LOD.[287]

bb) Unbalanced Living Paired Exchange

In unbalanced living paired exchanges, one donor-recipient pair in which the donor has blood type 0 is asked to participate in an exchange, even though he has the possibility to donate directly to his related recipient.[288] One donor-recipient pair does not have the possibility for a direct LOD, while the other pair does have this option and would not have to take part in the exchange to realise their intended LOD.[289] This type of LOD is illustrated in Fig. 1.

Those who argue against this type of LOD claim that the potential donor with blood type 0 might feel pressured to take part in the unbalanced living paired exchange.[290] To do so would not be in accord with his initial plan to donate directly to his relative.[291] Regarding the wish of recipient A, he could prefer to receive the kidney directly from his relative.[292] The mere request to participate in an unbalanced living paired exchange might be considered manipulative. The donor-recipient pair which could donate directly is forced to consider the well-being of another

[284] Cf. Ross et al., Vol. 336 The New England Journal of Medicine 1752, 1755 (1997).

[285] Gutmann/Schroth (2002), p. 118; Gutmann (2006), p. 32.

[286] Cf. Gutmann (2011), p. 4.

[287] Cf. Gutmann (2011), p. 4; Ross, for example, states that "[w]hile many individuals may have been disappointed that they could not help their potential intended recipient, some may have been relieved" (Ross, in Weimar/Bos/Busschbach (ed.) (2008), p. 184).

[288] Donors with blood type 0 are universal donors and can donate to their intended recipient, no matter whether he has blood type A, B or AB if there is no positive cross-match (Ross, Vol. 16 Kennedy Institute of Ethics Journal 151, 159 (2006)).

[289] Ross, Vol. 16 Kennedy Institute of Ethics Journal 151, 155 f.

[290] Bundestag printed paper 15/5050 (2005), p. 47; Fortin et al., in Weimar/Bos/Busschbach (ed.) (2011), p. 422 f.; Price (2010), p. 222.

[291] Bundestag printed paper 15/5050 (2005), p. 47.

[292] Ross, in Weimar/Bos/Busschbach (ed.) (2008), p.186.

Fig. 1 Unbalanced living paired exchange

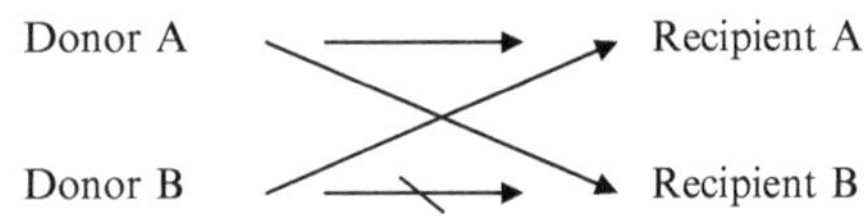

donor-recipient pair, even though they are not related to them.[293] They can refuse to participate in the exchange, but it is not uncommon for them to feel a certain obligation to agree to it. Therefore, some argue that the mere request already contains the danger of threatening the voluntariness of the one living organ donor.[294]

Is it ethically acceptable for the transplant team to ask a donor-recipient pair to donate to a stranger, even though a direct LOD is possible for them?[295] Does the mere request place the donor-recipient pair in an uncomfortable position?[296]

The consequence of these concerns would be that "[i]ndividuals should not be asked to participate in an exchange unless they cannot donate directly."[297]

Unbalanced living paired exchange has positive aspects as well. As will be discussed later in this chapter in detail, exchange programmes often disadvantage recipients with blood type 0. Unbalanced living paired exchange, in contrast, considers the difficulties blood type 0 recipients have.[298] In addition, it makes LOD possible for incompatible donor-recipient pairs that would otherwise not have the option of a LOD. As the critics themselves recognise, donor A has the possibility to say "No", he is not forced to take part in the exchange.[299] A mere question cannot be equated with force. It is also possible that donor A does not even want to decline the option, but is, in contrast, thrilled to save two lives instead of only one.

cc) Living Paired Cascade Exchange

Living paired cascade exchanges are LODs that consist of more than two donor-recipient pairs. Every intended recipient in the exchange gets an organ from a compatible donor who is also involved in the exchange.[300] The amount of involved couples and the amount of exchanges is variable. Figure 2 shows a living paired cascade exchange between three couples.

---

[293] Ross, Vol. 16 Kennedy Institute of Ethics Journal 151, 157 (2006).

[294] Ibid., p. 157 f.

[295] Fortin et al., in Weimar/Bos/Busschbach (ed.) (2011), p. 423; Ross, Vol. 16 Kennedy Institute of Ethics Journal 151, 157 f. (2006).

[296] Ross, in Weimar/Bos/Busschbach (ed.) (2008), p.186.

[297] Ross, Vol. 16 Kennedy Institute of Ethics Journal 151, 158 (2006); Ross, in Weimar/Bos/Busschbach (ed.) (2008), p. 186.

[298] Fortin et al., in Weimar/Bos/Busschbach (ed.) (2011), p. 422.

[299] Cf. Ross, Vol. 16 Kennedy Institute of Ethics Journal 151, 158 (2006).

[300] Ross, Vol. 16 Kennedy Institute of Ethics Journal 151, 154 (2006).

Fig. 2 Living paired cascade exchange

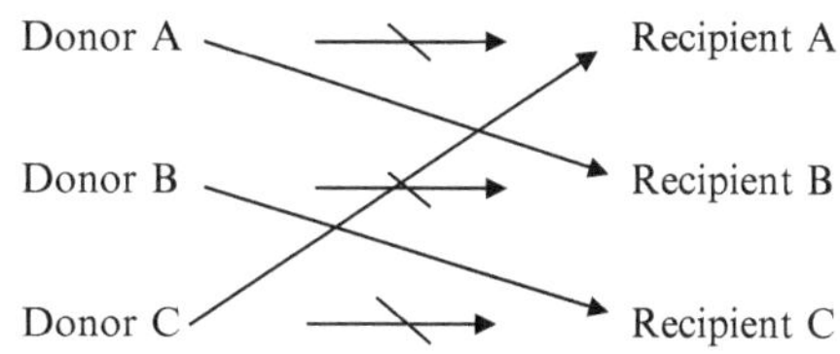

Since the amount of involved couples is variable, some fear that the comprehensibility may be lost.[301] Additionally, immunological compatibility could, at some point in the future, become unnecessary for organ transplantation. If this happens, living paired cascade exchanges would become useless.[302] The critics of this type of LOD are against the establishment of such programmes and, thus, regularly demand their prohibition.

Living paired cascade exchange is also supported. It has the same advantage cross-over LOD has, namely, to overcome the inability of a willing donor to actually donate to a specified recipient.[303] This helps patients who would otherwise have no therapeutic option.[304] With regard to cross-over LOD, a significant amount of pairs taking part in such a programme remain unmatched,[305] so that living paired cascade exchange seems advantageous. Proponents of living paired cascade exchange, furthermore, challenge whether comprehensibility can actually serve as the aim of the legislator and whether it is able to justify the interference with fundamental rights.[306] The reference to potential improvements in the future is also criticised. Predicted improvements related to LOD are uncertain. Above all, they do not help patients who are sick now and in need of help immediately.[307] In addition, the donor involved in a direct LOD intends to restore the health of a loved one. This, however, applies to the same extent in living paired cascade exchange. Consequently, if the donor's intention to help a relative is used as justification for direct LOD, this must apply to living paired cascade exchange as well.[308]

All types of LOD that involve more than one donor-recipient pair are criticised, but also supported. The main fear is that pressure towards potential living donors increases. The main argument in favour of those types of LOD is that it makes LODs possible in case a direct LOD is impossible because of incompatibility.

---

[301] Bundestag printed paper 15/5050 (2005), p. 42.

[302] Cf. Riedel, in Rittner/Paul (ed.) (2005), p. 75.

[303] Cronin, in Weimar/Bos/Busschbach (ed.) (2011), p. 401.

[304] Gutmann (2006), p. 37.

[305] Cf. Montgomery et al., Vol. 368 The Lancet 419, 419 (2006); Patel/Chadha/Papalois, Vol. 3 Experimental and Clinical Transplantation 181, 184 (2011); cf. Roodnat et al., Vol. 10 American Journal of Transplantation 821, 821 (2010); Zuidema et al. state with regard to the Dutch cross-over programme that approximately 50 % of the enrolled donor-recipient couples remain unmatched (Zuidema et al., in Weimar/Bos/Busschbach (ed.) (2008), p. 307).

[306] Gutmann (2006), p. 35 f.

[307] Ibid., p. 38.

[308] Patel/Chadha/Papalois, Vol. 3 Experimental and Clinical Transplantation 181, 184 (2011).

b) Living Organ Donations That Involve the Donation to an Anonymous Recipient or a Stranger

This section presents the types of LOD that involve an organ donation for the benefit of an unknown recipient. This applies (1) to unspecified LOD and (2) pool donation. (3) I introduce directed altruistic LOD and (4) analyse unspecified non-directed donation catalysing cascade exchanges.

aa) Unspecified Living Organ Donation

An unspecified LOD takes place if a person donates an organ to an anonymous and unspecified recipient,[309] in other words, to a stranger.[310]

The fear that unspecified LOD leads to organ trade and commercialisation has already been addressed.[311] The concern that unspecified LOD is connected to involuntariness has also already been considered.[312] Beyond this, several other arguments for and against unspecified LOD have already been brought forth in chapter "Comparative Analysis of European Transplant Laws Regarding Living Organ Donation."

Scepticism towards LOD for unknown recipients exists.[313] Experts fear that the decision to donate to a stranger is based on an excessive helper syndrome or on psychopathological reasons.[314] They claim that the decision might be inadvertent or premature.[315]

Many doubt that the amount of people actually performing such a donation would increase if unspecified LOD were allowed. They do not believe unspecified LOD would fulfil its purpose to increase the amount of donor organs to a satisfying extent.[316] Others fear that unspecified LOD would have a negative effect on direct LOD: the overall amount of LOD would not increase and may even possibly decrease.[317]

---

[309] Dor et al., 2011 Transplantation 1, 1 f.

[310] Cf. Henderson et al., Vol. 3 American Journal of Transplantation 203, 203 (2003).

[311] P. 176 f.

[312] P. 169.

[313] Massey et al., in Weimar/Bos/Busschbach (ed.) (2011), p. 370; Price (2010), p. 213; Taylor et al., in Gutmann et al. (ed.) (2004), p. 206.

[314] Bundestag printed paper 15/5050 (2005), p. 74.

[315] Ibid., p. 45.

[316] Bundestag printed paper 15/5050 (2005), p. 44; cf. Riedel, in Rittner/Paul (ed.) (2005), p. 73.

[317] Bundestag printed paper 15/5050 (2005), p. 44. This has been stated with regard to deceased donation as well, namely that promoting unspecified LOD, what would be necessary to actually gain a considerable amount of people to perform unspecified LOD, would have a negative effect on the acceptance and willingness with regard to deceased donation (Riedel, in Rittner/Paul (ed.) (2005), p. 73).

Counter-arguments assert that people have different motives for donating an organ and that a complete ban of unspecified LOD would lead to the nullification of several good intentions.[318] It would also disrespect the decisions of donor and recipient.[319] With respect to the scepticism towards unspecified LOD, empirical evidence shows that people who perform an unspecified LOD are "very self-directed and without psychopathology."[320] The mistrust many people have only arises because the majority does not nourish those altruistic principles themselves.[321] The argument that unspecified LOD does not need to be allowed, because it would not significantly raise the amount of donor organs available, can be countered by pointing out the false assumption that the legislator can only act if a benefit for the whole society is expected and not if the basic rights of individual persons are endangered.[322] Also, the opposite, namely that unspecified LOD might indeed make up a major source for donor organs in the future, is stated as well.[323] Studies show that it is more than unlikely that allowing unspecified LOD might decrease the overall amount of LODs.[324] Indeed, it seems very unrealistic to assume that a potential donor does not donate for the benefit of a relative,[325] or that a person becomes unwilling to donate after his death, just because unspecified LOD is allowed.

bb) Pool Donation

Pool donation means that a living donor gives an organ to a pool of organs and her relative receives a compatible organ from the pool in return.[326] This is illustrated in Fig. 3.

Some doubt that this would actually increase the amount of donor organs.[327] How big the pool would have to be to actually increase the amount of donor organs is also unclear. In addition, the potential increase of pressure on potential donors is referred to, because this type of LOD increases the possibility of finding a

[318] Gutmann/Schroth, in Oduncu/Schroth/Vossenkuhl (ed.) (2003), p. 276.

[319] Cf. Fortin/Dion-Labrie/Doucet, in Weimar/Bos/Busschbach (ed.) (2008), p. 353; cf. Glannon, Vol. 343 British Medical Journal (2011); Gutmann (2006), p. 40.

[320] Cf. Patel/Chadha/Papalois, Vol. 3 Experimental and Clinical Transplantation 181, 183 (2011); Price (2010), p. 214.

[321] Price (2010), p. 214.

[322] Gutmann (2006), p. 29 f.

[323] First, Vol. 29 Transplantation Proceedings 67, 68 (1997); Price (2000), p. 316.

[324] Cf. Danovitch, in Weimar/Bos/Busschbach (ed.) (2011), p. 391, who states as an experience from the United States that "[r]ates of related donation have remained steady while unrelated, non-commercially motivated donation has become a widely accepted modality."

[325] Gutmann (2006), p. 44.

[326] Ibid., p. 24.

[327] Bundestag printed paper 15/5050 (2005), p. 48.

Fig. 3 Pool donation

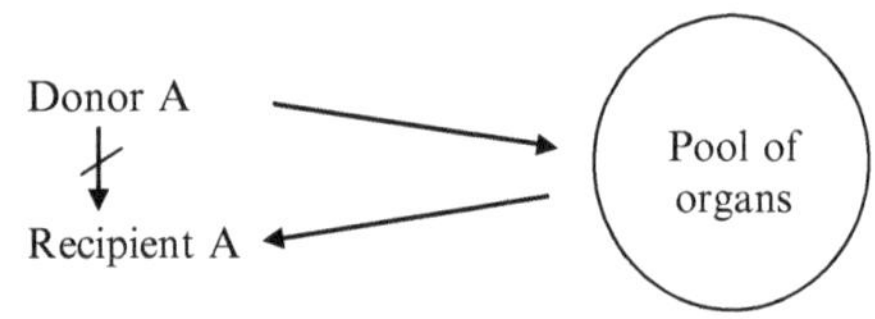

compatible organ for the intended recipient, which makes it difficult for a potential donor to decline.[328]

In favour of this method, one could claim that the amount of possible transplantations increases. People who are incompatible with their relative, but still want to help them, can realise this wish. It seems unfair for people who want to donate for a relative, but cannot because of incompatibility, to be at a disadvantage in comparison to compatible donor-recipient pairs.

cc) Directed Altruistic Living Organ Donation

The donor in directed altruistic LOD is a person who donates an organ to an anonymous recipient, but the recipient either has to meet specific criteria, e.g., regarding age, ethnicity, or is a specific individual, such as a famous person.[329]

This type of LOD is considered to be discriminatory,[330] and critics claim that it is not desirable for altruistic donors to be "perceived as tacitly endorsing such preferences."[331]

This kind of LOD usually comes up when a person hears about a patient, probably a public figure, in the media. According to the Committee of the Dutch Health Council, this could dangerously lead to the potential donor acting impulsively to help a celebrity. The Council also considers it unfair for a patient to receive an organ just because she is famous.[332]

Furthermore, a comparison to deceased donation is drawn. A person who consents to deceased donation cannot indicate a preference to donate to a specific group of persons or one particular person.[333] With regard to allocation, the Council recommends that the allocation of unspecified LODs should be based on impartial justice and equity, and special attention should be given to those with the greatest

---

[328] Bundestag printed paper 15/5050 (2005), p. 48; Riedel, in Rittner/Paul (ed.) (2005), p. 74 f.

[329] Dor et al., 2011 Transplantation 1, 3; cf. Glannon/Ross, Vol. 11 Cambridge Quarterly of Healthcare Ethics 153, 154 (2002); Hilhorst, Vol. 8 Ethical Theory and Moral Practice 197, 197 (2005); Wright et al., in Weimar/Bos/Busschbach (ed.) (2011), p. 413.

[330] Cf. Veatch, Vol. 23 Journal of Medicine and Philosophy 456, 456 (1998) (with regard to directed deceased donation).

[331] Ross, Vol. 30 The Journal of Law, Medicine and Ethics 440, 441 (2002); cf. Spital, Vol. 76 Transplantation 1252, 1253 (2003).

[332] The view of the Committee of the Dutch Health Council is stated in Hilhorst, Vol. 8 Ethical Theory and Moral Practice 197, 199 (2005).

[333] Ibid., p. 201.

medical need. Allowing directed altruistic LOD would endanger this scheme.[334] People in need could be treated unequally, which would violate the principle of justice.[335]

In addition, the Committee of the Dutch Health Council realistically states that it might be impossible to maintain anonymity between donor and recipient in case of a directed altruistic LOD.[336] This is connected to the concern that altruistic directed LOD makes the unwanted buying and selling of organs possible.[337] In short, the Council requires that an organ that is part of an unspecified LOD be given to the cadaveric pool without conditions.

However, why should the chance of a potential recipient to receive a life-saving organ be denied?[338] Prohibiting this type of LOD would prevent a sick person from getting the help she needs. This is an interference with the recipient's fundamental rights.[339] With regard to the potential donor, prohibiting altruistic directed LOD would disrespect her individual autonomy.[340] One must consider that the particular LOD might only happen because of a felt, special relationship to the group. Such significant, partial relationships govern human life and should be valued highly. They are part of the donor's moral identity. Respecting this identity provides the basis for forms of directed donation that are partial, but fair and justified.[341] The donor intends to include a certain person or a certain group of people, not, however, to exclude a particular person or group.[342] As a consequence, the particular LOD might not take place at all without the mentioned relationship.[343] This does not disadvantage anyone, not even the people waiting that would not be part of the group chosen anyway.[344] Allowing the LOD, on the other side, could even cause all

---

[334] Cf. Cronin, in Weimar/Bos/Busschbach (ed.) (2011), p. 404; cf. Roff, Vol. 33 Journal of Medical Ethics 437, 438 (2007); cf. Truog, Vol. 353 New England Journal of Medicine 444, 445 (2005).

[335] Cf. Ankeny, Vol. 10 Cambridge Quarterly of Healthcare Ethics 387, 395 (2001); cf. Kluge, Vol. 19 Hastings Center Report 10, 12 ff. (1989); cf. Spital, Vol. 76 Transplantation 1252, 1254 (2003); cf. Veatch, Vol. 23 Journal of Medicine and Philosophy 456, 461 (1998) (with regard to directed deceased donation).

[336] The view of the Committee of the Dutch Health Council is stated in Hilhorst, Vol. 8 Ethical Theory and Moral Practice 197, 202 (2005).

[337] Cf. Epstein, in Weimar/Bos/Busschbach (ed.) (2011), p. 139; Truog, Vol. 353 New England Journal of Medicine 444, 445 (2005).

[338] Hilhorst, Vol. 8 Ethical Theory and Moral Practice 197, 198 (2005).

[339] Gutmann (2006), p. 45.

[340] Cf. Ankeny, Vol. 10 Cambridge Quarterly of Healthcare Ethics 387, 393 (2001); cf. Kluge, Vol. 19 Hastings Center Report 10, 13 (1989); cf. Spital, Vol. 76 Transplantation 1252, 1255 (2003).

[341] Hilhorst, Vol. 8 Ethical Theory and Moral Practice 197, 205 f. (2005).

[342] Hilhorst, in Weimar/Bos/Busschbach (ed.) (2008), p. 138 f.

[343] Hilhorst, Vol. 8 Ethical Theory and Moral Practice 197, 205 (2005).

[344] Cf. Spital, Vol. 76 Transplantation 1252, 1255 (2003); Truog, Vol. 353 New England Journal of Medicine 444, 445 (2005); Veatch, Vol. 23 Journal of Medicine and Philosophy 456, 461 (1998) (with regard to directed deceased donation).

other patients on the waiting list to move up, because the selected person would be removed from the waiting list.[345] In the end, because of the severe shortage of donor organs, willing donors should never be turned away.[346]

The comparison of LOD and post-mortem organ donation is criticised as well, because the regular case for LOD, in contrast to deceased donation, is to be directed, and the close donor-recipient relationship is usually constitutive of the LOD.[347] Consequently, the following question is posed: "If the dispositional authority of some living donors entitles them to direct their gift to a loved one, relative or friend, one with whom they have a legitimate 'connectedness', why should others be prevented from executing their dispositional authority by directing their donation to a 'group of persons' with whom they might also have a legitimate 'connectedness'?"[348]

The fear that donor and recipient would not maintain anonymity in altruistic directed LOD can indeed not be dispelled. The rationale behind maintaining anonymity is "to avoid possible undesirable and adverse consequences, such as payment, claims and expectations."[349] Maintaining anonymity is not the only way to avoid these consequences, though. Some claim that it would be too radical to prohibit direct LOD just to prevent the risks that might arise if anonymity cannot be maintained. They compare it to banning sex just to avoid pregnancy. Such radical strategies are considered as being only justified in case no other, less radical, options exist.[350] The anonymity argument obviously has no validity; it cannot be used to support the prohibition of this type of LOD. The more desirable solution would be to address the unwanted consequences and to develop a strategy that aims at preventing those from happening.[351]

dd) Unspecified Non-Directed Donation Catalysing Cascade Exchanges

Unspecified non-directed donation catalysing cascade exchanges describe a living organ donor who performs an unspecified LOD and catalyses a cascade of LODs. It ends with a living donor donating an organ to the deceased donor waiting list.[352] It,

---

345 Veatch, Vol. 23 Journal of Medicine and Philosophy 456, 462 (1998) (with regard to directed deceased donation).

346 Cf. Spital, Vol. 76 Transplantation 1252, 1252 (2003).

347 Cf. Cronin et al., in Weimar/Bos/Busschbach (ed.) (2011), p. 186; Hilhorst, Vol. 8 Ethical Theory and Moral Practice 197, 201 (2005).

348 Cronin et al., in Weimar/Bos/Busschbach (ed.) (2011), p. 403.

349 Hilhorst, Vol. 8 Ethical Theory and Moral Practice 197, 202 (2005).

350 Ibid.

351 Ibid., p. 203.

352 Cronin, in Weimar/Bos/Busschbach (ed.) (2011), p. 402; Roodnat et al., Vol. 10 American Journal of Transplantation 821, 822 (2010); Ross, Vol. 16 Kennedy Institute of Ethics Journal 151, 166 (2006).

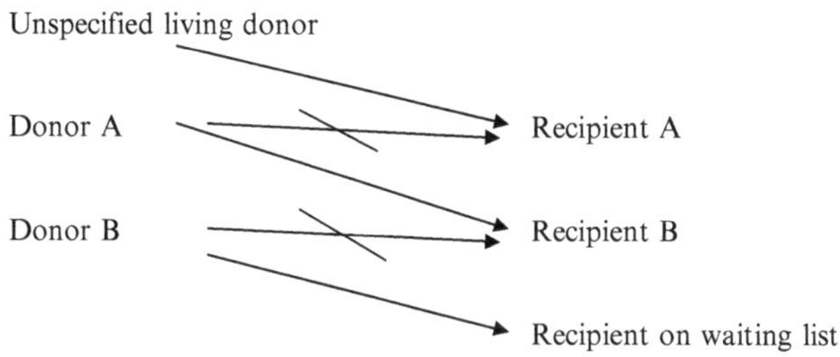

Fig. 4 Unspecified non-directed donation catalysing cascade exchanges

therefore, connects unspecified LOD and the participation of more than one donor-recipient pair. This is illustrated in Fig. 4.

Unspecified non-directed donation catalysing cascade exchanges are criticised with respect to blood group equity.[353] A study of the incompatible cross-over pairs shows that in more than 2/3 of those cases the recipient has blood type 0.[354] If the patient has blood type 0, he needs a donor with this blood type as well. If the intended donor does not have this blood type, an exchange is seldom, because donors with blood type 0 are universal donors.[355] Therefore, recipients with blood type 0 have to be protected and it is, for example, problematic if the unspecified living organ donor is blood type 0, while the cascade results in a blood type A organ going to the waiting list.[356] That leads to a further disadvantage for patients on the waiting list with blood type 0, who do not have a living organ donor who participates in the unspecified non-directed donation catalysing cascade exchanges.[357] That is inconsistent with the main idea of the Rawlsian *Theory of Justice* which permits inequity only in case it benefits those who are worst off.[358] It is consequently argued that an unspecified non-directed donation catalysing

[353] Ross, Vol. 16 Kennedy Institute of Ethics Journal 151, 166 (2006).

[354] Bundestag printed paper 15/5050 (2005), p. 48; Gutmann (2006), p. 39; Health Council of the Netherlands (2007); Price (2010), p. 224; cf. Ross/Zenios, Vol 4 American Journal of Transplantation 1553, 1553 (2004).

[355] Ackerman/Thistlethwaite/Ross, Vol. 6 American Journal of Transplantation 83, 83 (2006); Ross, Vol. 16 Kennedy Institute of Ethics Journal 151, 159 (2006). That causes that they can donate to their intended recipient, regardless of whether he has blood type A, B or AB unless there is a positive cross-match (Ross, Vol. 16 Kennedy Institute of Ethics Journal 151, 159 (2006)).

[356] Ross, Vol. 16 Kennedy Institute of Ethics Journal 151, 166 (2006).

[357] Cf. Bundestag printed paper 15/5050 (2005), p. 48; cf. Gutmann (2006), p. 39; cf. Health Council of the Netherlands (2007); Patel/Chadha/Papalois, Vol. 3 Experimental and Clinical Transplantation 181, 185 (2011); cf. Price (2010), p. 224; Ross, Vol. 16 Kennedy Institute of Ethics Journal 151, 161 f. (2006); cf. Ross/Zenios, Vol 4 American Journal of Transplantation 1553, 1553 (2004).

[358] Rawls (1971); Rawls (1975), p. 94; cf. Ross, in Weimar/Bos/Busschbach (ed.) (2008), p. 187; see also Ackerman/Thistlethwaite/Ross, Vol. 6 American Journal of Transplantation 83, 84 (2006) with regard to list-paired exchange, though.

cascade exchange "should be permitted only if it will not harm individuals [...] who are already worst off."[359]

Those in favour of such a programme argue that it makes a large number of transplantations possible.[360] The unspecified LOD does not only help one waiting person, but makes the best use of the LOD.[361] Thus, the potential disadvantage to patients on the waiting list with blood group 0 may be justified if the amount of transplantations that can be carried out as a result of the exchange is substantial.[362] According to *Weimar*, "[U]tilitarian arguments hold that benefits for an entire patient group can outweigh disadvantages for a subgroup."[363] Furthermore, this type of LOD is also advantageous for the patients remaining on the waiting list, because the number of patients on it is reduced, thus increasing the chances of others on the list to actually receive an organ.[364] It can also be assumed that donating within this scheme is in the interest of the unspecified living organ donor.[365] He acts with an altruistic motivation and such participation increases the probability of a good result.[366] Respecting the interests of unspecified living organ donors may increase unspecified LOD, which can be considered as being in the best interest of all patients suffering from organ shortage. With regard to the Rawlsian *Theory of Justice*, *Gutmann* argues that only duties to support disadvantaged groups can be deduced from this consideration. The Rawlsian *Theory of Justice* does not, however, create a duty to establish general paternalistic measures that prevent other people from exercising their right of self-determination in a sufficient way.[367] Applying *Gutmann's* argument to unspecified non-directed donation catalysing cascade exchanges would not speak in favour of a complete prohibition, but rather in favour of a duty to support patients with blood type 0.

As seen, arguments in favour of LODs that involve the donation to an anonymous recipient are brought forth, but there are also arguments which refute this type of LOD. Scepticism towards LOD for unknown recipients exists. Some fear that pressure on potential donors increases. In contrast, these types of LOD increase the overall amount of LODs, an outcome that is beneficial for all patients suffering from organ failure, given the current severe organ shortage.

---

[359] Ross, Vol. 16 Kennedy Institute of Ethics Journal 151, 166 f. (2006).

[360] Cronin, in Weimar/Bos/Busschbach (ed.) (2011), p. 402; cf. Ross, in Weimar/Bos/Busschbach (ed.) (2008), p. 187; Roodnat et al., in Weimar/Bos/Busschbach (ed.) (2011), p. 386; Wright et al., in Weimar/Bos/Busschbach (ed.) (2011), p. 413.

[361] Montgomery et al., Vol. 368 The Lancet 419, 419 ff. (2006).

[362] Patel/Chadha/Papalois, Vol. 3 Experimental and Clinical Transplantation 181, 185 (2011).

[363] Weimar et al., Vol. 368 The Lancet 987, 987 (2006).

[364] Cronin, in Weimar/Bos/Busschbach (ed.) (2011), p. 402.

[365] Cronin, in Weimar/Bos/Busschbach (ed.) (2011), p. 402; Roodnat et al., Vol. 10 American Journal of Transplantation 821, 825 (2010).

[366] Cronin, in Weimar/Bos/Busschbach (ed.) (2011), p. 402.

[367] Gutmann, in Schroth et al. (ed.) (2006), p. 249.

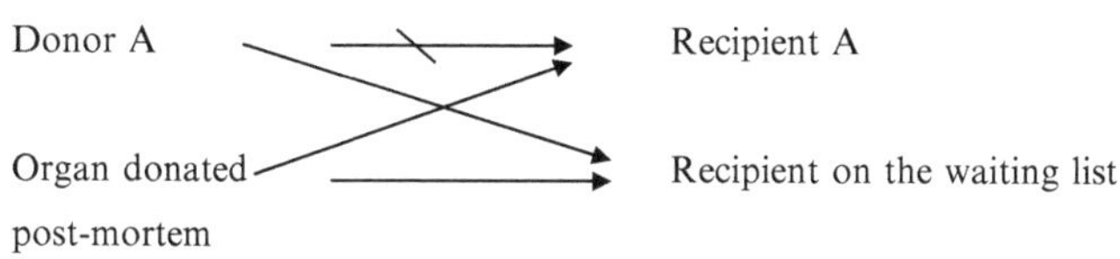

Fig. 5 List-paired exchange

c) Connection of Living Organ Donation and Post-Mortem Organ Donation: List-Paired Exchange

List-paired exchange involves a living donor-recipient pair, a deceased donor and a candidate on the waiting list. The living donor donates to a patient on the waiting list. Afterwards, the recipient related to the living donor is given the highest priority to receive a compatible organ donated post-mortem.[368] This can be seen in Fig. 5.

The main concern of this type of LOD is that it might be unfair. Again, many fear patients on the waiting list with blood type 0 are disadvantaged.[369] But, this is inconsistent with the Rawlsian *Theory of Justice*.[370]

---

[368] American Medical Association, Vol. 284 The Journal of the American Medical Association 2919, 2923 (2000); den Hartogh (2008), p. 119; Hilhorst in Weimar/Bos/Busschbach (ed.) (2008), p. 136; Morrissey, Vol. 6 American Journal of Transplantation 434, 434 (2006); Ross, Vol. 16 Kennedy Institute of Ethics Journal 151, 159 (2006); Veatch (2000), p. 187.

[369] As already mentioned, in more than 2/3 of the incompatible cross-over pairs, the recipient has blood type 0. If one of those pairs is considered for the list-paired-exchange, the recipient with blood type 0 is given priority for the next matching blood-type 0 organ donated post-mortem. As a consequence, the organs donated post-mortem that are available to other patients on the waiting list with this blood type decrease. Like unspecified non-directed donation catalysing cascade exchanges, this also leads to a disadvantage for patients on the waiting list with blood type 0, who do not have a living organ donor that participates in the list-paired exchange (American Medical Association, Vol. 284 The Journal of the American Medical Association 2919, 2923 (2000); Bundestag printed paper 15/5050 (2005), p. 48; Dondorp, in Weimar/Bos/Busschbach (ed.) (2008), p.112 f.; Gutmann (2006), p. 39; Health Council of the Netherlands (2007); Price (2010), p. 224; Ross, Vol. 16 Kennedy Institute of Ethics Journal 151, 161 f. (2006); Ross/Zenios, Vol 4 American Journal of Transplantation 1553, 1553 (2004); cf. Gentry/Segev/Montgomery, Vol. 5 American Journal of Transplantation 1914, 1914 f. (2005).

[370] Rawls (1971); cf. Ackerman/Thistlethwaite/Ross, Vol. 6 American Journal of Transplantation 83, 84 (2006). It is claimed that it has to be distinguished between ABO-compatible and ABO-incompatible list-paired exchanges, because these different types of exchanges have different ethical consequences. In cases of ABO-compatibility between potential donor and intended recipient, they have the same blood type, but a different crossmatch that prevents a direct LOD. An exchange can then be considered as a simple replace of a deceased donor kidney with the relevant blood type with a kidney from a living donor with the same blood type. Since a kidney donated by a living person has several advantages in comparison to a kidney donated post-mortem, it can be assumed that patients on the waiting-list would prefer a living kidney donation. That is considered as moral justification for ABO-compatible list-paired exchanges. In ABO-incompatible exchanges, in contrast, the potential donor will usually be of blood group A, B or C and therefore donate a corresponding organ to the waiting list, while the recipient will regularly be of blood type 0 and consequently removes a corresponding organ from the waiting list. This will typically cause

Some claim that the order of the waiting list is disregarded (so called queue jumping). The person who has an incompatible willing living donor moves to the top of the waiting list,[371] so that the person who was on top of the waiting list before moves back.[372] Consequently, persons who have a living organ donor are prioritised.[373] It also causes the allocation of organs donated post-mortem to be, in part, based on more than the medical criteria alone.[374]

As stated, simple cross-over LOD might put the potential donor under severe psychological pressure. It is feared that this pressure grows in list-paired exchange.[375]

But should list-paired exchange be prohibited? With regard to the argument that it disadvantages recipients with blood type 0, who do not have a willing living donor,[376] some suggest that this type of LOD only needs to be prohibited if blood type 0 recipients are involved.[377]

Those in favour of list-paired exchanges state that list-paired exchanges would make LOD available when the willing donor is not able to donate directly to his relative and no match can be found in the cross-over programme.[378] Or one could view list-paired exchanges positively insofar as both parties involved, the relative of the willing living donor and the patient on the waiting list, profit from the donation.[379] From a utilitarian point of view, this maximises the overall utility.[380] The overall amount of donor organs available would be increased,[381] and instead of declaring one LOD impossible, two are made possible.

---

a disadvantage for patients with blood type 0 who are on the waiting list (Ackerman/Thistlethwaite/Ross, Vol. 6 American Journal of Transplantation 83, 83 (2006); Ross/Woodle, Vol. 69 Transplantation 1539, 1540 f. (2000)).

[371] Gentry/Segev/Montgomery, Vol. 5 American Journal of Transplantation 1914, 1920 (2005); den Hartogh (2008), p. 83; Price (2010), p. 224.

[372] den Hartogh (2008), p. 83; Price (2010), p. 224.

[373] Health Council of the Netherlands (2007); Price (2010), p. 224.

[374] Cf. Dondorp, in Weimar/Bos/Busschbach (ed.) (2008), p.115; Nickel/Preisigke, Vol. 22 Medizinrecht 307, 309 (2004); Riedel, in Rittner/Paul (ed.) (2005), p. 75.

[375] Bundestag printed paper 15/5050 (2005), p. 48.

[376] Cf. Gutmann/Schroth (2002), p. 120; Ross/Zenios, Vol 4 American Journal of Transplantation 1553, 1554 (2004).

[377] Gutmann/Schroth (2002), p. 120; Gutmann (2006), p. 39; Ross/Zenios, Vol 4 American Journal of Transplantation 1553, 1554 (2004); Veatch (2000), p. 200. The experts who distinguish between ABO-compatible and ABO-incompatible exchanges suggest allowing list-paired exchanges only in cases of ABO-compatibility (Ackerman/Thistlethwaite/Ross, Vol. 6 American Journal of Transplantation 83, 84 (2006)).

[378] Dondorp, in Weimar/Bos/Busschbach (ed.) (2008), p. 112.

[379] Cf. Dondorp, in Weimar/Bos/Busschbach (ed.) (2008), p. 112; cf. Hilhorst in Weimar/Bos/Busschbach (ed.) (2008), p. 136.

[380] Cf. Dondorp, in Weimar/Bos/Busschbach (ed.) (2008), p. 113 f.; cf. Hilhorst in Weimar/Bos/Busschbach (ed.) (2008), p. 136.

[381] Ackerman/Thistlethwaite/Ross, Vol. 6 American Journal of Transplantation 83, 87 (2006).

The diverse critiques and defences of different types of LOD became apparent. Especially apparent is the danger of increasing pressure on the potential donor, a recurrent argument against almost all of the more unusual types of LOD. That a donor could be pressured or could feel pressured cannot be denied. But there is no empirical evidence which proves that pressure is more intense in cases of special types of LOD in comparison to direct LOD. However, against the existence of pressure in general, one could argue that the decision to donate is mostly due to the donor's own initiative.[382] The donor is not merely a passive object that must be protected from pressure.[383] If one agrees with the principle of respect for autonomy, the donor's wishes, as long as he is competent to consent, has done so voluntarily, has been informed properly and understands what he has been told, should be respected.[384] That all Member States of the European Union consider informed consent as a fundamental principle has been clarified in chapter "Comparative Analysis of European Transplant Laws Regarding Living Organ Donation" and can be confirmed by its inclusion in the *CFREU*. The *CFREU* puts the right of self-determination of one's own body in concrete terms[385] and provides it with constitutional status.[386] Apart from this, the implementation of several types of LOD grants more people the opportunity to become a living organ donor. More people can realise their wish to help a person by making an autonomous decision in favour of the LOD and by putting this decision into practice.[387] In times of severe organ shortage, (unnecessary) restrictions of LOD cannot remain unquestioned.[388] Organ shortage, in the worst case, causes the death of waiting patients. Keeping the *CFREU* in mind, which protects the right to life[389] and the right to the integrity of a person,[390] one should assume that the momentary situation cannot be endorsed. One should also presume that patients who suffer from organ failure also support methods that increase the amount of donor organs.[391]

---

[382] Frade et al., Vol. 40 Transplantation Proceedings 677, 679 (2008); cf. Gutmann (2006), p. 29; cf. Lennerling et al., Vol. 19 Nephrology Dialysis Transplantation 1600, 1603 (2004); Schneewind/Sedlmayer, in Schroth et al. (ed.) (2006), p. 39 & p. 44; cf. Waterman et al., Vol. 16 Progress in Transplantation 17, 17 (2006).

[383] Gutmann (2011), p. 4.

[384] Gutmann (2006), p. 25 f.; Roff, Vol. 33 Journal of Medical Ethics 437, 440 (2007).

[385] Borowsky, in Meyer (ed.) (2011), Art. 3 at 43.

[386] Höfling, in Tettinger/Stern (ed.) (2006), Art. 3 at 17.

[387] Cf. Gutmann (2006), p. 26 ff.

[388] Cf. Evans, Vol. 15 Journal of Medical Ethics 17, 19 (1989); cf. Patel/Chadha/Papalois, Vol. 3 Experimental and Clinical Transplantation 181, 185 (2011).

[389] Art. 2 (1) Charter of Fundamental Rights of the European Union.

[390] Ibid., Art. 3 (1).

[391] Ackerman/Thistlethwaite/Ross, Vol. 6 American Journal of Transplantation 83, 86 (2006).

8) Procedures

The material issues of LOD have been discussed. Chapter "Comparative Analysis of European Transplant Laws Regarding Living Organ Donation" showed that it is not uncommon for countries to also establish procedural safeguards for LOD. Those that can be considered as restricting LOD to a certain extent will be considered here.

Switzerland does not restrict the donor-recipient relationship. Also, it has not established a state commission to review LOD. In Switzerland, the situation for physicians is exceptional insofar as there is a high level of trust in them; they believe their physicians will make good decisions.[392] However, the Swiss *Federal Act on the Transplantation of Organs, Tissues and Cells* stipulates that "[a]ny person who removes organs [...] must notify the Federal Office of this."[393] In addition, the hospitals and transplant centres themselves decide on specific procedures for LOD. In the United Kingdom, Sec. 33 *Human Tissue Act 2004* standardises a criminal prohibition with a reservation to grant permission.[394] The donor-recipient relationship is not restricted by law. In contrast to Switzerland, every LOD has to be approved by the Human Tissue Authority through an independent assessment process.[395] The exact procedure depends on whether donor and recipient are genetically or emotionally related or if neither is.[396] Thus, the types of LOD are treated differently in the United Kingdom.[397] Another difference between the procedure in Switzerland and the United Kingdom is that "[a]ny person that removes organs" is addressed in Switzerland, while the members of the Human Tissue Authority in the United Kingdom are appointed by the Secretary of State, the National Assembly for Wales and the relevant Northern Ireland department.[398] Hence, a state commission becomes involved in the United Kingdom. In Germany, the donor-recipient relationship is restricted by law. In addition, a commission has to give an opinion of whether there are substantiated reasons to assume that the donor's consent is not being given freely or that the organ is the object of prohibited organ trade.[399] It is governmental, but the commissions are "responsible according to Land legislation."[400]

All in all, several countries have not only established material restrictions for LOD, but also restrictions on a procedural level.

---

392 This has been discussed at the 2011 ELPAT-Working Group on Legal Boundaries Conference in Berlin.

393 Art. 14 1 Federal Act on the Transplantation of Organs, Tissues and Cells.

394 Fateh-Moghadam (2008), p. 165.

395 Sec. 13 ff. Human Tissue Act; information from D. Price.

396 Fateh-Moghadam (2008), p. 292.

397 This is criticised by Choudhry et al., Vol. 29 Journal of Medical Ethics 169, 2003.

398 1 (1) Human Tissue Act 2004, Schedule 2 – The Human Tissue Authority.

399 Sec. 8 III 2 Act on the donation, removal and transplantation of organs.

400 Ibid.

III. Summary

As seen, several arguments in favour of restricting LOD are brought forth, but those are also subject to severe criticism.

(1) LOD could be seen as a violation of the principle of nonmaleficence and the donor's risks are mentioned in favour of restricting the donor-recipient relationship. Those against restricting the donor-recipient relationship mention the principle of autonomy, causing that every person has a right of self-determination. (2) There is no common opinion on the connection between related LOD and (in)voluntariness, or unrelated LOD and (in)voluntariness. The best approach, therefore, is to judge each case individually. (3) Some are concerned that a general legalisation of LOD would increase the pressure on living individuals. The rationale of this—that it is more important for the state to protect its citizens from having to face hard choices than to give them the chance to choose whether to help or even save a person in need—is unsound. Also, permitting unspecified LOD cannot be equated with the duty to actually become an unspecified living organ donor. (4) Unspecified LOD and cross-over LOD are associated with organ trade, but those concerns have been dispelled. (5) Some question whether LOD is just and equal, since it is not available to every patient in need. The solution, to prohibit LOD in general, would not improve the situation of any of the patients, though. (6) Several arguments in favour of completely prohibiting minors and mentally incapacitated adults from acting as living organ donors exist. However, there are also valid arguments against such a complete prohibition. (7) The different types of LOD that are (medically) possible have the positive effect of overcoming incompatibility between the willing donor and the intended recipient, increasing the amount of donor organs. The main point of critique is that it is feared that the pressure inflicted upon the potential donor might increase. (8) Several countries do not restrict the donor-recipient relationship, but do have specific procedures that must be followed.

To summarise, with the exception of specific arguments concerning special models of LOD exchanges, none of the arguments brought forth for the necessity and the legitimacy of most of the legal restrictions proposed for LOD seems cogent.[401] There is no class of donors and no category of LOD that should be generally dismissed by law. Rather, the solution seems to be a procedural one. Careful case-by-case decisions should be made.[402]

What is the opinion of the European Union on LOD? And has it become active in the field of (living) organ donation yet? That will be worked out in the next chapter.

---

[401] Cf. Daar et al., Vol. 11 Transplantation Review 95, 99 (1997); cf. Evans, Vol. 15 Journal of Medical Ethics 17, 20 (1989); Gutmann/Schroth (2002), p. 61; Gutmann (2006), p. 47; cf. Hilhorst et al., Vol. 24 Transplant International 1164, 1168 (2011).

[402] Cf. Gutmann et al., Terasaki (ed.) (1995), p. 356; Land, Vol. 2 Transplant International 168, 173 (1989) with regard to spouse-to-spouse donation.

Actions of the European Union So Far

The European Parliament and the European Council recognises LOD as an excellent option to have, but also see that it contains severe risks as well. They state: "The extensive therapeutic use of organs for transplantation demands that their quality and safety should be such as to minimize any risks associated with the transmission of disease."[1]

The European Commission has also put organ donation and transplantation high on their policy agenda. It, together with transplant professionals, strongly advocates the use of LOD to overcome the problem of organ shortage, under the condition that a legal framework is in place and that safety for both the donor and recipient can be guaranteed.[2]

The European Union has, thus, already shown attention to (living) organ donation.[3] (1) The European Commission conducted a survey on the legal requirements for organ donation in 2003. (2) In 2007, the *Communication from the Commission to the European Parliament and the Council – Organ Donation and Transplantation: Policy Actions at EU Level* (*Communication from the Commission*) was completed. (3) The Europesn Parliament has established resolutions on LOD.

I. Survey by the European Commission in 2003

The European Commission conducted a survey in 2003 on the legal requirements regarding organ transplantation. It included the 25 Member States of the European Union and also Bulgaria, Norway, Romania and Turkey. The aim of the survey was "to collect information on the legal framework related to ethical, organisational and

[1] (1) & (2) Directive 2010/53/EU of the European Parliament and of the Council of 7 July 2010 on standards of quality and safety of human organs intended for transplantation.

[2] EULOD WP 2 (2012), DOW: Deliverable 4, p. 15.

[3] The rather recent actions of the European Union will be introduced in the following.

DOI 10.1007/978-3-642-33799-4_5,

technical aspects in the field of organ transplantation."[4] The survey questionnaire contained questions regarding the general legal framework, and its organisatorial and technical aspects.[5] With regard to all three categories of questions it became apparent that the countries do share similar rules, but that they are not identical and exhibit individual differences. The report disclosed the differences with respect to the quality and safety requirements.[6] However, it did not make any suggestions for how to improve the current situation or for how to harmonise the different legal regulations in the Member States of the European Union. The report also has not led to any action on the part of the European Union.

II. *Communication from the Commission to the European Parliament and the Council – Organ Donation and Transplantation: Policy Actions at EU Level*

In 2007, The European Union took extensive action on the matter of organ donation when it released the *Communication from the Commission.* (1) The content of this document will be presented. (2) It is not sufficient to only examine the *Communication from the Commission.* In addition, the *Commission staff working document, Accompanying document to the Communication from the Commission to the European Parliament and the Council, Organ Donation and Transplantation: Policy Actions at EU Level, Impact Assessment* (*Commission staff working document, Accompanying document to the Communication*) will be considered. (3) Then, the consequences of this initiative will be presented in detail.

1) Content of the *Communication from the Commission to the European Parliament and the Council – Organ Donation and Transplantation: Policy Actions at EU Level*

The *Communication from the Commission* provides general information of organ donation. It highlights how the Member States deal differently with several aspects of organ donation.[7] This is in accord with the findings made in chapter "Comparative Analysis of European Transplant Laws Regarding Living Organ Donation." The *Communication from the Commission* points to the fact that the amount of

[4] European Commission (2003), p. 3.

[5] Ibid., p. 4.

[6] Council of the European Union (2007), p. 4. The detailed results of the report can be seen at http://ec.europa.eu/health/ph_threats/human_substance/documents/organ_survey.pdf.

[7] Commission of the European Communities (2007), p. 2.

organ transplantations has steadily increased over the past few decades, and that transplantations provide excellent results; it also mentions the problem of organ shortage.[8] Its purpose is to respond to the existing challenges, based on the mandate provided by Art. 168 IV (a) *Treaty on the Functioning of the European Union* (*TFEU*) (former Art. 152 IV (a) *Treaty establishing the European Community* (*EC*)).[9]

The *Communication from the Commission* mentions transplant risks, organ shortage and organ trafficking as some of the current challenges facing organ donation and transplantation.[10]

It then presents "the added value of EU action." The European Commission has actively made an effort to further organ transplantation.[11] In 2006, it launched an open consultation specifically for organ transplantion. It even proposed to make initiatives at Community level to address the existing challenges.[12] "Ensuring the quality and safety of human organs in Europe lies at the core of Community activities in this field, [...]." Organ shortage and organ trafficking were listed as the main priorities in this area.[13] They highlighted how certain aspects, such as quality and safety, have to be considered in the framework of supply and demand for organs.[14]

To address the challenges just introduced, the Commission found that "an integrated approach of three actions" was necessary. (1) A "[f]ramework for quality and safety of organ donation and transplantation" must be established. (2) The "[c] ooperation between Member States" is recommended. This section of the *Communication from the Commission* explicitly mentions LOD; it states that it might be important to promote unspecified LOD to expand the donor pool. Safeguards to protect the living organ donor and to prevent organ trafficking have to be guaranteed, though. It also lists the factors that lead to the increase of LOD. The increase of LOD is a result of, inter alia, the shortage of organs donated post-mortem, surgical advances, favourable results of transplantations and low risks for donors. (3) "Fighting organ trafficking" is also mentioned.[15]

The *Communication from the Commission* concludes with a list of suggested follow-up actions. "Improving quality and safety", "[i]ncreasing organ availability" and "[m]aking transplantation systems more efficient and accessible" are considered as the main areas of action in the field of organ donation and transplantation. As a consequence, the Commission proposes to first pass an "[a]ction plan on

---

[8] Ibid.

[9] Ibid.

[10] Ibid., p. 3 f.

[11] For an overview about EU funded projects see Council of the European Union (2007), p. 7.

[12] Council of the European Union (2007), p. 4.

[13] Commission of the European Communities (2007), p. 4–8; cf. also Council of the European Union (2007), p. 45.

[14] Commission of the European Communities (2007), p. 4–8.

[15] Ibid.

strengthened cooperation between Member States" and second, an "EU legal instrument on quality and safety of organ donation and transplantation."[16]

2) Content of the *Commission Staff Working Document, Accompanying Document to the Communication from the Commission to the European Parliament and the Council, Organ Donation and Transplantation: Policy Actions at EU Level, Impact Assessment*

Beyond the *Communication from the Commission*, the Council of the European Union also created the *Commission staff working document, Accompanying document to the Communication.*

The introduction also includes general information about organ transplantation. It emphasises the need to solve the problem of organ shortage, as well as aspects concerning quality and safety.[17] In the conclusion, several previous actions with respect to organ donation are listed, as well as EC-funded projects in this field.[18]

This document, like the *Communication from the Commission* itself, recognises that organ transplantation contains several problems. The risks connected to the aspect of quality and safety are presented in a rather detailed manner.[19] It is repeated that organ shortage is the main challenge in the field of organ donation.[20] Furthermore, it is stated that the organisational structure has a key role in the organ donation systems.[21]

The main objective of the European Union is to "ensure high standards of quality and safety for human organs." It is suggested to connect this aim to the problem of organ shortage. Another important objective, without doubt, is to avoid organ trafficking.[22]

The European Union must find a way to approach these challenges in the field of organ donation and transplantation. The following three options for possible European Union actions are stated and compared in the following. (1) The first level is to "[f]urther work under existing Community programmes." (2) The second level is to achieve "[a]ctive coordination between Member States on organ quality, safety and availability." (3) The third level combines several measures, namely "[s]econd level + Minimum harmonization on quality and safety + Initiative on organ

[16] Ibid., p. 9 ff.

[17] Council of the European Union (2007), p. 4.

[18] Ibid., p. 5 ff.

[19] Ibid., p. 11 f.

[20] Ibid., p. 21 f.

[21] Ibid., p. 24.

[22] Ibid., p. 45.

trafficking." Advantages as well as disadvantages have been brought forth for all three levels. For the first level, the disadvantages clearly outweigh the advantages. This is different with respect to the second level and the third level. Especially with respect to the duties established by Art. 168 *TFEU*, the third level is preferable to the second level. Establishing "[a]n appropriate and flexible European legal framework" seems appropriate to deal with the demands of quality and safety. A shared legal framework is also the only way to establish the same standards of quality and safety in the European Union, thus ensuring an equal level of health protection.[23] It has been concluded that, "[u]ndoubtly and EU initiative on organ donation and transplantation has an added value."[24]

The *Communication from the Commission* and the corresponding *Commission staff working document, Accompanying document to the Communication* show that the European Union can produce reasonable actions in the field of organ donation. The suggestions made did not remain without consequences, as will be presented next.

3) Consequences of the *Communication from the Commission and the Commission Staff Working Document, Accompanying Document to the Communication*

Both the *Communication from the Commission* and the *Commission staff working document, Accompanying document to the Communication* provide general information about the status quo of organ donation and transplantation. In addition, they also suggest actions at the Community and Member State level that (might) increase the amount of donor organs available in the European Union and help ensure that the necessary procedures are in accord with a sufficient level of quality and safety.[25] The *Communication from the Commission* was intended to serve as "a dual mechanism of action: an Action Plan enhancing active coordination and cooperation between Member States complemented by a legal instrument containing the basic quality and safety principles." In the following, I will therefore (1) describe the *Action plan on Organ Donation and Transplantation* (2009–2015) (*Action plan*) and (2) the *Directive 2010/53/EU*.

[23] Council of the European Union (2007), p. 59. For a detailed list with the advantages and disadvantages of the three policy options see Council of the European Union (2007), p. 57 ff.

[24] Council of the European Union (2007), p. 60.

[25] Commission of the European Communities (2008), p. 2.

a) *Action Plan on Organ Donation and Transplantation* (2009–2015)

The *Action plan* is the first component of the "dual mechanism". It has been established with "the aim of strengthening cooperation between Member States in terms of organ donation and transplantation." This is in accord with the competence to coordinate according to Art. 168 II *TFEU*.[26]

In general, an action plan is passed to put the programmes to make law and the general objectives from the Treaties of the European Union in concrete terms.[27]

The *Action plan* contains ten priority actions. These can be grouped under the following three challenges: (1) It intends to increase organ availability. (2) It aims at improving the transplantation systems with regard to efficiency and accessibility. (3) It intends to raise the standards of quality and safety.[28] The Member States can decide what actions and measures they should take to achieve the objectives,[29] but the *Action plan* suggests including the measures in their Sets of National Priority Actions. Those should "serve as a platform for discussion, exchange of expertise, and identification of best practices in the framework of this Action Plan." The actions of the countries should be country-specific and should consider the specific situation of each Member State.[30]

(1) To increase organ availability, the *Action plan* clarifies that LOD is complementary to post-mortem organ donation. The *Action plan* should be deployed by the Member States "to promote the exchange of best practices on living donation programmes (Priority Action 3)."[31] For LOD, an exchange of good practice is encouraged and altruistic donation programmes are approved.[32] Living donors should be protected and organ trafficking has to be prevented. To evaluate and to ensure the living donors' health and safety, the *Action plan* recommends developing registers for living donors.[33]
(2) To enhance the efficiency and accessibility of transplantation systems, the *Action plan* refers to the "interchange of organs between Member States." The *Action plan* contains the goal to increase the exchange of organs with regard to urgent patients and those who are difficult to treat.[34]
(3) To improve quality and safety, it suggests collecting information in registers. The registers would contain post-transplant results that could be later evaluated. The registers are meant to "help to develop good medical practices in organ

---

[26] Wollenschläger, in Müller-Graff/Schmahl/Skouris (ed.) (2011), p. 457.
[27] Borchardt (2010), at 552.
[28] Commission of the European Communities (2008), p. 2.
[29] Wollenschläger, in Müller-Graff/Schmahl/Skouris (ed.) (2011), p. 457.
[30] Commission of the European Communities (2008), p. 3.
[31] Ibid., p. 4.
[32] Ibid.
[33] Ibid., p. 10.
[34] Ibid., p. 6.

donation and transplantation." The *Action plan* also focuses on increasing the amount of donor organs. It considers the use of expanded donors,[35] and it recommends establishing "common definitions of terms and methodology to help determine acceptable levels of risk in the use of expanded donors."[36]

Since the *Action plan* focuses on strengthening the cooperation between the Member States in the field of organ donation and transplantation, it refers to former Art. 152 II *EC*, now Art. 168 II *TFEU*. This provision determines that the European Union "shall encourage cooperation between the Member States in the areas referred to in this Article and, if necessary, lend support to their action." It, therefore, seems beneficial for the Member States to share experience and expertise in the field of organ donation and transplantation. The methods of the *Action plan* to achieve this aim are "the identification and development of common objectives and guidelines, jointly-agreed indicators and benchmarks, regular reporting, and identification and sharing of best practices."[37]

Hence, the *Action plan* involves three important challenges for organ donation: (1) to increase organ availability, (2) to improve the transplantation systems with regard to efficiency and accessibility and (3) to raise the standards of quality and safety. It intends to stimulate the Member States to become active, but it is not binding. In addition, the *Action plan* focuses on strengthening the cooperation between the Member States in the field of organ donation and transplantation.

b) ***Directive 2010/53/EU of the European Parliament and of the Council of 7 July 2010 on Standards of Quality and Safety of Human Organs Intended for Transplantation***

The *Directive 2010/53/EU* is the latest and the most extensive action of the European Union with regard to LOD to date.

The proposal for the *Directive 2010/53/EU* was already adopted by the Commission in 2008.[38] The *Directive 2010/53/EU* entered into force on the 20th day after its publication in the Official Journal of the European Union (6 August, 2010).[39] The *Directive 2010/53/EU* must be transposed by 27 August, 2012.[40]

---

[35] Expanded donors means donors that, from a medical point of view, can only be considered for specific recipients under specific circumstances (Commission of the European Communities (2008), p.7).

[36] Commission of the European Communities (2008), p.7.

[37] Ibid., p.2.

[38] The European Data Protection Supervisor, Vol. 52 Official Journal of the European Union 6, 7 (2009).

[39] Art. 32 Directive 2010/53/EU of the European Parliament and of the Council of 7 July 2010 on standards of quality and safety of human organs intended for transplantation.

[40] Ibid., Art. 31, 1.

During this time period the Member States are allowed to maintain or introduce more stringent rules if they comply with the *TFEU*.[41]

The aim of the proposal for the *Directive 2010/53/EU* was to ensure high standards of quality and safety for human organs intended for transplantation, in order to guarantee a high level of human health protection.[42] Organ donation is a very important subject matter of health care policy and several important aspects are connected to it, e.g., the requirements to justify organ donation. With the coming into effect of the *Directive 2010/53/EU*, this subject matter came to the fore of the European Union Law.[43] It became part of the general approach to set common standards for different areas of the national healthcare services. Furthermore, it complements the already existing legal framework with regard to the use of human material, which has focussed on blood and blood products[44] and on human tissues and cells[45] so far.[46] The *Directive 2004/23/EC of the European Parliament and of the Council of 31 March 2004 on setting high standards of quality and safety for the donation, procurement, testing, processing, preservation, storage and distribution of human tissues and cells* has been used as a model for *Directive 2010/53/EU*.[47]

(1) A description of the procedure to pass the *Directive 2010/53/EU* is provided. (2) I will then regard the content of the *Directive 2010/53/EU* with respect to LOD. (3) In the end, I will draw an interim result.

aa) Procedure to Pass the *Directive 2010/53/EU of the European Parliament and of the Council of 7 July 2010 on Standards of Quality and Safety of Human Organs Intended for Transplantation*

The legislative procedure stipulated to pass the *Directive 2010/53/EU* has been followed.[48] (1) First, the *Directive 2010/53/EU* is based on Art. 168 IV (a) *TFEU*. (2) That the proposal from the European Commission has been followed has already been described. (3) The European Economic and Social Committee was consulted by the European Council on January 21, 2009. It, overall, appreciated the proposal for the intended *Directive 2010/53/EU*. It was satisfied with the

[41] Art. 31, 2. Directive 2010/53/EU of the European Parliament and of the Council of 7 July 2010 on standards of quality and safety of human organs intended for transplantation.

[42] The European Data Protection Supervisor, Vol. 52 Official Journal of the European Union 6, 7 (2009).

[43] Wollenschläger, in Müller-Graff/Schmahl/Skouris (ed.) (2011), p. 448 f.

[44] Directive 2002/98/EC; Directive 2004/33/EC; Directive 2005/61/EC; Directive 2005/62/EC.

[45] Directive 2004/23/EC; Directive 2006/17/EC; Directive 2006/86/EC.

[46] The European Data Protection Supervisor, Vol. 52 Official Journal of the European Union 6, 7 (2009).

[47] Council of the European Union (2008), p. 30.

[48] Directive 2010/53/EU of the European Parliament and of the Council of 7 July 2010 on standards of quality and safety of human organs intended for transplantation.

Directive's aim to increase the protection of health of the citizens of the European Union. However, it made a few remarks on the provisions of the proposal. Art. 1 of the proposal clarifies that its aim is to guarantee high levels of quality and safety for organs and high levels of health protection. The European Economic and Social Committee does not consider this as adequate; the word "high" is too vague. Therefore, the following phrasing of Art. 1 was suggested: "to ensure the necessary standards of quality and safety for organs of human origin intended for transplantation to the human body, in order to ensure the highest level of human health protection." Specifically with regard to LOD, the Committee criticises Art. 15, 2. of the proposal which addresses the "[p]rotection of the living donor." The proposal states in Art. 15, 2. : "Member States shall ensure that living donors are selected on the basis of their health and medical history, including a psychological evaluation if deemed necessary, by qualified and trained professionals. Such assessments may provide for the exclusion of persons whose donation could present a health risk to others, such as the possibility of transmitting diseases, or a serious risk to themselves." The Committee suggested reformulating the second sentence accordingly: "[...] may provide for the exclusion of persons, whose donation could present a serious risk to themselves." However, on the whole, the European Economic and Social Committee agrees with the proposal.[49] (4) The Commitee of the Regions has been involved as well. (5) The opinion of the European Data Protection Supervisor has been obtained. It generally supports the idea, but also emphasises that a well-coordinated and uniform data protection is necessary.[50] (6) The acting was in accordance with the ordinary legislative procedure. The opinion of the European Parliament corresponds to the final legislative act.[51] The Council of the European Union adopted the *Directive 2010/53/EU* without debate.[52]

In addition to the requirements stated in Art. 168 IV (a) *TFEU*, general principles of the competences of the European Union are relevant as well. The principle of subsidiarity must have been followed. The *Directive 2010/53/EU* states that the Member States cannot sufficiently lay down quality and safety standards for organs intended for transplantation and that these objectives can be better achieved at European Union level. That is explained by reference to the scale of the action. Consequently, the *Directive 2010/53/EU* considers the principle of subsidiarity.[53] The Commission of the European Communities also considers the principle of subsidiarity. It states that: "[1)] The European Community has a clear opportunity and obligation to implement binding measures laying down high standards of

[49] The President of the European Economic and Social Committee, 2009 Official Journal of the European Union 64, 64 ff. (2009).

[50] The European Data Protection Supervisor, Vol. 52 Official Journal of the European Union 6, 7 (2009).

[51] European Parliament (2010).

[52] Council of the European Union (2010).

[53] (28) Directive 2010/53/EU of the European Parliament and of the Council of 7 July 2010 on standards of quality and safety of human organs intended for transplantation.

quality and safety for the use of blood, organs, and substances of human origin [2)] European Community action is likely to contribute to public value by providing a platform for implementation and mutual learning which combines standardization of reporting with diversity of service."[54] Furthermore, the principle of proportionality has to be taken into account. A directive, in comparison to other legal acts the European Union may pass, can be considered as the best legal act for quality and safety measures with respect to organs, because it leaves the most freedom to act for the Member States.[55]

bb) Content of the *Directive 2010/53/EU of the European Parliament and of the Council of 7 July 2010 on Standards of Quality and Safety of Human Organs Intended for Transplantation*

After describing the procedure to pass the *Directive 2010/53/EU*, the content is focused on.

The *Directive 2010/53/EU* starts by clarifying that organ donation has a great value and several advantages, but contains risks as well. As a consequence, the quality and safety of organs used for transplantation "should be such as to minimise any risks associated with the transmission of diseases."[56] To harmonise the quality and safety would not only guarantee a high standard of quality and safety, but would make an organ exchange between the Member States easier.[57] This can be considered as a helpful step to reduce the problem of organ shortage.[58] One method of organ donation that cannot be accepted is organ trade. The *Directive 2010/53/EU* thus also indirectly fights organ trade.[59]

Reason (8) of the *Directive 2010/53/EU* refers to Art. 168 VII *TFEU*, which states that the action of the European Union should not affect national provisions on the donation of organs. One objective of the *Directive 2010/53/EU* is to reduce the dangers connected to organ transplantation.[60] Consequently, it might be necessary to include provisions for the transplantation in the *Directive 2010/53/EU*. This

[54] Council of the European Union (2008), p. 25.

[55] Cf. Berg, in Schwarze (ed.) (2009), Art. 152 at 22.

[56] (1) & (2) Directive 2010/53/EU of the European Parliament and of the Council of 7 July 2010 on standards of quality and safety of human organs intended for transplantation.

[57] Council of the European Union (2010); Spielberg, Vol. 107 (22) Deutsches Ärzteblatt A-1102/B-973/C961 (2010); cf. Wollenschläger, in Müller-Graff/Schmahl/Skouris (ed.) (2011), p. 457.

[58] Cf. Council of the European Union (2010); cf. Eurotransplant (2012).

[59] Council of the European Union (2010); (7) Directive 2010/53/EU of the European Parliament and of the Council of 7 July 2010 on standards of quality and safety of human organs intended for transplantation; Spielberg, Vol. 107 (22) Deutsches Ärzteblatt A-1102/B-973/C961 (2010).

[60] Council of the European Union (2010).

especially applies to unexpected situations that might occur during the transplantation process and which might affect quality and safety of the organ concerned.[61]

To reduce the risks of organ transplantation and to maximise the benefits of transplantations, it is considered necessary for the Member States to make good use of a framework for quality and safety of organ donation.[62]

Art. 2 I *Directive 2010/53/EU* clarifies that "[t]his Directive applies to the donation, testing, characterisation, procurement, preservation, transport and transplantation of organs intended for transplantation." The *Directive 2010/53/EU* applies to post-mortem organ donation and to LOD.

Chapter III, which deals with "[d]onor and recipient protection and donor selection and evaluation", is the most relevant chapter of the *Directive 2010/53/EU* for this thesis. The relevant provisions of this chapter are (1) Art. 13, (2) Art. 14 and (3) Art. 15 *Directive 2010/53/EU.*

(1) Art. 13 *Directive 2010/53/EU of the European Parliament and of the Council of 7 July 2010 on Standards of Quality and Safety of Human Organs Intended for Transplantation*

LOD (as well as deceased donations) must be voluntary.[63] The principle of non-payment governs organ donation,[64] although compensation is allowed and may be granted under circumstances that are defined by the Member States.[65]

The *Directive 2010/53/EU* considers the principles of voluntariness and non-payment as aspects that concern the quality and safety of organ donation.[66] The European Union argues that the quality of the organ donation might be jeopardised if the organ donation is not voluntary and/or undertaken with a view to financial gain.[67] This argumentation is objected to by legal experts, claiming that adequate

[61] (8) Directive 2010/53/EU of the European Parliament and of the Council of 7 July 2010 on standards of quality and safety of human organs intended for transplantation.

[62] Art. 4, 1. Directive 2010/53/EU of the European Parliament and of the Council of 7 July 2010 on standards of quality and safety of human organs intended for transplantation; Council of the European Union (2010); Europa (2012).

[63] Art. 13 I Directive 2010/53/EU of the European Parliament and of the Council of 7 July 2010 on standards of quality and safety of human organs intended for transplantation.

[64] Art. 13 II Directive 2010/53/EU of the European Parliament and of the Council of 7 July 2010 on standards of quality and safety of human organs intended for transplantation; Spielberg, Vol. 107 (22) Deutsches Ärzteblatt A-1102/B-973/C961 (2010); cf. Wollenschläger, in Müller-Graff/Schmahl/Skouris (ed.) (2011), p. 458.

[65] Art. 13 II Directive 2010/53/EU of the European Parliament and of the Council of 7 July 2010 on standards of quality and safety of human organs intended for transplantation; cf. Wollenschläger, in Müller-Graff/Schmahl/Skouris (ed.) (2011), p. 458.

[66] (19) Directive 2010/53/EU of the European Parliament and of the Council of 7 July 2010 on standards of quality and safety of human organs intended for transplantation; Wollenschläger, in Müller-Graff/Schmahl/Skouris (ed.) (2011), p. 460.

[67] (19) Directive 2010/53/EU of the European Parliament and of the Council of 7 July 2010 on standards of quality and safety of human organs intended for transplantation.

standards for quality and safety would also be able to solve this problem.[68] The European Union replied, stating that "[e]ven if the process is developed in accordance with appropriate quality standards, a clinical history obtained from either a potential living donor or the relatives of a potential deceased donor who are seeking financial gain or are subjected to any kind of coercion might not be sufficiently accurate in terms of conditions and/or diseases potentially transmissible from donor to recipient."[69] The opponents doubt this argumentation. In their opinion, it is wrong to think that every donation that happens under dubious circumstances is also a danger for the recipient.[70] The argumentation of the European Union is understandable. It is not implausible that chances of financial gain or plight hold certain dangers. This also concerns aspects of quality and safety so that the European Union is considered to be competent. Art. 168 VII *TFEU* clarifies that the competence for the donation of organs remains with the Member States, a strict examination must be carried out to assess whether a regulation actually addresses aspects of quality and safety.[71] That Art. 13 *Directive 2010/53/EU* actually aims at ensuring a high level of quality and safety cannot be doubted, though.

(2) Art. 14 *Directive 2010/53/EU of the European Parliament and of the Council of 7 July 2010 on Standards of Quality and Safety of Human Organs Intended for Transplantation*

Art. 14 *Directive 2010/53/EU* states that the requirements of the national laws with regard to the patient's consent have to be met.[72] Thus, the European Union does not curtail these competences of the Member States, and it especially does not influence the national models of consent for post-mortem organ donation.[73]

(3) Art. 15 *Directive 2010/53/EU of the European Parliament and of the Council of 7 July 2010 on Standards of Quality and Safety of Human Organs Intended for Transplantation*

Art. 15 *Directive 2010/53/EU* directly addresses "[q]uality and safety aspects of living donation."[74] According to Art. 15, the aim is to ensure the highest possible

---

[68] Cf. Pühler et al., Vol. 25 Medizinrecht 584, 585 f. (2007); Wollenschläger, in Müller-Graff/Schmahl/Skouris (ed.) (2011), p. 460.

[69] (19) Directive 2010/53/EU of the European Parliament and of the Council of 7 July 2010 on standards of quality and safety of human organs intended for transplantation.

[70] König, Vol. 23 Medizinrecht 22, 26 (2005) (His argumentation is with regard to the Directive 2004/23/EC of the European Parliament and of the Council of 31 March 2004 on setting standards of quality and safety for the donation, procurement, testing, processing, preservation, storage and distribution of human tissues and cells).

[71] Wollenschläger, in Müller-Graff/Schmahl/Skouris (ed.) (2011), p. 461. Art. 168 Treaty on the Functioning of the European Union will be examined in detail below.

[72] Art. 14 Directive 2010/53/EU of the European Parliament and of the Council of 7 July 2010 on standards of quality and safety of human organs intended for transplantation.

[73] Wollenschläger, in Müller-Graff/Schmahl/Skouris (ed.) (2011), p. 458.

[74] Art. 15 I Directive 2010/53/EU of the European Parliament and of the Council of 7 July 2010 on standards of quality and safety of human organs intended for transplantation; Jackson (2012).

protection of living donors to guarantee a certain level of quality and safety for the organs used for transplantation.[75] It also stipulates that living donors should be selected on the basis of their health and medical history and that they should be excluded if the donation could present unacceptable risks for the potential donor's health.[76] In addition, it is regulated that a register or record of living organ donors must be kept.[77]

A register is considered a key instrument for the health and safety of living organ donors.[78] Furthermore, a register serves in the fight against organ trafficking, because it makes it possible for the "Member States to monitor and evaluate the practice of living donation within the EU and across borders."[79] Additionally, a follow-up of living donors should be established and the Member States should create a system in accordance with their national law that recognises, reports and manages all events that might be connected to the quality and safety of the donated organ. The register is related to the safety of the recipient and to all serious adverse reactions that might be the result of the LOD.[80]

Art. 15 *Directive 2010/53/EU* is the only article in the directive that solely addresses LOD. It does not harmonise the requirements for LOD and does thus not interfere with Art. 168 VII *TFEU*. It establishes rules to guarantee the quality and safety of LOD. This also includes requirements with respect to the selection of the living organ donor and the post-care.[81]

cc) Interim Result

The *Directive 2010/53/EU* concentrates on aspects of quality and safety of LOD. This is done with regard to post-mortem organ donation and also with regard to LOD. It is unavoidable to also touch upon the requirements for LOD, but the *Directive 2010/53/EU* never (completely) stipulates the requirements for LOD. This competence remains with the Member States.

The Member States of the European Union have to transpose the *Directive 2010/53/EU* by 27 August 2012. The implementation of the *Directive 2010/53/EU* leaves a lot of flexibility to the Member States, making it possible for them to retain

[75] Art. 15 I Directive 2010/53/EU of the European Parliament and of the Council of 7 July 2010 on standards of quality and safety of human organs intended for transplantation.

[76] Ibid., Art. 15 II.

[77] Ibid., Art. 15 III.

[78] Solomon/Pavlou (2011).

[79] Ibid.

[80] Art. 15 IV Directive 2010/53/EU of the European Parliament and of the Council of 7 July 2010 on standards of quality and safety of human organs intended for transplantation.

[81] Wollenschläger, in Müller-Graff/Schmahl/Skouris (ed.) (2011), p. 458.

particular features of their national health care systems.[82] Also, the *Directive 2010/53/EU* only establishes a minimum standard,[83] meaning that the Member States may implement stronger regulations.[84]

Germany transposed the *Directive 2010/53/EU* by passing a law that changed the German *Act on the donation, removal, and transplantation of organs* on May 25, 2012.[85] The German Federal Government stated that the transposition of the *Directive 2010/53/EU* does not require a fundamental change of the structures of the current law on transplantation, because the *Directive*, in consideration of the national transplant systems, leaves a broad scope for the countries to act within.[86] The changes mainly focus on post-mortem organ donation. Germany used to apply the opt-in system for post-mortem organ donation.[87] This approach remains in principle,[88] but is intended to be changed insofar as every citizen will be asked to decide in advance whether he will become an organ donor after his death (so called *Entscheidungslösung*[89]).[90] Every citizen should be able to come to an informed and independant decision[91]; however, a duty to come to a decision does not exist.[92] The aim is to improve the amount of donor organs and to make the population more aware of the topic of organ donation.[93] With respect to LOD, changes are intended as well. The changes are meant to improve the situation of living organ donors with respect to their claims against the recipient's health care insurance and continued remuneration.[94]

Beyond the (rather mild) effect the *Directive 2010/53/EU* has right now, it might produce a further dynamism with regard to a Europeanising of the transplantation law in future. The extent of this cannot be foreseen.[95] The Europeanising of this field continues to move forward. The quantity of legal acts of the European Union in the field of health care policy increases. One of those acts is the *Directive 2010/53/EU*.

---

[82] Bundesverband der Organtransplantierten (2012); di Bella, Vol. 9 Rechtsdepesche für das Gesundheitswesen 38, 38 (2012).

[83] Art. 31, 1. Directive 2010/53/EU of the European Parliament and of the Council of 7 July 2010 on standards of quality and safety of human organs intended for transplantation; Spielberg, Vol. 107 (22) Deutsches Ärzteblatt A-1102/B-973/C961 (2010).

[84] Spielberg, Vol. 107 (22) Deutsches Ärzteblatt A-1102/B-973/C961 (2010).

[85] Cf. Bundestag printed paper 17/7376 (2011); cf. Bundestag printed paper 17/9773 (2012).

[86] Bundestag printed paper 17/9773 (2012), p. 1.

[87] Bundestag (2012); Bundestag printed paper 17/9030 (2012), p. 4.

[88] Bundestag (2012).

[89] Bundesrat printed paper 322/12 (2012); Bundestag printed paper 17/9030 (2012), p. 4.

[90] Bundestag printed paper 17/9030 (2012), p. 4.

[91] Ibid.

[92] Bundestag (2012).

[93] Bundesrat (2012); Bundestag (2012).

[94] See p. 136 f.

[95] Cf. Wollenschläger, in Müller-Graff/Schmahl/Skouris (ed.) (2011), p. 462.

Consequently, the European Union does not regulate the requirements for LOD in the *Directive 2010/53/EU*. The effect of the *Directive 2010/53/EU* cannot be considered as being very extensive. However, it might cause a further Europeanising of this subject matter and is probably not the last legal act of the European Union in this field.

III. Resolutions from the European Parliament

The European Commission is not the only European authority that has become active in the field of LOD. (1) After the European Commission adopted the *Communication from the Commission*, the European Parliament reacted by adopting the *resolution on organ donation and transplantation: Policy actions at EU Level (2007/2210 (INI))* (*resolution on organ donation and transplantation*)in 2008.[96] (2) About 2 years later, in 2010, the European Parliament adopted another resolution on the Commissions *Communication: Action plan on Organ Donation and Transplantation (2009–2015): Strengthened Cooperation between Member States (2009/2014 (INI))* (*resolution on the Action plan*).

1) European Parliament *Resolution of 22 April 2008 on Organ Donation and Transplantation: Policy Actions at EU Level (2007/2210 (INI))*

Following the *Communication from the* Commission, the European Parliament adopted a resolution on it in April 2008.[97]

The *resolution on organ donation and transplantation* addresses the problems of organ shortage and organ trade. It also emphasises the advantages of organ transplantation in comparison to alternative treatments and mentions the organ exchange between Member States.[98]

The *resolution on organ donation and transplantation* welcomes the *Commission Communication to the European Parliament and the Council on Organ Donation and Transplantation: Policy Actions at EU level.* It expects the European Commission to work out a proposal for a Directive. This should lay "down quality and safety requirements for organ donation, procurement, testing, preservation,

[96] Council of the European Union (2008), p. 6; Joint Council of Europe/United Nations Study (2009), p. 43.

[97] Council of the European Union (2008), p. 6; Joint Council of Europe/United Nations Study (2009), p. 43.

[98] European Parliament resolution of 22 April 2008 on organ donation and transplantation: Policy actions at EU level (2007/2210 (INI)), passim.

transport and allocation across the EU and the resources necessary to meet those requirements."[99] It suggests that this directive "should complement and reinforce efforts made by Member States to achieve an active and efficient method of coordination without preventing the introduction or maintenance of more stringent measures."[100]

It addresses the three issues the *Action plan* primarily considers as well (the increase of organ availability, the efficiency and accessibility of transplantation systems and the improvement of quality and safety). In addition, it specifically mentions the "[c]o-operation between Member States," "[i]ncreasing public awareness" and "[o]rgan trafficking."[101]

With regard to LOD in particular, the *resolution on organ donation and transplantation* asks the Member States to consider quality and safety while allowing LOD and emphasises that LOD should only be seen as complementary to post-mortem organ donation.[102] It supports measures with the aim to protect living organ donors. Member States that allow unspecified LOD are asked to ensure anonymity.[103] It additionally recommends the Member States, with regard to unspecified LOD, "to adopt or maintain strict legal provisions." It suggests only allowing this type of LOD "under the conditions defined in national law and following authorisation by a suitable independent body".[104]

All in all, the *resolution on organ donation and transplantation* is very extensive. It addresses several issues of organ donation in general, but also contains suggestions solely for post-mortem organ donation and LOD in particular.

2) European Parliament *Resolution of 19 May 2010 on the Commission Communication: Action Plan on Organ Donation and Transplantation (2009–2015): Strengthened Cooperation Between Member States (2009/2104(INI))*

The *resolution on the Action plan* starts by mentioning the problem of organ shortage, more precisely, that "there are currently 56,000 patients waiting for a suitable organ donor in the EU, and it is estimated that every day 12 people die while waiting for a solid organ transplant."[105] Afterwards, it addresses several

[99] Ibid., number 2.

[100] Ibid., number 3.

[101] Ibid., passim.

[102] Ibid., number 15.

[103] Ibid., number 22.

[104] Ibid., number 23.

[105] European Parliament resolution of 19 May 2010 on the Commission Communication: Action plan on Organ Donation and Transplantation (2009–2015): Strengthened Cooperation between Member States (2009/2104(INI)), letter A.

issues concerning both post-mortem organ donation and LOD: "[O]rgan transplantation provides the possibility of saving lives, offers a better quality of life and (in the case of kidney transplantation) has the best cost/benefit ratio when compared with other replacement therapies as well as increasing the possibilities for patients to participate in social and working life."[106]

With respect to the relationship between LOD and post-mortem organ donation, it explains that LOD can be a helpful additional measure for patients who cannot receive an organ donated post-mortem,[107] and it states that LOD should be complementary to post-mortem organ donation.[108]

It also contains considerations solely with regard to LOD. LOD is considered legal only "if any illegal activity and payment for the donation has been ruled out."[109] The potential donor has to give free and informed consent, he has to be informed before the intervention about the purpose, the nature, the consequences and the risks of it and that he may freely withdraw his consent.[110]

The European Parliament has also expressed its opinion on the *Action plan*. It welcomes the *Action plan* and makes several suggestions to improve organ donation. It claims that "the establishment of common binding standards of quality and safety will be the only mechanism that can ensure a high level of health protection throughout the EU."[111] The European Parliament also mentions that measures must be approved to protect the living organ donor and to ensure that LOD is performed altruistically and voluntarily and without any payment beyond compensation.[112] Furthermore, it stresses that living organ donors should be treated with the highest medical standards in case medical problems are caused by the transplantation process,[113] and it recognises the importance of post-care.[114] The European Parliament also calls on the European Commission to evaluate whether living organ donors are sufficiently insured in all Member States.[115] The selection of living organ donors is addressed as well. Living donors should be chosen "on the basis of their health and medical history, including a psychological evaluation if deemed necessary, by qualified or trained and competent professionals."[116]

Thus, the *resolution on the Action plan* is very extensive and mainly approves of the *Action plan*.

[106] Ibid., letter F.
[107] Ibid., letter N.
[108] Ibid., letter Q.
[109] Ibid., letter N.
[110] Ibid., letter O.
[111] Ibid., number 11.
[112] Ibid., number 15.
[113] Ibid., number 28.
[114] Ibid., number 46.
[115] Ibid., number 16.
[116] Ibid., number 17.

IV. Summary

The European Union has already passed legal acts on organ donation, which also address LOD. (1) The European Commission made an extensive survey in 2003. This report disclosed that the countries differ with respect to quality and safety requirements for organ donation, but did not cause any actions of the European Union, though. (2) The *Communication from the Commission* lead to an *Action plan* and the *Directive 2010/53/EU*. Thus, the actions by the European Union that have happened so far are mostly non-binding, and a binding directive that considers LOD has been passed as well. This, however, does not completely regulate LOD, but only addresses aspects regarding quality and safety of organ donation. (3) In addition, the European parliament passed resolutions on organ donation and transplantation.

Whether a further harmonisation of LOD by the European Union would be possible with respect to the competences of the European Union will be considered in the next chapter.

Possibilities of Harmonisation

The medical and social problems connected to LOD, especially that too few donor organs are available to help all patients that suffer from organ failure, are very similar in the countries considered.[1] However, as seen in chapter "Comparative Analysis of European Transplant Laws Regarding Living Organ Donation," the legal rules regarding the issues of LOD in need of regulation are only partly similar; significant differences exist as well. This is problematic given the fact that the borders in Europe are very open and it is not uncommon for patients to engage in cross-border health care.[2]

The *TEU*, as a consequence of the *Treaty of Lisbon*, clarifies that it "makes a new stage in the process of creating an ever closer union among the peoples of Europe, in which decisions are taken as openly as possible and as closely as possible to the citizen."[3] With respect to LOD in particular, the relevance and advantages of harmonising the requirements for LOD, meaning "a process of replacing diverse national rules with common rules,"[4] were already mentioned. One particular cause for concern is individuals who try to make use of the regulations in other countries if they are not able to receive a donor organ in their home country.[5] As seen in the previous chapter, the European Union already acted in the field of (living) organ donation. It even passed a directive regulating quality and safety aspects of (living) organ donation. The European Union did not, however, harmonise the rules for LOD.

(1) Why is the European Union focused on when considering a harmonisation of legislation regarding LOD? (2) What are the general competences of the European Union to make law? (3) How can these be applied to the field of LOD? (4) What is *soft law*, and (5) how can the option to pass soft law be applied to the field of LOD?

---

[1] Hohmann (2003), p. 191.

[2] Hendriks, in Weimar/Bos/Busschbach (ed.) (2011), p. 34.

[3] Art. 1 Treaty on European Union.

[4] Weatherill (2010), p. 574.

[5] Hendriks, in Weimar/Bos/Busschbach (ed.) (2011), p. 32.

DOI 10.1007/978-3-642-33799-4_6,

I. European Union as Subject to Harmonise Legislation Regarding Living Organ Donation

Different legal possibilities exist to harmonise national legislation. This also applies to legislation for LOD. One possibility is to conclude a bilateral treaty.[6] This is an agreement under national law between two parties only.[7] This thesis focuses on the harmonisation of the Member States of the European Union, therefore making a bilateral treaty insufficient. Another option is to conclude collective treaties, which can either be regional or continental or can even exceed one continent. Agreements that are this extensive require a very high degree of comparative preparation.[8] A collective treaty seems to be an effective way to harmonise the legislation for LOD in the countries considered. But why is the European Union considered as possible initiator of such a collective treaty?

The final aim of this dissertation is to create a best practise proposal for LOD for the Member States of the European Union (and Moldova, Norway and Switzerland). Therefore, only organs and institutions that are able to act European Union-wide are considered.[9]

1) WHO

The WHO could act. It has already passed papers on the topic of organ donation.[10] The WHO is the authority within the system of the United Nations that concentrates on health matters.[11] It has no relation to the European Union, though. Since this thesis focuses on the European Union, the WHO is not taken into consideration.

2) Council of Europe

The Council of Europe could act. This international organisation includes almost all European States, and with it all Member States of the European Union. It is independent of the European Union and its organs.[12] The Council of Europe has been very active in the field of organ transplantation in the past.[13] In 1978, it already

[6] Schnitzer (1961), p. 81.
[7] Creifelds/Weber (ed.) (2011), p. 214.
[8] Schnitzer (1961), p. 81.
[9] World Health Organization (2011).
[10] Norba (2009), p. 352.
[11] World Health Organization (2011).
[12] Cf. Council of Europe; cf. Haratsch/König/Pechstein (2010), p. 17.
[13] Hendriks, in Weimar/Bos/Busschbach (ed.) (2011), p. 35; Price (2010), p. 18 f.

passed the *Resolution (78) 29 on Harmonisation of Legislations of Member States Relating to Removal, Grafting and Transplantation of Human Substances*. This *Resolution* is not binding to the Member States; it just contains recommendations.[14] The Council recently became active again by passing the *Convention on Human Rights and Biomedicine* and the *Additional Protocol on Transplantation of Organs*. Both documents consider LOD. The *Convention on Human Rights and Biomedicine* states that LOD "may be carried out solely for the therapeutic benefit of the recipient."[15] Furthermore, it lays down that LOD is subsidiary to post-mortem organ donation and alternative therapies.[16] It also explicitly requires consent to be "given expressly and specifically either in written form or before an official body."[17] The *Convention on Human Rights and Biomedicine* especially protects persons who are not able to give consent to the organ removal. People who are not able to give valid consent are not allowed to perform a LOD. Exceptions are only provided for regenerative tissue.[18] The *Additional Protocol on Transplantation of Organs* contains even more rules about LOD. It addresses the donor-recipient relationship, the evaluation of potential risks for the donor, the disclosure of the donor, consent of the donor and the protection of persons who are not able to consent to a LOD.[19] Neither the *Convention on Human Rights and Biomedicine* nor the *Additional Protocol on the Transplantation of Organs* is automatically in force in the Member States of the European Council. Both only become effective in Member States after they are ratified, because the structure of the Council of Europe is intergovernmental.[20] Consequently, all conventions passed by the Council of Europe are not directly binding in its Member States, but each Member State has to agree to each convention to enforce it.[21] Since the Council of Europe has no competence to pass binding legislation without the states concerned explicitly agreeing to it, it cannot have competences to harmonise legislation regarding LOD.

[14] Cf. Resolution (78) 29 on Harmonisation of Legislations of Member States Relating to Removal, Grafting and Transplantation of Human Substances.

[15] Art. 19, 1. Convention for the Protection of Human Rights and Dignity of the Human Being with regard to the Application of Biology and Medicine.

[16] Ibid.

[17] Ibid., Art. 19, 2.

[18] Ibid., Art. 20.

[19] Chapter 6 Additional Protocol on the Convention on Human Rights and Biomedicine concerning Transplantation of Organs and Tissues of Human Origin.

[20] Haratsch/König/Pechstein, p. 18 (2010); Hendriks, in Weimar/Bos/Busschbach (ed.) (2011), p. 39.

[21] Haratsch/König/Pechstein (2010), p. 18.

3) European Union

The European Union is, in contrast to the Council of Europe, supranational.[22] That means that decisions are directly binding for its Member States.[23] The supranational legislation is not dependant on the national legislator.[24] The European Union established a procedure which makes decisions binding for the Member States without requiring their approval of a particular decision.[25] Hence, the European Union has the competence to directly bind the Member States in defined areas. Therefore, and because this thesis focuses on the Member States of the European Union, the general competences of the European Union will be analysed and will be applied to LOD.

II. General Competences of the European Union to Make Law

Does the European Union have the competence to harmonise the legislation regarding LOD? To evaluate whether this is the case, an overview of the general competences of the European Union will first be given. The European Union is not a state,[26] but an international organisation with legal personality based on a treaty between the Member States.[27] Therefore, the Member States have the authority to decide which competences they confer to the European Union,[28] and the European Union does not have a general competence to make law.[29]

The Member States only conferred certain competences to the European Union. Consequently, the competences of the Member States and those of the European Union must be distinguished.[30] (1) This will be done by presenting the relevant, general principles. (2) In case the European Union is responsible for a certain subject matter, the details of the competences are not always similar. Therefore, the different types of competences will be described. (3) The competences make it possible for the European Union to pass different legal acts. Those will be explained last.

[22] Case 26/62 Van Gend en Loos v. Nederlandse Administratie der Belastingen [1963] ECR 1; Haratsch/König/Pechstein (2010), p. 28; Hendriks, in Weimar/Bos/Busschbach (ed.) (2011), p. 36; Herdegen (2011), p. 70; Oppermann/Classen/Nettesheim (2011), p. 60.

[23] Haratsch/König/Pechstein (2010), p. 28.

[24] Constantinesco (1972), p. 401; cf. Ehlers, in Großfeld et al. (ed.) (2006), p. 45.

[25] Haratsch/König/Pechstein (2010), p. 28. The direct application of European Union law in the Member States of the European Union has to be distinguished from directly being in force (Ehlers, Vol. 33 JURA 187, 187 (2011)).

[26] Nowak (2011), p. 117.

[27] Hobe (2011), p. 31 f.

[28] Herdegen (2011), p. 71 & 176; Moussis (2011).

[29] Nowak (2011), p. 117.

[30] Cf. Nowak (2011), p. 116; cf. Schmidt am Busch (2007), p. 2.

1) General Principles

The competences of the Member States and those of the European Union must be distinguished, Art. 5 *TEU*.[31] This rule contains (1) the principle of conferral, (2) the principle of subsidiarity and (3) the principle of proportionality. (4) In addition, the priority application of European Union Law has to be considered.

a) Principle of Conferral

Art. 5 *TEU* states that "[t]he limits of Union competences are governed by the principal of conferral." The Member States have to decide on the aspects that, according to their view, can be regulated in a more efficient manner by the European Union. They may then confer certain sovereign powers to the European Union. The European Union can only act based on the competences conferred upon it by its Member States,[32] and "competences not conferred upon the Union in the Treaties remain with the Member States."[33] The main purpose of the principle of conferral is to divide the competences between the European Union and the Member States.[34] Furthermore, it limits the competences of the European Union to a certain extent.[35]

In addition to the relation between the European Union and its Member States, the horizontal relationship between the organs of the European Union is noteworthy.[36] Art. 13, 2. *TEU* stipulates that "[e]ach institution shall act within the limits of the powers conferred to it in the Treaties." Therefore, the European Union cannot choose which of its organs acts; it is prescribed.

b) Principle of Subsidiarity

According to the principle of subsidiarity, "[I]n areas which do not fall within its exclusive competence, the Union shall act only if and insofar as the objectives of the proposed action cannot be sufficiently achieved by the Member States, either at central level or at regional and local level, but can rather, by reason of the scale or the effects of the proposed action, be better achieved at Union level." If an area does

[31] Calliess (2010), p. 186.

[32] Moussis (2011).

[33] Art. 4 Treaty on European Union.

[34] Biervert, in Schwarze (ed.) (2009), Art. 249 at 11.

[35] Cf. Oppermann/Classen/Nettesheim (2011), p. 61.

[36] Cf. Nowak (2011), p. 118.

not fall within the exclusive competence of the European Union, it is only necessary for the Union to act if the objectives of the action cannot be sufficiently achieved by the Member States. A second requirement for the Union acting is that the aim must be better achieved at European Union level. Those two requirements have to be fulfilled cumulatively.[37] In order to analyse whether an action of the European Union is in accordance with the principle of subsidiarity, it has to be regarded whether the subject matter concerned contains transnational aspects and whether an acting of the European Union would be significantly advantageous.[38]

c) Principle of Proportionality

The principle of proportionality stipulates that actions of the European Union "shall not exceed what is necessary to achieve the objectives of the Treaties." This principle does not address if the European Union may act at all, but *how* the European Union may act.[39] The European Union has a wide scope with regard to acting, because it has to face rather complex decisions concerning political, economical and social aspects while making law.[40]

d) Priority Application of European Union Law

Although it is not explicitly stated in the founding treaties of the European Union, the ECJ has clarified that the application of European Union Law has priority over the application of the national laws of the Member States.[41] The national law of a Member State may not be applied in case of a conflict with (primary or secondary) European Union Law.[42] However, the national law does not lose its legal force.[43] Granting priority application to the European Union Law guarantees a uniform application of it[44] and ensures the functioning of the European Union,[45] because the

[37] Geiger, in Geiger/Khan/Kotzur (2010), p. 35; Norba (2009), p. 338; Ludwigs (2004), p. 131.

[38] Norba (2009), p. 339.

[39] Ibid.

[40] Arndt/Fischer/Fetzer (2010), p. 59.

[41] Ehlers, Vol. 23 JURA 266, 267 f. (2001); Nowak (2011), p. 130; Schroeder, in Streinz (ed.) (2012), p. 2168; Terhechte, in Terhechte (ed.) (2011), p. 61 f.

[42] Biervert, in Schwarze (ed.) (2009), Art. 249 at. 6; Ehlers, Vol. 23 JURA 266, 268 (2001); Nowak (2011), p. 130.

[43] Biervert, in Schwarze (ed.) (2009), Art. 249 at 6; Ehlers, Vol. 33 JURA 187, 190 (2011); Terhechte, in Terhechte (ed.) (2011), p. 62.

[44] Nowak (2011), p. 130.

[45] Ehlers, Vol. 33 JURA 187, 187 (2011); Nowak (2011), p. 130.

supranational European Union would not be able to act effectively if the Member States themselves could choose whether they accept the European Union law or not.[46]

The relationship between the competences of the European Union and those that remain with the Member States is mainly based on the four principles just introduced.

2) *Types of Competences*

After the European Union Law was reformed by the *Treaty of Lisbon*, the distinction between the different types of competences of the European Union in Art. 2–6 *TFEU* became more apparent.[47] The relevant competences are (1) the exclusive competence of the European Union, (2) the shared competence between the European Union and the Member States and (3) the competence of the European Union to support, coordinate or supplement the Member States.[48]

a) Exclusive Competence of the European Union

Certain subject matters cannot be regulated by the Member States.[49] In those cases, "only the Union may legislate and adopt legally binding acts, the Member States being able to do so only if so empowered by the Union or for the implementation of Union acts."[50] This means, for example, that if the German legislator passes a law in the area of an exclusive competence of the European Union without being authorised to do so, the law is void because the German legislator has no competence to make law in this case.[51] The principle of subsidiarity does not apply if the European Union has the exclusive competence.[52] Art. 3 *TFEU* contains a concluding list of the areas in which the European Union has the exclusive competence.[53]

---

[46] Ehlers, Vol. 33 JURA 187, 187 (2011).

[47] Schwarze, in Schwarze/Hatje (ed.) (2009), p. 16.

[48] Cf. Nowak (2011), p. 119; cf. Wollenschläger, in Müller-Graff/Schmahl/Skouris (ed.) (2011), p. 449.

[49] Cf. Nowak (2011), p. 119.

[50] Art. 2 Treaty on the Functioning of the European Union.

[51] Ehlers, Vol. 33 JURA 187, 191 (2011).

[52] Art. 5 III Treaty on European Union.

[53] Calliess (2010), p. 189.

b) Shared Competence

Art. 4 I *TFEU* lays down that "[t]he Union shall share competence with the Member States where the Treaties confer on it a competence which does not relate to the areas referred to in Article 3 and 6." The main areas of that competence are listed in Art. 4 II *TFEU*. The list contains, inter alia, the internal market, consumer protection and common safety concerns in public health matters for the aspects defined in this *Treaty*.

The legal consequences of a shared competence are set down in Art. 2 I *TFEU*. The European Union as well as the Member States may regulate the particular areas, but they are not exactly ranked equally. The European Union takes precedence over the Member States, because the Member States can only act insofar as the European Union has not exercised competence or has ceased exercising its competence. If the European Union has regulated a certain area, the scope of this regulation must be defined, because it determines the areas in which the Member States must not act anymore. In response, the Member States clarified that only those aspects actually governed by an act of the European Union are covered and not the whole area.[54]

Additional areas of shared competence are listed in Art. 4 III *TFEU*, namely the "areas of research, technological development and space." In contrast to the areas enumerated in Art. 4 II *TFEU*, a regulation of the areas listed in Art. 4 III *TFEU* does not prevent the Member States from acting. This also applies to the "areas of development cooperation and humanitarian aid,"[55] named in Art. 4 IV *TFEU*.[56]

c) Competence of the European Union to Support, Coordinate or Supplement

Art. 6 *TFEU* contains a list of areas in which the European Union has the competence to support, coordinate or supplement acts of the Member States. This applies, e.g., to protection and improvement of human health. Actions of the European Union based on that provision "shall not entail harmonisation of Member States' laws or regulations."[57] It, however, entitles the European Union to pass legally binding acts.[58]

[54] European Union, Vol. 53 Official Journal of the European Union 1, 307 (2010).

[55] Art. 4, 4. Treaty on the Functioning of the European Union.

[56] Cf. Nowak (2011), p. 121 f.

[57] Art. 2, 5. Treaty on the Functioning of the European Union.

[58] Calliess (2010), p. 190.

3) *Types of Legal Acts of the European Union*

Art. 288 *TFEU* contains a part of the sources of European Union law, not all though. All sources of law together make up the European Union Law.[59] The different sources of European Union Law can be roughly divided into (1) primary legislation and (2) secondary legislation.[60]

a) Primary Legislation

Primary legislation comprises written and unwritten sources of law.[61] The (written) founding treaties are first and foremost the primary legislation in the European Union,[62] namely the *TEU*, the *TFEU* and the *Treaty establishing the European Atomic Energy Community* (*EAC*).[63] They were created by the Member States, partly in cooperation with the organs of the European Union.[64] They oblige all Member States.[65] In exceptional cases, they also directly establish rights and duties for individuals.[66]

Unwritten sources can also be primary legislation.[67] They contain general principles of law, customary law and European agreements relating to international law.[68] The unwritten sources have not addressed LOD so far and will hence be left out of consideration.

The primary legislation has a higher rank than the secondary legislation.[69] An authorisation in the primary legislation is necessary for the European Union to pass secondary legislation.[70] Thus, while judging the lawfulness of secondary legislation, primary legislation is the decisive measure.[71]

---

[59] Schroeder, in Streinz (ed.) (2012), p. 2158.

[60] Cf. Borchardt (2010), at 47; Ehlers, in Großfeld et al. (ed.) (2006), p. 45; Lenz/Erhard, in Lenz (ed.) (1994), p. 55; Streinz (2012), at 2.

[61] Cf. Ehlers, Vol. 23 JURA 266, 268 (2001); Lenz/Erhard, in Lenz (ed.) (1994), p. 57.

[62] Haratsch/Koenig/Pechstein (2010), p. 161; Lenz/Erhard, in Lenz (ed.) (1994), p. 57; Ruffert, in Calliess/Ruffert (ed.) (2011), Art. 288 at 8 AEUV; Streinz (2012), at 446 f.

[63] Cf. Borchardt (2010), at 76; cf. Hobe (2011), p. 88 f. cf. Lenz/Erhard, in Lenz (ed.) (1994), p. 57.

[64] Haratsch/Koenig/Pechstein (2010), p. 160.

[65] Hobe (2011), p. 87; Streinz (2012), at 448.

[66] Haratsch/Koenig/Pechstein (2010), p. 161 ff. Hobe (2011), p. 87; cf. Streinz (2012), at 448.

[67] Arndt/Fischer/Fetzer (2010), p. 43 f.; Haratsch/Koenig/Pechstein (2010), p. 161; Streinz (2012), at 453.

[68] Hobe (2011), p. 87.

[69] Haratsch/Koenig/Pechstein (2010), p. 160; Schroeder, in Streinz (ed.) (2012), p. 2161.

[70] Hobe (2011), p. 89; cf. Schroeder, in Streinz (ed.) (2012), p. 2157.

[71] Haratsch/Koenig/Pechstein (2010), p. 160; Ruffert, in Calliess/Ruffert (ed.) (2011), Art. 288 at 8 AEUV.

b) Secondary Legislation

Art. 288 *TFEU* lists secondary legislation,[72] but this enumeration is not concluding.[73]

The organs of the European Union are usually free in deciding what legal act they use.[74] While exercising this freedom, the responsible organ has to act after a due assessment of the circumstances and has to adhere to the principle of proportionality.[75] In exceptional cases, however, the treaties prescribe the kind of legal act the European Union has to use.[76]

The legal acts listed in Art. 288 *TFEU* will be introduced in the following. Within the secondary legislation, the binding legal acts, namely (1) regulations, (2) directives and (3) decisions in contrast to (4) recommendations and opinions, which are not binding, must be distinguished. (5) In addition, "incentive measures" will be explained.

aa) Regulation

According to Art. 288 *TFEU,* a regulation has general application, is entirely binding and directly applicable in all Member States. General application means that it applies to an undetermined amount of cases, and thus has the quality of a legal rule.[77] It is "binding" in such a way that the rights and duties stipulated in the regulation are mandatory for its addressees,[78] and it is "directly applicable" insofar as the regulation is effective in all Member States without any national act of transposition.[79] Consequently, the addressees are not only the Member States and the organs of the European Union, but the citizens of the European Union as well.[80]

---

[72] Arndt/Fischer/Fetzer (2010), p. 44; Borchardt (2010), at 81; Haratsch/Koenig/Pechstein (2010), p. 160; Hobe (2011), p. 91.

[73] Borchardt (2010), at 81; Ruffert, in Calliess/Ruffert (ed.) (2011), Art. 288 at 98 AEUV; Schroeder, in Streinz (ed.) (2012), p. 2155.

[74] Hobe (2011), p. 91.

[75] Biervert, in Schwarze (ed.) (2009), Art. 249 at 12.

[76] Hobe (2011), p. 91; Schroeder, in Streinz (ed.) (2012), p. 2157.

[77] Streinz (2012), at 468.

[78] Biervert, in Schwarze (ed.) (2009), Art. 249 at 19.

[79] von Arnauld, in Riesenhuber/Takayama (ed.) (2006), p. 251; Haratsch/Koenig/Pechstein (2010), p. 160; Hobe (2011), p. 92; Ruffert, in Calliess/Ruffert (ed.) (2011), Art. 288 at 20 AEUV; Streinz (2012), at 470.

[80] Arndt/Fischer/Fetzer (2010), p. 45; Biervert, in Schwarze (ed.) (2009), Art. 249 at 21.

bb) Directive

According to Art. 288 *TFEU*, “A directive shall be binding, as to the result to be achieved, upon each Member State to which it is addressed, but shall leave to the national authorities the choice of form and methods.” Unlike the regulation which is entirely binding, the directive is only binding with regard to its aim.[81] Another difference between the directive and the regulation is that the directive is not directly effective in the Member States, but has to be put into national legislation.[82] This transposition has to take place entirely and within the stipulated period of time.[83] The character of the directive can be described as a compromise between demanding uniform law in the European Union on the one hand and keeping national characteristics on the other hand. Consequently, the European Union usually has the authority to pass a directive with regard to subject matters where a harmonisation of the national laws is intended, not a unification, though.[84]

cc) Decision

A decision is binding, but “[a] decision which specifies those to whom it is addressed shall be binding only on them.”[85] Decisions, thus, cause legal effects.[86] Before the *Consolidated Version of the TFEU* was passed, the decision always only applied to a particular case.[87] The addressee could be a Member State or an individual[88] and was either named explicitly or it was at least possible to individualise the addressee.[89] The new version contains a dichotomy in European decisions — there is no addressee and it is not directed toward a certain addressee.[90] In case the decision is directed to a certain addressee, though, this can still be either a Member State or an individual.[91]

[81] Haratsch/Koenig/Pechstein (2010), p. 167; Hobe (2011), p. 92; Streinz (2012), at 474.

[82] Streinz (2012), at 476.

[83] Haratsch/Koenig/Pechstein (2010), p. 168; Hobe (2011), p. 93.

[84] Biervert, in Schwarze (ed.) (2009), Art. 249 at 22; Streinz (2012), at 475.

[85] Art. 288 Treaty on the Functioning of the European Union.

[86] Ruffert, in Calliess/Ruffert (ed.) (2011), Art. 288 at 86–90 AEUV.

[87] Arndt/Fischer/Fetzer (2010), p. 54; Haratsch/Koenig/Pechstein (2010), p. 177; Ruffert, in Calliess/Ruffert (ed.) (2011), Art. 288 at 87AEUV.

[88] Arndt/Fischer/Fetzer (2010), p. 54; Bievert, in Schwarze (ed.) (2009), Art. 249 at 31; Haratsch/Koenig/Pechstein (2010), p. 177; Hobe (2011), p. 95.

[89] Bievert, in Schwarze (ed.) (2009), Art. 249 at 31; Haratsch/Koenig/Pechstein (2010), p. 177.

[90] Schroeder, in Streinz (ed.) (2012), Art. 288 at 133 ff. AEUV.

[91] Ruffert, in Calliess/Ruffert (ed.) (2011), Art. 288 at 90 ff. AEUV.

dd) Recommendations and Opinions

Recommendations and opinions have, in contrast to regulations, directives and decisions, no binding force.[92] They, nevertheless, have political[93] and legal effects[94] and the addressees usually comply with them.[95] The intention of these legal acts is to suggest a certain behaviour to the addressee without binding him.[96] The addressee can be a Member State, but also an individual.[97] A recommendation can be described as a statement of an organ of the European Union due to its own initiative, while an opinion exists if an organ acts because of the initiative of another organ.[98]

ee) Incentive Measures

Art. 288 *TFEU* explains several legal acts of the European Union, but does not contain an explanation of "incentive measures." This term is frequently used in the treaties, but is nowhere defined.[99] It is described as being an unlabelled or atypical action.[100] They can be roughly defined as actions by the European Union that either supplement measures of the Member States concerning the content or support those.[101] Incentive measures are usually used if the European Union does not have the explicit competence to regulate a certain subject matter and does not intend to harmonise the national laws of the Member States.[102] It usually allows the use of all legal acts listed in Art. 288 *TFEU*,[103] but it is noted that they need to be non-binding if they concern contents, while they may have binding character if they regard organisational aspects.[104] In case they are supposed to be non-binding, they are usually carried out with actions plans.[105]

---

[92] Art. 288 Treaty on the Functioning of the European Union.

[93] Arndt/Fischer/Fetzer (2010), p. 54; Streinz (2012), at 517.

[94] European Court of Justice [Europäischer Gerichtshof], 13 December, 1989, Case C-322/88 at 17 f.; Stefan, Vol. 14 European Law Journal 753, 753 (2008) (only with regard to recommendations).

[95] Arndt/Fischer/Fetzer (2010), p. 54.

[96] Haratsch/Koenig/Pechstein (2010), p. 179; Streinz (2012), at 518.

[97] Bievert, in Schwarze (ed.) (2009), Art. 249 at 35; Haratsch/Koenig/Pechstein (2010), p. 179.

[98] Hobe (2011), p. 96.

[99] Bonde (2003).

[100] Cf. Berg, in Schwarze (ed.) (2009), Art. 152 EGV at 32 with respect to the former regulation; Kingreen, in Calliess/Ruffert (ed.) (2011), Art. 168 at 15; Lurger, in Streinz (ed.) (2012), Art. 168 at 44.

[101] Norba (2009), p. 343 f.

[102] Bonde (2003).

[103] Cf. Berg, in Schwarze (ed.) (2009), Art. 152 EGV at 32 with respect to the former regulation; Lurger, in Streinz (ed.) (2012), Art. 168 at 44.

[104] Cf. Berg, in Schwarze (ed.) (2009), Art. 152 EGV at 33 f. with respect to the former regulation; Lurger, in Streinz (ed.) (2012), Art. 168 at 44.

[105] Cf. Lurger, in Streinz (ed.) (2012), Art. 168 at 44; Norba (2009), p. 344.

Hence, the competence of the European Union has to be evaluated on a case-by-case basis. That also applies to the question of *what kind* of competence the European Union has and what kinds of (legal) acts it might pass. It, however, does not only apply to the question of whether the European Union has the competence to act at all.

III. Competences in the Field of Living Organ Donation

The general competences of the European Union have to be applied to the field of LOD. The primary legislation is decisive to evaluate whether the European Union has the competence to pass secondary legislation with regard to LOD. Therefore, the relevant provisions of the primary legislation of the European Union will be discussed in detail.

(1) Art. 168 *TFEU* contains details about the competences of the European Union in the field of public health and can be denominated as the heart of the European Union health care policy.[106] (2) Art. 114 *TFEU* and (3) Art. 9 *TFEU* also address human health. (4) In addition, the *CFREU* will be considered. (5) A conclusion will be drawn.

1) Art. 168 *Treaty on the Functioning of the European Union*

Art. 168 *TFEU* establishes several tasks and competences for the European Union in the field of health policy.[107] The *Treaty of Lisbon* extended the competences of the European Union in the field of health care policy.[108] This does not mean, however, that the competences for the European Union resulting from Art. 168 *TFEU* are all-embracing. In fact, Art. 168 *TFEU* mainly concerns competences for precautionary measures and protective measures with respect to health, but does not affect the public health systems of the Member States.[109] (1) Art. 168 I, (2) Art. 168 IV and (3) Art. 168 V *TFEU* might be relevant for regulating the requirements of LOD.

[106] Schmidt am Busch, in Grabitz/Hilf/Nettesheim (ed.) (2011), Art. 168 at 6.
[107] Herdegen (2011), p. 408; Kingreen, in Calliess/Ruffert (ed.) (2011), Art. 168 at 1 AEUV.
[108] Herdegen (2011), p. 408.
[109] Kingreen, in Calliess/Ruffert (ed.) (2011), Art. 168 at 1 AEUV.

a) Art. 168 I *Treaty on the Functioning of the European Union*

(1) Art. 168 I 1 *TFEU* demands the European Union and the Member States to ensure "[a] high level of human health protection [...] in the definition and implementation of all Union policies and activities." (2) Art. 168 I 2, 3 *TFEU* lists several aims and tasks of the European Union in the field of health policy, but is not concluding.[110] (3) Art. 168 I 4 *TFEU* particularly addresses drugs-related health damage, which is irrelevant for LOD. Art. 168 I 1 *TFEU* and Art. 168 I 2, 3 *TFEU* might be relevant for LOD and will be analyed in the following.

aa) Art. 168 I 1 *Treaty on the Functioning of the European Union*

Art. 168 I 1 *TFEU* demands a high level of human health protection. It neither provides a competence for the commission,[111] nor can be used to evade the prohibition to harmonise stipulated in Art. 168 VII *TFEU*.[112] It, however, creates an obligation to guarantee human health protection when performing other competences.[113] Thus, when pursuing the other aims established in the treaties of the European Union, the European Union should also have the secondary aim to establish a high level of human health protection.[114] Human health protection should not be the main purpose for creating measures other than Art. 168 *TFEU*, though, because this article solely determines the limits for the European Union to make measures that mainly concern health policy.[115]

Hence, Art. 168 I 1 *TFEU* does not provide any competence to the European Union. No regulations can be based on it. Therefore, the article does not, under any circumstance, provide the European Union with a competence to regulate LOD.

bb) Art. 168 I 2, 3 *Treaty on the Functioning of the European Union*

Art. 168 I 2, 3 *TFEU* lists areas of responsibility of the European Union.[116] The list is not concluding,[117] but contains very important tasks of the European Union.[118]

---

[110] Ibid., at 11 AEUV.

[111] Kingreen, in Calliess/Ruffert (ed.) (2011), Art. 168 at 2 AEUV.

[112] Cf. European Court of Justice [Europäischer Gerichtshof], October, 2000, Case C-376/98, at 79; Kingreen, in Calliess/Ruffert (ed.) (2011), Art. 168 at 9 AEUV.

[113] Council of the European Union (2008), p. 2; Kingreen, in Calliess/Ruffert (ed.) (2011), Art. 168 at 2 AEUV; Lurger, in Streinz (ed.) (2012), Art. 168 at 24; Schmidt am Busch, in Grabitz/Hilf/Nettesheim (ed.) (2011), Art. 168 at 84.

[114] Kingreen, in Calliess/Ruffert (ed.) (2011), Art. 168 at 9 AEUV.

[115] Lurger, in Streinz (ed.) (2012), Art. 168 at 26.

[116] Frenz/Götzkes, Vol. 28 Medizinrecht 613, 614 (2010); Kingreen, in Calliess/Ruffert (ed.) (2011), Art. 168 at 2 AEUV; cf. Lurger, in Streinz (ed.) (2012), Art. 168 at 14.

[117] Kingreen, in Calliess/Ruffert (ed.) (2011), Art. 168 at 11 AEUV; Lurger, in Streinz (ed.) (2012), Art. 168 at 14; Schmidt am Busch, in Grabitz/Hilf (ed.) (2011), Art. 168 AEUV at 21.

[118] Lurger, in Streinz (ed.) (2012), Art. 168 at 14.

It explicitly states that the activities of the European Union are intended to improve the human health of the general public.[119]

The general distribution of competences between the European Union and the Member States needs to be considered when working out the competences evolving for the European Union from Art. 168 I 2, 3 *TFEU*. With respect to the protection and improvement of human health, Art. 6 (a) *TFEU*, with reference to Art. 2 V *TFEU*, is relevant.[120] As a consequence, "[t]he Union shall have competence to carry out actions to support, coordinate or supplement the actions of the Member States."[121] A harmonisation of the national legislations in the concerned areas is barred.[122] The European Union can consequently not use Art. 168 I 2, 3 *TFEU* to harmonise the national legislation for LOD.

The provision does not only address the relationship between the European Union and the Member States,[123] but also refers to coordination between the Member States. The European Union should encourage the cooperation between the Member States with regard to the concerned areas. Cooperation is particularly important "to improve the complementarity of their health services in cross-border areas."[124] This might make sense with respect to LOD, but does not provide the European Union with the competence to create the requirements for LOD.

b) Art. 168 IV (a) *Treaty on the Functioning of the European Union*

Art. 168 IV (a) *TFEU* explicitly states that the European Parliament and the Council shall adopt "measures setting high standards of quality and safety of organs [...]."

The preceding provision of Art. 168 IV (a) *TFEU*, namely Art. 152 IV (a) *TEC*, was established because of the scandal in several Member States, especially in France and Germany, when AIDS was transmitted due to contaminated infusions with blood in 1993.[125] The main aim was to strengthen the populations trust in biological substances again.[126] Before Art. 152 *TEC* came into effect, measures on the quality of blood products were based on the general competence resulting from the internal market. By establishing a specific competence, it became clear that those products are not equalised with conventional commodity.[127] The competence

---

[119] Cf. Berg, in Schwarze (ed.) (2009), Art. 152 at 8 with respect to the former regulation.

[120] Kingreen, in Calliess/Ruffert (ed.) (2011), Art. 168 at 4 AEUV.

[121] Art. 6 S. 2 (a) Treaty on the Functioning of the European Union.

[122] Kingreen, in Calliess/Ruffert (ed.) (2011), Art. 168 at 4 AEUV.

[123] Cf. Kingreen, in Calliess/Ruffert (ed.) (2011), Art. 168 at 2 AEUV.

[124] Art. 168, 2. Treaty on the Functioning of the European Union.

[125] Frenz/Götzkes, Vol. 28 Medizinrecht 613, 616 (2010); Kingreen, in Calliess/Ruffert (ed.) (2011), Art. 168 at 19 AEUV; Lurger, in Streinz (ed.) (2012), Art. 168 at 40; Norba (2009), p. 341; Schmidt am Busch (2007), p. 346; cf. Spiegel (1993).

[126] Schmidt am Busch, in Grabitz/Hilf/Nettesheim (ed.) (2011), Art. 168 at 43.

[127] Schmidt am Busch (2007), p. 346.

excludes organs, substances of human origin, blood and blood derivates from the "logic of the internal market."[128]

This is the only provision that explicitly mentions organs and will therefore be discussed in detail. (1) The type of competence that results from Art. 168 IV (a) *TFEU*, (2) the legal acts the European Union may use with respect to Art, 168 IV (a) *TFEU* and (3) the scope of the acting of the European Union will be presented.

aa) Type of Competence: Shared Competence

To determine the exact competences of the European Union that result from Art. 168 IV (a) *TFEU,* one must cross-reference Art. 4, II (k) *TFEU*.[129] This article determines a shared competence between the European Union and the Member States for the area of "common safety concerns in public health matters, for the aspects defined in this Treaty." Consequently, a shared competence exists in the field of Art. 168 IV (a) *TFEU*.[130]

Since the European Union does not have the exclusive competence in the field of health care, the principle of subsidiarity must be applied.[131] To explain measures by the European Union based on Art. 168 IV (a) *TFEU*, their function to supplement must be proven.[132] The European Union shall "act only if and insofar as the objectives of the proposed action cannot be sufficiently achieved by the Member States, either at central level or at regional and local level, but can rather, by reason of the scale or the effects of the proposed action, be better achieved at Union level."

This leads one to question whether a common standard for LOD would be useful in the countries considered. We have already seen how the differences within national transplant laws are quite problematic. Hence, the creation of equal rules for LOD in the countries seems to make sense. It cannot be expected, though, that the countries will create bilateral contracts for LOD. A harmonisation by the European Union seems to reach this goal more sufficiently. The European Union may thus act in the field of LOD and has already done so by passing the *Directive 2010/53/EU*.[133]

---

[128] Hendriks, in Weimar/Bos/Busschbach (ed.) (2011), p. 36; Norba (2009), p. 340; Schmidt am Busch (2007), p. 346.

[129] Nowak (2011), p. 120 f.

[130] Lurger, in Streinz (ed.) (2012), Art. 168 at 2.

[131] Cf. ibid., at 38.

[132] Ibid., at 39.

[133] Lurger, in Streinz (ed.) (2012), Art. 168 at 39; Schmidt am Busch, in Grabitz/Hilf/Nettesheim (ed.) (2011), Art. 168 at 48.

bb) Legal Acts the European Union May Use

What legal acts may the European Union pass to set "high standards of quality and safety of organs?" Art. 168 IV (a) *TFEU* only refers to "measures." However, legal acts that are based on Art. 168 IV (a) *TFEU* need to be passed in accordance with the "ordinary legislative procedure." Art. 289 I *TFEU* states that this procedure is only possible for regulations, directives or decisions.[134]

In addition, according to Art. 168 VI *TFEU* the Council may also adopt recommendations.[135]

No additional guidelines are given. The European Union can thus choose the exact legal act among those that have just been listed. The most suitable legal act must be freely decided after a due assessment of the circumstances, and the principle of proportionality has to be taken into account.[136]

cc) Scope of the Competence Resulting from Art. 168 IV (a) *Treaty on the Functioning of the European Union*

Art. 168 IV (a) *TFEU* vests the European Union with a strong competence to make law in the field of public health.[137] It enables the European Union to pursue a health policy which is independent from the health policies of the Member States.[138] As just seen, the term "measure" even makes it possible to pass regulations and directives.[139]

Art. 168 IV (a) *TFEU*, however, only contains part of the competences connected to the issue of LOD.[140] For everything that is not included, the European Union only has a competence to support, coordinate or supplement the actions of the Member States or it has the weaker competences of Art. 168 II *TFEU* and Art. 168 V *TFEU*.[141]

To determine the exact competences that result from Art. 168 IV (a) *TFEU* it is not sufficient to solely examine this paragraph. Art. 168 VII *TFEU* restricts the competences of the European Union,[142] and is also relevant for the competences

---

[134] Cf. Schmidt am Busch, in Grabitz/Hilf/Nettesheim (ed.) (2011), Art. 168 at 41.

[135] Cf. Frenz/Götzkes, Vol. 28 Medizinrecht 613, 616 (2010); cf. Wollenschläger, in Müller-Graff/Schmahl/Skouris (ed.) (2011), p. 454.

[136] Berg, in Schwarze (ed.) (2009), p. 1437.

[137] Lurger, in Streinz (ed.) (2012), Art. 168 at 39.

[138] Frenz/Götzkes, Vol. 28 Medizinrecht 613, 614 (2010).

[139] Cf. Lurger, in Streinz (ed.) (2012), Art. 168 at 39.

[140] Wollenschläger, in Müller-Graff/Schmahl/Skouris (ed.) (2011), p. 455.

[141] Cf. Lurger, in Streinz (ed.) (2012), Art. 168 at 1; Wollenschläger, in Müller-Graff/Schmahl/Skouris (ed.) (2011), p. 455.

[142] Schmidt am Busch, in Grabitz/Hilf/Nettesheim (ed.) (2011), Art. 168 at 10.

that result from Art. 168 IV (a) *TFEU*.[143] Consequently, (1) Art. 168 IV TFEU assigns competences to the European Union, while (2) Art. 168 VII TFEU sets limits.[144] (3) An overlap between Art. 168 IV (a) *TFEU* and Art. 168 VII *TFEU* is possible. (4) An interim result will be drawn last.

(1) Positive: Possibilities That Result from Art. 168 IV (a) *Treaty on the Functioning of the European Union*

Art. 168 IV (a) *TFEU* authorises the European Union to adopt "measures setting high standards of quality and safety of organs." Even if the European Union actually acts, this does not prevent the Member States from keeping their stricter rules or from implementing stricter regulations afterwards.[145] Rules passed by the European Union thus create a minimum standard.[146]

The European Union is granted with the competence to adopt legal rules with regard to the safety and quality of organ donation.[147] This includes determining the criteria for the constitution of organs,[148] for example, the criterion that an organ must be free of disease-causing agents.[149] It also includes the organ removal, laboratory tests and storage.[150] Furthermore, disclosure about the optimal usage of biological substances and the determination of basic criteria for the inspection and the instruction of the inspectors are included.[151] The European Union may also determine that LOD must be nonpaid.[152]

[143] Cf. Lurger, in Streinz (ed.) (2012), Art. 168 at 21.

[144] Cf. Schmidt am Busch, in Grabitz/Hilf/Nettesheim (ed.) (2011), Art. 168 at 10.

[145] Art. 168, IV (a) Treaty on the Functioning of the European Union.

[146] Schmidt am Busch, in Grabitz/Hilf/Nettesheim (ed.) (2011), Art. 168 at 46; Weatherill (2010), p. 584. With respect to that, Art. 168 Treaty on the Functioning of the European Union can even be considered as a downgrading of the competences of the European Union, because it only grants the European Union to establish minimum standards, while Art. 95 Treaty on European Union that had been applied before Art. 168 Treaty on the Functioning of the European Union came into force only allowed to depart from the measures based on Art. 95 Treaty on European Union under very tight circumstances (Schmidt am Busch, in Grabitz/Hilf/Nettesheim (ed.) (2011), Art. 168 at 47).

[147] Kingreen, in Calliess/Ruffert (ed.) (2011), Art. 168 at 18 AEUV; Lurger, in Streinz (ed.) (2012), Art. 168 at 21; Schmidt am Busch, in Grabitz/Hilf/Nettesheim (ed.) (2011), Art. 168 at 45.

[148] Berg, in Schwarze (ed.) (2009), p. 1437; Schmidt am Busch, in Grabitz/Hilf/Nettesheim (ed.) (2011), Art. 168 at 45.

[149] Wollenschläger, in Müller-Graff/Schmahl/Skouris (ed.) (2011), p. 455.

[150] Cf. Berg, in Schwarze (ed.) (2009), Art. 152 at 26 with respect to the former regulation; Lurger, in Streinz (ed.) (2012), Art. 168 at 21; Schmidt am Busch, in Grabitz/Hilf/Nettesheim (ed.) (2011), Art. 168 at 6.

[151] Schmidt am Busch, in Grabitz/Hilf/Nettesheim (ed.) (2011), Art. 168 at 45.

[152] Ibid.

(2) Negative: Limits Established by Art. 168 VII *Treaty on the Functioning of the European Union*

Art. 168 VII *TFEU* clarifies that the European Union is not authorised to "affect national provisions on the donation or medical use of organs and blood." The Member States regulate all issues of the organ donation that are not relevant for its quality.[153]

The competence of the European Union does not include the use of organs and the regulation of patients' rights.[154] The use of organs includes the allocation of organs, which (indirectly) also includes the selection of the recipient.[155] The European Union is consequently not competent to regulate what organs may be donated and what organs may not be donated.[156] The European Union is also not granted with the competence to regulate the requirements concerning the patient's consent[157] and other requirements regarding the legitimacy of LOD.[158] Regulations with respect to basic ethical principles of organ donation and the protection of the donor are also not covered by the competences of the European Union.[159]

(3) Overlap of Art. 168 IV (a) *Treaty on the Functioning of the European Union* and Art. 168 VII *Treaty on the Functioning of the European Union*

Apart from the (positive) scope and the (negative) limits of the provisions that have just been presented, the competences that actually result from Art. 168 IV (a) *TFEU* for the European Union are not always as clear.[160] Art. 168 VII *TFEU* restricts the scope of Art. 168 IV (a) *TFEU*, but the extent of the restriction is only partly clear, making it necessary to interpret the provision.[161] In particular, the competence of the European Union concerning quality and safety measures for organs[162] and the reservation that the actions by the European Union may not affect provisions by the Member States on the donation or medical use of organs[163] may overlap.[164]

---

153 Lurger, in Streinz (ed.) (2012), Art. 168 at 21.

154 Cf. Berg, in Schwarze (ed.) (2009), Art. 152 at 26 with respect to the former regulation; Frenz/Götzkes, Vol. 28 Medizinrecht 613, 615 (2010); Lurger, in Streinz (ed.) (2012), Art. 168 at 21.

155 Norba (2009), p. 341.

156 Ibid., p. 342.

157 Wollenschläger, in Müller-Graff/Schmahl/Skouris (ed.) (2011), p. 456.

158 Gutmann/Schroth (2002), p. 43 with regard to the former Art. 152 IV.

159 Schmidt am Busch, in Grabitz/Hilf/Nettesheim (ed.) (2011), Art. 168 at 45.

160 Cf. ibid.

161 Cf. Norba (2009), p. 340 ff.

162 Art. 168 IV a) Treaty on the Functioning of the European Union.

163 Ibid., Art. 168 VII.

164 Cf. Norba (2009), p. 341; Wollenschläger, in Müller-Graff/Schmahl/Skouris (ed.) (2011), p. 455.

Requirements to donate an organ might also be relevant with regard to the quality and safety of organs.[165]

The initial purpose of this provision must first be discussed before the overlap can be considered. The provision initially focussed primarily on screening-tests and criteria to evaluate the quality of organs and human substances.[166] In particular, protection against communicable diseases was addressed.[167] The European Union has only aimed at ensuring high standards of quality and safety of organs, not at establishing requirements for LOD. Therefore, it only has the competence to ensure such high standards, not to pass laws that would regulate the requirements for LOD. The European Union must try to avoid implementing rules to ensure high standards, but which also simultaneously stipulate requirements for LOD. If that is not possible and a regulation that intends to ensure high standards concurrently regulates a requirement for LOD, such action must take place in a very cautious manner. The same result can be achieved when interpreting the wording of Art. 168 IV (a) *TFEU* and Art. 168 VII *TFEU*. Art. 168 IV (a) *TFEU* explicitly mentions "high standards of quality and safety of organs." It does not, however, provide the European Union with the competence to harmonise the requirements for LOD. Art. 168 VII *TFEU* even explicitly prohibits such harmonisation. The phrasing of the regulations must, consequently, concern the quality and safety of organs and not harmonise the requirements of LOD. If regulating the quality and safety of organs affects the requirements of LOD, then the affects should be as minimal as possible.

c) Art. 168 V *Treaty on the Functioning of the European Union*

Art. 168 V *TFEU* contains the competence to "adopt incentive measures designed to protect and improve human health." Severe organ shortage is a problem in all countries considered. The life quality of people that suffer from organ failure, but do not receive a healing donor organ, is impaired. The organ shortage even causes death every single day. A measure that aims at improving this situation could definitely be considered as protecting and improving human health. If it would actually be able to increase the amount of donor organs, it would most likely be able to improve the quality of life of many patients and would even save lives.

However, apart from Art. 168 IV (a) *TFEU*, the other paragraphs of Art. 168 *TFEU* only provide the European Union with the competence to support, coordinate or supplement actions of the Member States (Art. 6 (a) *TFEU*).[168] According to Art. 2 V *TFEU*, measures adopted on the basis of this competence may not harmonise

[165] Wollenschläger, in Müller-Graff/Schmahl/Skouris (ed.) (2011), p. 455.

[166] Norba (2009), p. 341; cf. Schmidt am Busch (2007), p. 347.

[167] Norba (2009), p. 341.

[168] Kingreen, in Calliess/Ruffert (ed.) (2011), Art. 168 at 14 AEUV; Lurger, in Streinz (ed.) (2012), Art. 168 at 43.

the laws or the regulations of the Member States.[169] The European Union can hence base actions for LOD on Art. 168 V *TFEU*, but cannot harmonise the national laws for LOD under Art. 168 V *TFEU*.[170]

Art. 168 V *TFEU* does not prohibit a harmonising measure, passed on the basis of another rule of the *TFEU*, from having an impact on human health protection.[171] The judgement of the ECJ on Articles 3 and 4 of *Directive 2003/33/EC of the European Parliament and of the Council of 26 May 2003 on the approximation of laws, regulations and administrative provisions of the Member States relating to the advertising and sponsorship of tobacco products* affirms this. *Directive 2003/33/EC* aims at regulating "advertising and sponsorship in respect of tobacco products in media other than television."[172] Germany brought a legal action against this *Directive*. It claims that the Court should annul Art. 3 and Art. 4 of it,[173] because Art. 95 *EC*, which aims at the approximation of laws (after the *Treaty of Lisbon*, basically the same content can be found in Art. 114 *TFEU*),[174] is no suitable legal basis for the *Directive*.[175] It furthermore stated that this *Directive* breaches Art. 152 IV (c) *EC* (That is the basis for Art. 168 V *TFEU* which even goes beyond Art. 152 IV c *EC*[176]).[177] With respect to basing this *Directive* on Art. 95 *EC*, the Court determined that the national rules on the advertisement and sponsorship for tobacco products differed, justifying an intervention by the Community legislature.[178] The Court found that, "those differences also meant that there was an appreciable risk of distortions of competition."[179] This justified the establishment of this *Directive* based on Art. 95 *EC*.[180] But are Art. 3 and Art. 4 actually "designed to eliminate or prevent obstacles to the free movement of goods or the freedom to provide services or to remove distortions of competition?"[181] The Court concluded

---

[169] Cf. Kingreen, in Calliess/Ruffert (ed.) (2011), Art. 168 at 14 AEUV; Lurger, in Streinz (ed.) (2012), Art. 168 at 43.

[170] Cf. European Court of Justice [Europäischer Gerichtshof], 5 October, 2000, Case C-376/98, at 79, in Vol. 41 Zeitschrift für Wirtschaftsrecht 1848, 1853 (2000); Weatherill (2010), p. 39.

[171] Cf. European Court of Justice [Europäischer Gerichtshof], 5 October, 2000, Case C-376/98, at 88, in Vol. 41 Zeitschrift für Wirtschaftsrecht 1848, 1854 (2000).

[172] European Court of Justice [Europäischer Gerichtshof], 12 December, 2006, Case C-380/03, at 3.

[173] Ibid., at 12.

[174] Kahl, in Calliess/Ruffert (ed.) (2011), Art. 114 at 4.

[175] European Court of Justice [Europäischer Gerichtshof], 12 December, 2006, Case C-380/03, at 16.

[176] Kingreen, in Calliess/Ruffert (ed.) (2011), Art. 168 at 14; Lurger, in Streinz (ed.) (2012), Art. 168, at. 9.

[177] European Court of Justice [Europäischer Gerichtshof], 12 December, 2006, Case C-380/03, at 16.

[178] Ibid., at 51.

[179] Ibid., at 66.

[180] Ibid., at 68.

[181] Ibid., at 69.

that this *Directive* actually aims at the improvement of the conditions for a functioning internal market, meaning that Art. 95 *EC* is an adequate legal basis.[182] The second claim that this *Directive* violates Art. 152 IV c *EC* remains. The applicant doubts that the real purpose of Art. 3 and Art. 4 of this *Directive* is to improve the functioning of the internal market, causing it to be contrary to the prohibition to harmonise the national laws and regulations in the field of public health according to Art. 152 IV c *EC*.[183] The Court replied that the prohibition of Art. 152 IV c *EC* does not mean that the harmonising measures, which are based on a different article of the *EC*, may not impact the protection of public health in any way.[184] This, for example, also applies to the *Regulation (EC) No 1024/2006 of the European Parliament and of the Council of 20 December 2006 on nutrition and health claims made on foods*. The primary purpose of this *Regulation* is to harmonise the national legal rules, regulations or administrative actions of the Member States for information on nutrition and health claims. It aims at ensuring the "effective functioning of the internal market whilst providing a high level of consumer protection."[185] When relating this *Regulation* to health care matters, it is important to consider Art. 10 I of the *Regulation*, which states, "Health claims shall be prohibited unless they comply with the general requirements in Chapter II and the specific requirements in [Chapter IV] and are authorised in accordance with this Regulation and included in the lists of authorised claims provided for in Articles 13 and 14." Art. 10, 2. of the *Regulation* lists information that must be included in the labelling, or if there is no labelling in the presentation and advertising in order to allow health claims. Thus, a reference to health care matters is only allowed under specific conditions, but it is generally possible. This *Regulation* refers to Art. 95 *EC*. It, however, also concerns matters of health care policy.[186] The ECJ declared this to be allowed. This must be kept in mind when considering a harmonisation of LOD.

d) Interim Result

Art. 168 *TFEU* is relevant for the competences of the European Union regarding LOD. This provision does not grant the European Union with an all-embracing competence in the field of LOD, though. The harmonisation of the national rules is no "competence *per se*" in the European Union.[187]

---

[182] Ibid., at 78.

[183] Ibid., at 90.

[184] European Court of Justice [Europäischer Gerichtshof], 12 December, 2006, Case C-380/03, at 95; Schmidt am Busch, in Grabitz/Hilf/Nettesheim (ed.) (2011), Art. 168 at 106.

[185] Art. 1, 1. Regulation (EC) No 1024/2006 of the European Parliament and of the Council of 20 December 2006 on nutrition and health claims made on foods.

[186] Cf. Leible, Vol. 3 Zeitschrift für das gesamte Lebensmittelrecht 285, 288 (2010).

[187] Weatherill (2010), p. 43.

Art. 168 IV (a) *TFEU* is the only paragraph of this article that grants shared competences to the European Union, as opposed to the other articles that only allow the European Union to support, coordinate or supplement the actions of the Member States. The European Union may use the competence resulting from Art. 168 IV (a) *TFEU* to pass rules ensuring high standards of quality and safety of organs. It must not harmonise the requirements for LOD, though. In case the European Union wants to take initiatives beyond the scope of powers granted, it depends on the willingness of its Member States.[188] The main responsibility with respect to public health and also for LOD hence remains with the Member States.[189]

2) Art. 114 *Treaty on the Functioning of the European Union*

Measures regarding health protection can also be passed apart from Art. 168 *TFEU*. In this case, the prohibition to harmonise laws or legislations of the Member States does not apply, but the other regulation should not be used to avoid this prohibition.[190]

Art. 114 III *TFEU* mentions health protection. It clarifies that the commission takes a high level of health protection as a starting point for proposals in this field.[191] This is a statutory duty.[192] Art. 168 *TFEU* and Art. 114 *TFEU* play striking roles since measures to harmonise laws and legislation of Member States are mostly prohibited by them.[193] Art. 114 *TFEU* grants the European Union with a competence to harmonise the national laws, but such harmonisation "must have as its object the establishment and functioning of the internal market."[194] Furthermore, Art. 168 *TFEU* is lex specialis towards Art. 114 *TFEU*, and thus enjoys priority.[195] A harmonisation of the requirements for LOD therefore can not be based on Art. 114 *TFEU*.

---

188 Hendriks, in Weimar/Bos/Busschbach (ed.) (2011), p. 39.

189 Cf. Berg, in Schwarze (ed.) (2009), Art. 152 at 11 with respect to the former regulation; Lurger, in Streinz (ed.) (2012), Art. 168 at 38; Schmidt am Busch, in Grabitz/Hilf/Nettesheim (ed.) (2011), Art. 168 at 24.

190 Cf. European Court of Justice [Europäischer Gerichtshof], 5 October, 2000, Case C-376/98, at 79, in Vol. 41 Zeitschrift für Wirtschaftsrecht 1848, 1853 (2000).

191 Kahl, in Calliess/Ruffert (ed.) (2011), Art. 114 at 33 AEUV.

192 Kahl, in Calliess/Ruffert (ed.) (2011), Art. 114 at 33 AEUV; Leible/Schröder, in Streinz (ed.) (2012), Art. 114 at 73.

193 Frenz/Götzkes, Vol. 28 Medizinrecht 613, 615 (2010).

194 Weatherill (2010), p. 37.

195 Frenz/Götzkes, Vol. 28 Medizinrecht 613, 618 (2010); Fischer, in Lenz/Borchardt (ed.), EU-Verträge, 5 Auflage 2010, Art. 168 AEUV, Rn. 24; cf. Wollenschläger, in Müller-Graff/Schmahl/Skouris (ed.) (2011), p. 455.

3) Art. 9 *Treaty on the Functioning of the European Union*

Art. 9 *TFEU* obliges the European Union to keep a high level of "protection of human health."[196] It, however, is no legal basis to base secondary legislation on.[197] Thus, the European Union cannot base a harmonisation of the legislation on LOD on this article.

4) *Charter of Fundamental Rights of the European Union*

The *CFREU* came into effect on 1 December 2009.[198] It is legally binding[199] and its main purpose is "to strengthen the protection of fundamental rights [. . .]."[200] The possibility of LOD, or the prohibition of it affects fundamental rights, such as the life of the patient suffering from organ failure and the right of self-determination of the potential donor. (1) The relevant articles of the *CFREU* and (2) the consequences of protecting those rights by the *CFREU* will be presented.

a) Relevant Articles of the *Charter of Fundamental Rights of the European Union*

LOD is often the only life-saving option for many patients suffering from organ failure. It could also bring about at least improvements of many patients' physical (and maybe also mental) well-being.[201] Therefore, the existence of unjustified legal barriers and misguided rationales, or a complete prohibition of LOD, might conflict with Art. 2 (1), Art. 3 (1) and Art. 35 *CFREU*. In addition, the right of self-determination (of the donor) needs to be kept in mind. Consequently, it will be proven whether unjustified legal rules and a complete ban of LOD would conflict with these fundamental rights, causing the CFREU to be relevant for the European Union in the field of LOD.

Art. 2 (1) *CFREU* guarantees that "[e]veryone has the right to life." This fundamental right contains a right of non-intervention.[202] Art. 2 (1) *CFREU* is infringed upon if an addressee of this regulation causes the death of someone

---

[196] Cf. Wollenschläger, in Müller-Graff/Schmahl/Skouris (ed.) (2011), p. 449.

[197] Schmidt am Busch, in Grabitz/Hilf/Nettesheim (ed.) (2011), Art. 168 at 7.

[198] Jarass (ed.) (2010), p. 5.

[199] Art. 6 I Treaty on European Union.

[200] Preamble of the Charter of Fundamental Rights of the European Union.

[201] Wagner/Fateh-Moghadam, Vol. 56 Soziale Welt 73, 77 (2005).

[202] Borowksy, in Meyer (ed.) (2011), Art. 2 at 34; Höfling, in Tettinger/Stern (ed.) (2006), Art. 2 at 30 f.

else.[203] Completely prohibiting LOD would deprive several sick patients of this option and would consequently likely cause the death of (some of) those. This result would not be the intent of the regulation, though, and must therefore be classified as being indirect. However, all measures that interfere with Art. 2 (1) *CFREU* are in conflict with this regulation, which implies that indirect interventions are also in conflict with this regulation. This includes every adverse impact caused by governmental behavior with regard to fundamental rights of individuals.[204] Thus, a complete prohibition of LOD would cause an impairment of Art. 2 (1) *CFREU* with regard to patients suffering from organ failure. This also applies to the milder case, where LOD is not completely prohibited, but is restricted by law in an unjustified way, because such restrictions most likely prevent life saving or at least life quality enhancing LODs. However, a complete prohibition of LOD or at least a restriction would guarantee that no potential donor would be exposed to the risk of dying as a result of acting as living organ donor. Even though the mortality risk is very low, it exists. However, just because LOD is not completely prohibited does not mean that potential donors are always exposed to danger. It does not force people to become living organ donors. It rather makes it possible for potential donors to come to their own decision. The *CFREU* even protects the right of self-determination about one's own body.[205] Therefore, a justified admissibility of LOD under certain circumstances would not be against Art. 2 (1) *CFREU*, while a complete prohibition of LOD or an unjustified restriction of LOD would be in conflict with this fundamental right.

Art. 3 (1) *CFREU* states that "[e]veryone has the right to respect for his or her physical and mental integrity." An impairment of Art. 3 (1) *CFREU* exists if a person, who is engaged in fundamental rights, intervenes on a person's physical integrity in a positive way.[206] The mere establishment of a concrete danger can already be qualified as legal injury.[207] From the potential recipient's point of view, completely prohibiting LOD or restricting it in an unjustified manner would be a disadvantage for his physical integrity and could therefore be considered as being in conflict with Art. 3 (1) *CFREU*.[208] The potential donor's physical integrity has to be considered as well. Becoming a living organ donor might be an impairment to his health. Just making it legally possible for living people to donate organs under understandable circumstances does not mean that anyone has to do so, though. Also, the potential donor's right of self-determination about his body has to be considered once again. Thus, a complete prohibition of LOD, or a pointless restriction of it, would also be in conflict with Art. 3 (1) *CFREU*.

---

[203] Cf. Borowksy, in Meyer (ed.) (2011), Art. 2 at 34; Jarass (ed.) (2010), Art. 2 at 7.

[204] von Danwitz, in Tettinger/Stern (ed.) (2006), Art. 52 at 32.

[205] Borowksy, in Meyer (ed.) (2011), Art. 2 at 35.

[206] Jarass (ed.) (2010), Art. 3 at 7.

[207] Höfling, in Tettinger/Stern (ed.) (2006), Art. 3 at 10.

[208] That indirect interventions are also included by the Charter of Fundamental Rights of the European Union has already been clarified.

In general, with regard to Art. 2 (1) and Art. 3 (1) of the *CFREU*, the same principle can be derived from what the German Constitutional Court decided regarding LOD, a decision based on the basic right in the German *Constitution*: "Whenever legal regulations entail that a citizen in need of treatment is denied a therapy which, according to the state of medical research, could be provided to him or her, and which would entail a prolonging of life or at least a substantial mitigation of the ailment, the constitutionally guaranteed basic negative right to life and health is infringed upon."[209]

The obligations that arise from Art. 2 (1) and 3 (1) *CFREU* are closely connected to Art. 35 *CFREU*.[210] This article demands that "[e]veryone has the right of access to preventive health care and the right to benefit from medical treatment under the conditions established by national laws and practices. A high level of human health protection shall be ensured in the definition and implementation of all the Union's policies and activities." A high level of human health protection is demanded, not the highest, though.[211] Withholding medical treatment from sick people certainly does not ensure a high level of human health protection. The requirements for LOD must, therefore, make a reasonable system of LOD possible.

The complete prohibition of LOD, or the creation of unjustified legal barriers and misguided rationales, is not in accord with the *CFREU*.

b) Consequences

As just seen, several articles of the *CFREU* become relevant when thinking about LOD. As a consequence, prohibiting LOD or passing unjustified restrictions for LOD conflicts with the *CFREU*. The *CFREU* does not grant competences to the European Union, though.[212] It is only effective within the competences of the European Union.[213] The European Union thus has to consider the *CFREU*, but it cannot base any law regulating LOD on it, meaning that the *CFREU* does not provide any legal basis to act.

---

[209] German Federal Constitutional Court [Bundesverfassungsgericht], 11 August, 1999, Case No. 1 BvR 218/98, in Vol. 46 Neue Juristische Wochenschrift 3399, 3400 (1999).

[210] Borowksy, in Meyer (ed.) (2011), Art. 3 at 38.

[211] Nußberger, in Tettinger/Stern (ed.) (2006), Art. 35 at 50.

[212] Art. 51 II Charter of Fundamental Rights of the European Union; Wollenschläger, in Müller-Graff/Schmahl/Skouris (ed.) (2011), p. 460.

[213] Wollenschläger, in Müller-Graff/Schmahl/Skouris (ed.) (2011), p. 460.

5) Treaty of Lisbon

The Treaty establishing a Constitution for Europea was rejected.[214] The *Treaty of Lisbon* has then been established.[215] Its consequences are not as extensive as those that were planned for the European Constitution, but the *Treaty of Lisbon* nevertheless caused consequences.[216] The European Union and the European Community subsequently merged into one single European Union,[217] and the European Union was granted with legal personality.[218] Furthermore, the original treaties of the European Union were amended.[219] The *TEU* and the *TFEU* are currently in force.[220] In addition, the *EAC* is in force.[221] The amendments were a further step of the European process of integration.[222] The *TEU* states that it "makes a new stage in the process of creating an ever closer union among the peoples of Europe, in which decisions are taken as openly as possible and as closely as possible to the citizen."[223] This proves that the European Union is increasingly growing together, which has already been named as a reason to harmonise the requirements of LOD.

Does the *Treaty of Lisbon* extend the competences of the European Union? And does that apply to the field of health care, which includes LOD? The *Treaty of Lisbon* establishes a new legal basis with respect to several areas for the European Union.[224] This, for example, applies to the public health system.[225] It does not annul the principle of conferral,[226] though, meaning that the European Union can

---

[214] Calliess (2010), p. 40; Ehlers, in Ehlers (ed.) (2009), p. 455; Jarass (ed.) (2010), p. 4.

[215] Cf. Calliess (2010), p. 40 ff.; Ehlers, in Ehlers (ed.) (2009), p. 455.

[216] Haratsch/Koenig/Pechstein (2010), p. 14 ff.

[217] Borchardt (2010), at 33; Cremer, Vol. 32 JURA 296, 298 (2010); Ehlers, in Ehlers (ed.) (2009), p. 443; German Federal Constitutional Court [Bundesverfassungsgericht], 6 June, 2009, Case No. 2 BvE 2/08 u. a. in Vol. 62 Neue Juristische Wochenschrift 2266, 2267 (2009); cf. Herrmann, in Terhechte (ed.) (2011), p. 620 f.; Terhechte, Vol. 43 Zeitschrift Europarecht 143, 146 (2008).

[218] Art. 46 A Treaty of Lisbon, in Vol. 50 Official Journal of the European Union 1, 38 (2007). See also Cremer, Vol. 32 JURA 296, 298 (2010); Terhechte, Vol. 43 Zeitschrift Europarecht 143, 146 (2008).

[219] Fischer (2010), p. 93; Nowak (2011), p. 51.

[220] Fischer (2010), p. 93; Oppermann/Classen/Nettesheim (2011), p. 18; German Federal Constitutional Court [Bundesverfassungsgericht], 30 June, 2009, Case No. 2 BvE 2/08 u. a. in Vol. 62 Neue Juristische Wochenschrift 2267, 2276 (2009); Terhechte, Vol. 43 Zeitschrift Europarecht 143, 148 (2008); Weatherill (2010), p. 3.

[221] Terhechte, Vol. 43 Zeitschrift Europarecht 143, 148 (2008).

[222] Nowak (2011), p. 78.

[223] Art. 1 Treaty on European Union.

[224] Fischer (2010), p. 98.

[225] Ibid.

[226] Cf. Cremer, Vol. 32 JURA 296, 302 f. (2010); Nowak (2011), p. 116 ff.; German Federal Constitutional Court [Bundesverfassungsgericht], 30 June, 2009, Case No. 2 BvE 2/08 u. a. in Vol. 62 Neue Juristische Wochenschrift 2267, 2271 (2009).

only act within the competences conferred upon it by its Member States.[227] The regulations that grant competences to the European Union in the field of health care have already been analysed.

The German Federal Constitutional Court's judgement on the *Treaty of Lisbon* serves as a model for how the competences of the European Union in the field of health care and those of the Member States can be divided (after the *Treaty of Lisbon* came into force). First of all, the judgement of the German Federal Constitutional Court can be considered as clarifying that the German *Constitution* has a future, even though the *Treaty of Lisbon* has been passed.[228] More detailed, the German Federal Constitutional Court lists the decisions it considers to be especially relevant for a state to be able to organise itself in a democratic way.[229] This, inter alia, applies to the structure of living conditions with respect to social welfare.[230] As a consequence, the German legislator must come to his own decisions with respect to the issues that are substantial for social politic.[231] The judgment of the German Federal Constitutional Court does not explicitly mention health care. Because Germany is a social federal state,[232] its duty is to create a just social order.[233] Does this include aspects of health care? The German Federal Constitutional Court clarifies that the state has an extensive scope to act when fulfilling the duty to create a just social order; it must establish the minimum conditions to make a human existence of its citizens possible.[234] It can be assumed that both social welfare and social politics are broad in scope. The state must provide a subsistence level of health.[235] Health plays a central role in the well-being of a human being. It

---

[227] Moussis (2011).

[228] Cf. Laas, in von Arnauld/Hufeld (ed.) (2011), p. 397; Wittreck, in Wittreck (ed.) (2010), p. 5.

[229] German Federal Constitutional Court [Bundesverfassungsgericht], 30 June, 2009, Case No. 2 BvE 2/08 u. a. in Vol. 62 Neue Juristische Wochenschrift 2276, 2274 (2009).

[230] Cf. Cremer, Vol. 32 JURA 296, 301 f. (2010); German Federal Constitutional Court [Bundesverfassungsgericht], 30 June, 2009, Case No. 2 BvE 2/08 u. a. in Vol. 62 Neue Juristische Wochenschrift 2276, 2275 (2009); cf. Kingreen, Vol. 54 Neue Juristische Wochenschrift 3382, 3382 (2001); cf. Laas, in von Arnauld/Hufeld (ed.) (2011), p. 414; cf. Musil, in von Arnauld/Musil (ed.) (2009), p. 22.

[231] Cf. Cremer, Vol. 32 JURA 296, 302 ff. (2010); German Federal Constitutional Court [Bundesverfassungsgericht], 30 June, 2009, Case No. 2 BvE 2/08 u. a. in Vol. 62 Neue Juristische Wochenschrift 2276, 2275 (2009).

[232] Art. 20 German Constitution.

[233] Cf. German Federal Constitutional Court [Bundesverfassungsgericht], 17 August, 1956, Case 1 BvB 2/51, in BVerfGE 5, 85 at 206; Creifelds/Weber (ed.) (2011), p. 1099; cf. Musil, in von Arnauld/Musil (ed.) (2009), p. 24.

[234] Jarass, in Jarass/Pieroth (ed.) (2011), Art. 20 at 124 f.; Laas, in von Arnauld/Hufeld (ed.) (2011), p. 414; Sommermann, in von Mangoldt/Klein/Starck (ed.) (2010), Art. 20 at 166 ff.; German Federal Constitutional Court [Bundesverfassungsgericht], 30 June, 2009, Case No. 2 BvE 2/08, in Vol. 43 Neue Juristische Wochenschrift 2869, 2870 (1990).

[235] German Federal Constitutional Court [Bundesverfassungsgericht], 9 February, 2010, Case No. 1 BvL 1/09, at 135; Musil, in von Arnauld/Musil (ed.) (2009), p. 28; Sommermann, in von Mangoldt/Klein/Starck (ed.) (2010), Art. 20 at 120; Steiner, in Spickhoff (ed.) (2011), p. 686 f.

has a great influence on a person's life and living conditions. If a person is ill, it is substantial to him to be helped, making health care policy essential for him. Thus, the social welfare state has the duty to provide medical care to its citizens.[236] This right is neither part of the German *Constitution*, nor of a more specific law, but it can be derived from Art. 2 II German *Constitution*, which guarantees the right to life and physical integrity.[237] The same argumentation can also be applied if the focus is on the entire nation, and individual consideration is insufficient, because the health of each individual has an influence on the condition of the whole nation. The competences for health care policy therefore must lie with the Member States. This area of competence is one that has a formative influence on a country being a sovereign state (being independent of a foreign power[238]), which means that the European Union has no competence in this field.

The same conclusion can be drawn if one looks beyond general health care and at organ transplantation in particular. Art. 20 I German *Constitution* is very abstract with respect to the duties resulting from the fact that Germany is a social federal state.[239] One particular task of the social federal state can be found in the catalogue of competences of Art. 74 German *Constitution*,[240] where it explicitly mentions organ transplantation[241] and confirms the (social federal) state is competent in this field. The German legislator has already made use of this competence by basing the *Act on the donation, removal and transplantation of organs* passed in 1997 on this rule.[242]

6) Result

The competences of the European Union in the field health care are restricted, and the European Union is not able to pursue an entirely autonomous health care policy.[243] In fact, the Member States must use discretion when protecting public health.[244] This field of competence can be considered necessary for a country to be

[236] Cf. Däubler, in Iliopoulos-Strangas (ed.) (2010), p. 132.

[237] Däubler, in Iliopoulos-Strangas (ed.) (2010), p. 132.

[238] Posener (ed.) (1909), p. 526.

[239] Cf. Herzog (1980), in Maunz/Dürig (ed.) (2011), Art. 20 at 5; cf. Schnapp, in von Münch/Kunig (ed.) (2012), Art. 20 at 55; cf. Sommermann, in von Mangoldt/Klein/Starck (ed.) (2010), Art. 20 at 120.

[240] Cf. Sommermann, in von Mangoldt/Klein/Starck (ed.) (2010), Art. 20 at 102.

[241] Art. 74 I No. 26 German Constitution.

[242] Müller-Terpitz, in Spickhoff (ed.) (2011), p. 691.

[243] Berg, in Schwarze (ed.) (2009), Art. 152 at 11 with respect to the former regulation; Frenz/Götzkes, Vol. 28 Medizinrecht 613, 615 (2010); cf. Oppermann/Classen/Nettesheim (2011), p. 620; Schmidt am Busch, in Grabitz/Hilf/Nettesheim (ed.) (2011), Art. 168 at 6.

[244] Weatherill (2010), p. 348.

a sovereign state, which means that the European Union has no all-embracing competence in this field. That also applies to the field of (living) organ donation.[245]

When evaluating the competences the European Union has in the field of health care, Art. 168 *TFEU* is especially relevant. The European Union has competences to pass regulations for LOD, but this does not include the whole area of LOD. To create a best practise proposal with regard to LOD, a decisive matter would be to determine whether the European Union has the competence to regulate the requirements for LOD. The European Union is not competent to regulate all aspects that arise in this field; it is only able to pass regulations "setting high standards of quality and safety of organs,"[246] and it must "not affect national provisions on the donation or medical use of organs."[247] The European Union is therefore not in a position to establish binding requirements for LOD.[248] But, this does not mean that it has absolutely no possibility to act. The option to establish soft law remains,[249] which will be explained in the following.

IV. Soft Law

The European Union wants to regulate the requirements for LOD, but it is not competent to pass hard law. However, the possibility to pass soft law remains. (1) This option will be described in general, and (2) it will also be applied to the field of LOD.

1) Soft Law in General

The possibility to use soft law is not new for the European Union. The European Commission already used it in 1962[250] and it started to appear in the public law literature to a great extent in the beginning of the 1970s.[251] Meanwhile, there is a general trend towards international soft law.[252] This also applies to the European Union, where the use of soft law has increased.[253]

---

[245] Norba (2009), p. 343.

[246] Art. 168 IV (a) Treaty on the Functioning of the European Union.

[247] Ibid., Art. 168 VII.

[248] Gutmann/Schroth, p. 43 with regard to the former Art. 152, 4.; Spielberg, Vol. 107 (22) Deutsches Ärzteblatt A-1102/B-973/C961 (2010).

[249] Cf. Cosma/Whish, Vol. 14 European Business Law Review 25, 34 (2003); cf. Klabbers, Vol. 65 Nordic Journal of International Law 167, 171 (1996).

[250] Senden/Prechal, in de Witte/Hanf/Vos (ed.) (2001), p. 185.

[251] Cini (2000), p. 4.

[252] Boyle, Vol. 48 International and Comparative Law Quarterly 901, 901 (1999); Chinkin, Vol. 38 International and Comparative Law Quarterly 850, 860 (1989).

[253] Cini (2000), p. 3; Cosma/Whish, Vol. 14 European Business Law Review 25, 25 (2003), Dawson (2009), abstract.

I will introduce soft law in general in the following. (1) I will give a general description of soft law and, afterwards, (2) I will focus on the impact of soft law. Additionally, special attention will be given to the European Union's use of soft law.

a) Description of Soft Law

In every agreement, the parties choose whether it should be binding or not.[254] An agreement is only binding if the parties intend it to be, thus, parties can also choose to *only* establish soft law.[255] (1) Nevertheless, the legitimacy of soft law is questioned. (2) Afterwards, soft law will be explained in detail. (3) The location of soft law in the framework of international law will be examined, and (4) its relation to other regulative mechanisms, namely hard law and mere commitments, will be clarified. (5) The advantages and disadvantages of soft law will be named. (6) Last, the possible content of soft law will be analysed.

aa) Legitimacy of Soft Law

It is stated that "[i]nternational legal scholars and practitioners are accustomed to the existence of soft law because it is such an integral part of the international legal system."[256] This is not undisputed. Positivist legal scholars, for example, deny the existence of soft law. They claim that law cannot be "more or less binding."[257]

It is a matter of fact that soft law is used.[258] This also applies to the European Union. It already adopted several soft law measures, for example, "'action programmes' to promote equality between women and men in the workplace, based on a Council decision establishing a 4-year programme on gender equality for 2001–2005 [Decision 2001/51 [2001]])."[259] Therefore, soft law will be considered in the following.

bb) Explanation of Soft Law

(1) A definition of soft law is indispensable to understanding its elaboration here. (2) Afterwards, the different instruments that are considered as soft law will be introduced. (3) Different organs of the European Union have the ability to develop soft.

[254] Guzman, Vol. 16 European Journal of International Law 579, 580 (2005).

[255] Schachter, Vol. 71 The American Journal of International Law 296, 296 f. (1977).

[256] Guzman/Meyer, Vol. 2 Journal of Legal Analysis 171, 176 (2010).

[257] Cf. Cini (2000), p. 4; Klabbers, Vol. 65 Nordic Journal of International Law 167, 181 (1996).

[258] Cf. Knauff, Vol. 46 Zeitschrift Europarecht 735, 735 (2011).

[259] Eurofound (2011).

(1) Definition of the Term Soft Law

Several different definitions of *soft law* exist,[260] which makes it impossible to refer to one universal definition. The most relevant definitions will hence be presented in the following.

First, several scholars use the term soft law for agreements that are not binding.[261] It is stated that soft law does not create enforceable rights and duties.[262] However, several of those scholars admit that soft law, nevertheless, has practical and legal effects.[263]

Second, regarding the European Union, soft law is considered as "the voluntary adherence of the Member States to social, political or economical norms that were agreed upon among official representatives."[264]

Third, soft law is sometimes defined by delimiting it to hard law. The creation of an obligation, the precision of rules and the delegation to a third-party decision maker are considered as the three main characteristics of hard law.[265] If those features are combined, it gives a 'hard' character to a law.[266] Consequently, as soon as a legal agreement is weakened with regard to one or more dimensions of the three just listed, it is considered as soft law.[267]

Fourth, rules that do not comply with constitutional and other formalities are called soft law.[268] As a consequence of this noncompliance, such rules are not legally binding.[269]

Fifth, agreements that contain imprecise obligations are described as soft law.[270]

As seen, several different definitions are brought forth, but it is nevertheless possible to deduce a few core elements from those to clarify the meaning of soft

---

[260] Boyle, Vol. 48 International and Comparative Law Quarterly 901, 901 f. (1999); Cosma/Whish, Vol. 14 European Business Law Review 25, 28 f. (2003); Guzman/Meyer, Vol. 2 Journal of Legal Analysis 171, 174 (2010); Meyer, Vol. 32 Fordham International Law Journal 888, 905 (2009); Shaffer/Pollack, Vol. 94 Minnesota Law Review 706, 712 (2010).

[261] Baxter, Vol. 29 International and Comparative Law Quarterly 549, 549 (1980); Borgen, Vol. 37 George Washington International Law Rev 573, 642 (2005); Boyle, Vol. 48 International and Comparative Law Quarterly 901, 901 f. (1999); Meyer, Vol. 32 Fordham International Law Journal 888, 889 (2009); Snyder, in Daintith (ed.) (1995), p. 64.

[262] Baxter, Vol. 29 International and Comparative Law Quarterly 549, 549 (1980).

[263] Meyer, Vol. 32 Fordham International Law Journal 888, 890 (2009); Snyder, in Daintith (ed.) (1995), p. 64.

[264] Christians, Vol. 25 Wisconsin International Law Journal 1, 7 (2007).

[265] Abbot et al., Vol. 54 International Organization 401, 401 (2000); Shaffer/Pollack, Vol. 94 Minnesota Law Review 706, 715 (2010).

[266] Abbot et al., Vol. 54 International Organization 401, 401 (2000).

[267] Abbot/Snidal, Vol. 54 International Organization 421, 422 (2000).

[268] Gersen/Posner, Vol. 61 Stanford Law Review 573, 577 (2008).

[269] Ibid.

[270] Abbot/Snidal, Vol. 54 International Organization 421, 422 (2000); Weil, Vol. 77 American Journal International Law 413, 414 (1983).

law.[271] A significant amount of legal scholars states that soft law is not legally binding. This is a main characteristic of soft law. Furthermore, it is common for soft law to be contrasted with hard law. Both are (often) contrasted with respect to the issue of legalisation. Legalisation refers to certain characteristics, namely the creation of an obligation, the precision of rules and the delegation to a third-party decision maker.[272] As a consequence, when all three characteristics just listed are maximised, hard law exists. If that is not the case, hard law does not exist, but rather different forms of soft legalisation.[273] So, a rule can either be classified as hard law or as soft law; both possibilities are distinguished.

(2) Instruments Considered as Soft Law

Several different instruments are denominated as soft law.[274] Treaties that only include soft obligations are included.[275] Non-binding resolutions and codes of conduct formulated and accepted by international and regional organisations can also be soft law.[276] In addition, statements prepared by individuals in a non-governmental capacity, which purport to lay down international principles, might form soft law.[277]

Thus, soft law can be established with various regulative instruments.[278] With respect to the European Union, the following list enumerates the soft law measures used by its organs: "resolutions; recommendations; declarations; conclusions; green-and white papers; reports; guidelines; action papers; programmes; codes of conduct; communications; memoranda; opinions."[279]

(3) Contributors to Develop Soft Law in the European Union

The European Commission is the main contributor to the development of soft law in the European Union. Beyond the European Commission, the European Court of Justice (ECJ) and the Court of First Instance (CFI) have also been making important contributions to the further development of soft law, since they are empowered to

---

[271] Cf. Cosma/Whish, Vol. 14 European Business Law Review 25, 29 (2003).

[272] Abbot et al., Vol. 54 International Organization 401, 401 (2000); Shaffer/Pollack, Vol. 94 Minnesota Law Review 706, 715 (2010).

[273] Ibid.

[274] Chinkin, Vol. 38 International and Comparative Law Quarterly 850, 851 (1989); Guzman, Vol. 16 European Journal of International Law 579, 580 (2005); Guzman/Meyer, Vol. 2 Journal of Legal Analysis 171, 173 (2010).

[275] Chinkin, Vol. 38 International and Comparative Law Quarterly 850, 851 (1989).

[276] Ibid.

[277] Ibid.

[278] Chinkin, Vol. 38 International and Comparative Law Quarterly 850, 852 (1989); cf. Stefan, Vol. 14 European Law Journal 753, 758 (2008).

[279] Cosma/Whish, Vol. 14 European Business Law Review 25, 46 (2003).

review the decisions of the European Commission. The courts consequently have an impact on the policy of the European Commission.[280]

As seen, not only a number of different definitions of soft law exist, but several different types as well. Different organs of the European Union may establish soft law.

cc) Location of Soft Law in International Law

The location of soft law in the framework of international law is not clear.[281]

It is the prevailing view that soft law is not binding. So, it is (often) located outside the boundaries of the legal discourse.[282] The borders of legal discourse are permeable, though. Such border lines are crossed if (first) a particular conflict is verbalised and if (second) these communications are published.[283]

It also has to be considered that an agreement, even though it is only soft law, implies that a certain arrangement exists.[284] It cannot be put on a level with having no agreement at all. This applies to soft law in the European Union as well. Soft law is significant in the European Union, even though it is not part of the European order of competences and cannot be equalized with hard law.[285] Soft law is *less* than hard law.[286] According to the middle-of-the-road strategy, it is less than "full-blown international law," but still more than the absence of any commitment.[287] Soft law is thus regarded as being in a "gray zone in between."[288]

As a consequence, soft law creates its own category within the framework of international law. It can be placed in between mere commitments and hard law.[289] Therefore, its relation to hard law and to mere commitments will be clarified in the following.

dd) Relation of Soft Law to Other Regulative Mechanisms

Soft law has to be distinguished from other regulative mechanisms. As seen, its place is in between (1) mere commitments and (2) hard law.

[280] Ibid., p. 37.
[281] Guzman, Vol. 16 European Journal of International Law 579, 583 f. (2005).
[282] Calliess/Renner, Vol. 22 Ratio Juris p. 20 (2009).
[283] Ibid.
[284] Guzman/Meyer, Vol. 2 Journal of Legal Analysis 171, 180 (2010).
[285] Knauff, Vol. 46 Zeitschrift Europarecht 735, 738 (2011).
[286] Guzman, Vol. 16 European Journal of International Law 579, 583 f. (2005).
[287] Guzman/Meyer, Vol. 2 Journal of Legal Analysis 171, 180 (2010).
[288] Klabbers, Vol. 65 Nordic Journal of International Law 167, 167 (1996).
[289] Cf. Cini (2000), p. 4; cf. Guzman/Meyer, Vol. 2 Journal of Legal Analysis 171, 173 (2010).

(1) Soft Law Versus Mere Commitments

Soft law is not legally binding and neither are mere commitments. Nevertheless, soft law cannot be put on a level with mere commitments, such as political or moral commitments.[290]

Soft law, in comparison to pure commitments, can be the basis of hard law or might affect the interpretation or implementation of hard law.[291] Mere commitments, in contrast, are not legal at all.[292]

(2) Soft Law Versus Hard Law

Soft law is frequently contrasted with hard law.[293] Hence, a short elaboration of hard law, and how it differs from soft law, is necessary.

One main feature of hard law is that it is legally binding.[294] So, it is subject to the rules that are responsible for international treaties, for example, the regulations of the *Vienna Convention on the Law of Treaties*.[295] It is expected that the people concerned comply with hard law.[296] It makes it possible for states to "assert legal claims, engage in legal discourse, invoke legal procedures, and resort to legal remedies."[297]

Since a rule can either be legally binding or not, it has to be clearly distinguished as either soft or hard law. But, these two instruments are also related to each other. On the one hand, the validity of soft law depends on its conformity with binding treaties. On the other hand, soft law might have an impact on hard law.[298] Even if it needs to be clear whether a rule is legally binding or not, this status can change.[299] In the past, non-binding arrangements have become binding, or, in other words, soft law was turned into hard law.[300]

Soft law and hard law are different with regard to the aspect of being legally binding. Their implications, consequently, differ as well. A development from soft law to hard law is possible, though.

---

[290] Klabbers, Vol. 65 Nordic Journal of International Law 167, 168 (1996).

[291] Meyer, Vol. 32 Fordham International Law Journal 888, 906 f. (2009).

[292] Klabbers, Vol. 65 Nordic Journal of International Law 167, 168 (1996).

[293] Boyle, Vol. 48 International and Comparative Law Quarterly 901, 901(1999); Stefan, Vol. 14 European Law Journal 753, 754 (2008).

[294] Meyer, Vol. 32 Fordham International Law Journal 888, 906 (2009); Senden (2004), p. 45; Stefan, Vol. 14 European Law Journal 753, 755 (2008).

[295] Meyer, Vol. 32 Fordham International Law Journal 888, 906 (2009).

[296] Baxter, Vol. 29 International and Comparative Law Quarterly 549, 549 (1980).

[297] Abbot et al., Vol. 54 International Organization 401, 411 (2000).

[298] Stefan, Vol. 14 European Law Journal 753, 768 (2008).

[299] Abbot/Snidal, Vol. 54 International Organization 421, 423 (2000); Hillgenberg, Vol. 10 European Journal of International Law 499, 501 (1999).

[300] Cini (2000), p. 6; Hillgenberg, Vol. 10 European Journal of International Law 499, 501 (1999).

ee) Advantages and Disadvantages of Soft Law

Soft law has (1) advantages as well as (2) disadvantages.

(1) Advantages

Soft law has several advantages.

First, soft law is considered as useful if it is the only possibility to come to an agreement at all.[301] Hard (international) law might be rejected if countries do not voluntarily give up more sovereignty than necessary. Hard law imposes high sovereignty costs on states,[302] whereas soft law imposes lower sovereignty costs.[303] Apart from that, it is not always easy to come to a compromise. Soft law can cope well with diversity[304] and is easier to negotiate than hard law.[305] That might make it possible to come to a compromise between parties with different interests, values, and degrees of power.[306] Consequently, if an agreement is strived for, but it is impossible to establish hard law, soft law might be the only option besides renunciating a regulation. In comparison to having no agreement at all, soft law is (usually) preferable.[307] Soft law can therefore be considered as a compromise between having no regulation at all and passing hard law.[308] Besides, cases exist, in which a party is not empowered to conclude hard (international) law.[309] This, however, does not necessarily apply to soft law, so the possibility to pass soft law might nevertheless be available to them,[310] making it possible for the parties concerned to remain inactive or to pass soft law. This applies to the European Union's possibilities to harmonise the regulations for LOD.

Second, it might be easier and faster to establish soft law than hard law.[311] The formal requirements necessary for hard law are not necessary for soft law[312] and the

---

301 Hillgenberg, Vol. 10 European Journal of International Law 499, 501 (1999).

302 Cf. Abbot/Snidal, Vol. 54 International Organization 421, 422 (2000).

303 Abbot/Snidal, Vol. 54 International Organization 421, 439 (2000); Shaffer/Pollack, Vol. 94 Minnesota Law Review 706, 719 (2010).

304 Cf. Christians, Vol. 25 Wisconsin International Law Journal 1, 8 (2007); Dawson (2009), p. 3; Shaffer/Pollack, Vol. 94 Minnesota Law Review 706, 719 (2010).

305 Abbot/Snidal, Vol. 54 International Organization 421, 444 f. (2000); Shaffer/Pollack, Vol. 94 Minnesota Law Review 706, 719 (2010).

306 Abbot/Snidal, Vol. 54 International Organization 421, 423 (2000).

307 Cf. Chinkin, Vol. 38 International and Comparative Law Quarterly 850, 861 (1989).

308 Chinkin, Vol. 38 International and Comparative Law Quarterly 850, 861 (1989).

309 Hillgenberg, Vol. 10 European Journal of International Law 499, 501 (1999).

310 Cf.Dawson (2009), p. 3; Hillgenberg, Vol. 10 European Journal of International Law 499, 501 (1999).

311 Cosma/Whish, Vol. 14 European Business Law Review 25, 33 (2003).

312 Gersen/Posner, Vol. 61 Stanford Law Review 573, 580 (2008).

required procedures are weaker for soft law.[313] This, especially, applies to hard international law, because it has to be transformed into the domestic legal systems in order to come into effect. Soft law, in contrast, can avoid the domestic approval procedures,[314] which makes the creation of soft law cheaper,[315] in particular, because the bureaucratic transaction costs[316] and the contracting costs are lower.[317]

Third, cases exist in which the aim of the intended regulation can be achieved better by using soft law, causing soft law to be sufficient or even more desirable than hard law.[318] This occurs, for example, if a strict regulation is not intended and the only aim is to convey information about congressional views.[319] The same applies if coordinated national legislation is the aim,[320] or if soft law serves as a coordinating device.[321] In addition, soft law makes it possible for governments to obtain policy goals without resorting to law.[322]

Fourth, soft law is very flexible,[323] which is advantageous if a development is still in progress or if the development is uncertain.[324] In such cases, the lasting development can be stimulated by soft law,[325] and the created rules can be adjusted easier to further development.[326] Thus, soft law makes it possible to create a preliminary, flexible regime which can consider the development in stages.[327] In addition, even if a development seems to be completed, the flexibility of soft law

---

[313] Cosma/Whish, Vol. 14 European Business Law Review 25, 33 (2003); Hillgenberg, Vol. 10 European Journal of International Law 499, 501 (1999).

[314] Boyle, Vol. 48 International and Comparative Law Quarterly 901, 902 f. (1999); Guzman, Vol. 16 European Journal of International Law 579, 592 (2005); Hillgenberg, Vol. 10 European Journal of International Law 499, 501 (1999).

[315] Christians, Vol. 25 Wisconsin International Law Journal 1, 8 (2007); Cosma/Whish, Vol. 14 European Business Law Review 25, 33 (2003); Gersen/Posner, Vol. 61 Stanford Law Review 573, 587 (2008).

[316] Guzman/Meyer, Vol. 2 Journal of Legal Analysis 171, 176 f. (2010).

[317] Abbot/Snidal, Vol. 54 International Organization 421, 434 (2000).

[318] Gersen/Posner, Vol. 61 Stanford Law Review 573, 594 ff. (2008).

[319] Ibid., p. 594.

[320] Hillgenberg, Vol. 10 European Journal of International Law 499, 501 (1999).

[321] Guzman/Meyer, Vol. 2 Journal of Legal Analysis 171, 188 ff. (2010).

[322] Cf. Christians, Vol. 25 Wisconsin International Law Journal 1, 9 f. (2007);

[323] Christians, Vol. 25 Wisconsin International Law Journal 1, 8 (2007); Cini (2000), p. 5; Guzman, Vol. 16 European Journal of International Law 579, 591 (2005); Meyer, Vol. 32 Fordham International Law Journal 888, 900 ff. (2009); Shaffer/Pollack, Vol. 94 Minnesota Law Review 706, 719 (2010).

[324] Hillgenberg, Vol. 10 European Journal of International Law 499, 501 (1999); Meyer, Vol. 32 Fordham International Law Journal 888, 908 (2009).

[325] Hillgenberg, Vol. 10 European Journal of International Law 499, 501 (1999).

[326] Cf. Shaffer/Pollack, Vol. 94 Minnesota Law Review 706, 719 (2010).

[327] Hillgenberg, Vol. 10 European Journal of International Law 499, 501 (1999).

might be valuable, because it is not as fixed as hard law and is consequently easier to amend or to replace.[328]

Fifth, the potential penalities for violating soft law are usually more lenient than in cases where hard law is violated.[329] This is due to the parties who explicitly decide against passing hard law, which also excludes the application of treaty law. The principle of *pacta sunt servanda* cannot be applied,[330] and the consequences that would arise from this principle also cannot be applied.[331] States are risk adverse and prefer to lower the costs of avoiding their obligations.[332] It, consequently, seems probable that states engage in deeper cooperation than they would if they had to worry about a stricter enforcement and more severe consequences in case of a violation.[333] Furthermore, hard law might cause the risk of failure and deterioration in relations. This may overburden international relations.[334]

As seen, soft law has several positive features. This comparison of soft and hard law has shown that soft law offers many advantages. But, there are disadvantages as well, and these will be presented next.

(2) Disadvantages

In contrast to the advantages of soft law, several disadvantages are stated as well.

First, soft law is an option for states or nonstate organisations to act in cases where they lack the power to pass hard law. Since there are usually reasons for a lack of powers, it is feared that this *backdoor* causes the abuse of powers.[335]

Second, soft law does not produce legally binding rules.[336] A decision is consequently less binding for the parties involved,[337] but, nevertheless, has consequences. It is, therefore, criticised that the consequences of soft law are vague and uncertain.[338] Not only the consequences of soft law are criticised as being unclear, but soft law, or rather its content in general, is also characterised in

328 Boyle, Vol. 48 International and Comparative Law Quarterly 901, 902 f. (1999); cf. Guzman, Vol. 16 European Journal of International Law 579, 591 (2005).

329 Guzman/Meyer, Vol. 2 Journal of Legal Analysis 171, 184 (2010); Hillgenberg, Vol. 10 European Journal of International Law 499, 504 (1999).

330 Hillgenberg, Vol. 10 European Journal of International Law 499, 513 (1999); Schachter, Vol. 71 The American Journal of International Law 296, 301 (1977).

331 Hillgenberg, Vol. 10 European Journal of International Law 499, 513 (1999).

332 Guzman/Meyer, Vol. 2 Journal of Legal Analysis 171, 184 (2010).

333 Shaffer/Pollack, Vol. 94 Minnesota Law Review 706, 719 (2010).

334 Hillgenberg, Vol. 10 European Journal of International Law 499, 501 (1999).

335 Cosma/Whish, Vol. 14 European Business Law Review 25, 34 (2003).

336 Cf. Cini (2000), p. 5; Gersen/Posner, Vol. 61 Stanford Law Review 573, 575 ff. (2008).

337 Guzman, Vol. 16 European Journal of International Law 579, 591 (2005).

338 Cosma/Whish, Vol. 14 European Business Law Review 25, 34 (2003).

this way.[339] This uncertainty is considered as causing more confusion than clarification.[340]

Third, different people perceive soft law in different ways. Some perceive it better than others. People that do not perceive it very well might find themselves in a bad position to control their lives.[341] This disadvantage is connected to the second aspect mentioned, namely that soft law is uncertain to a certain extent.

Fourth, it is necessary to differentiate soft law from hard law. This is not always easy. Those two instruments might become indistinguishable,[342] and the threshold between rules that are legally binding and those that are not might be blushed.[343] That can lead to confusion and is contrary to the intent to provide legal clarity.

Fifth, that soft law does not have to fulfil specific formalities is criticised. It is stated that a certain degree of formalism protects the addressees from arbitrariness. When no specific formalities are required, this might cause or further arbitrariness.[344]

Sixth, violations of soft law obligations might increase given its lower penalties.[345] The likelihood of deviating from the commitment increases for all parties concerned, raising the risks for all parties.[346] The mild punishments in cases of violation probably leads to low compliance rates. Therefore, while deciding on whether soft law makes sense or not, a cost-benefit calculation has to be carried out[347]: "The delegation theory suggests that soft law will be used when the expected benefits from the violations with the latter motivation — to unilaterally amend suboptimal legal rules — are greater than the expected costs from both the violations with the former motivation and opportinustic efforts at unilateral amendment."[348] Hard law, in contrast, is enforceable.[349] It encourages all parties involved to comply with it.[350] All partners can rely on the agreement to a greater extent and can resolve common problems in a cooperative way.[351] This raises the credibility of the parties involved in the particular agreement.[352]

---

[339] Cf. Cini (2000), p. 5.

[340] Cosma/Whish, Vol. 14 European Business Law Review 25, 33 (2003).

[341] Gersen/Posner, Vol. 61 Stanford Law Review 573, 597 (2008).

[342] Cf. Klabbers, Vol. 65 Nordic Journal of International Law 167–182 (1996).

[343] See for further reference Cosma/Whish, Vol. 14 European Business Law Review 25, 32 (2003).

[344] Klabbers, Vol. 67 Nordic Journal of International Law 381, 391 (1998).

[345] Guzman/Meyer, Vol. 2 Journal of Legal Analysis 171, 178 (2010).

[346] Ibid., p. 184.

[347] Cf. Guzman/Meyer, Vol. 2 Journal of Legal Analysis 171, 177 (2010); cf. Meyer, Vol. 32 Fordham International Law Journal 888, 909 f. (2009).

[348] Guzman/Meyer, Vol. 2 Journal of Legal Analysis 171, 178 (2010).

[349] Ibid., p. 182.

[350] Ibid.

[351] Ibid.

[352] Abbot/Snidal, Vol. 54 International Organization 421, 427 ff. (2000); Meyer, Vol. 32 Fordham International Law Journal 888, 900 (2009); Shaffer/Pollack, Vol. 94 Minnesota Law Review 706, 717 f. (2010).

As seen, soft law has several advantages, but numerous disadvantages as well. This applies to hard law as well. Which of these two mechanisms is preferable cannot be answered in a general way. Rather, each instrument offers particular advantages for different contexts.[353] Thus, different forms of legislation will be preferred in different subject areas.[354] When evaluating the regulative mechanism for an individual case, several factors have to be considered in order to decide which mechanism fits best. Therefore, several scholars advocate a pragmatic approach. According to this approach, the characteristics of the issue and the negotiating and institutional context in question should be taken into consideration while deciding on whether to apply hard- or soft law instruments.[355]

It needs to be considered that soft law is the only option for the European Union to regulate the requirements of LOD, because it does not have the competence to pass hard law in these cases. It cannot decide which type of law to use from a consideration of the advantages and disadvantages. The European Union can hence either remain inactive with respect to the requirements for LOD or it can *regulate* the requirements with soft law.

ff) Possible Content of Soft Law

The subject matter concerned must be able to become the content of soft law.

When states enter into an agreement, they can decide what they want to include.[356] The range of the possible content of soft law is rather broad. It can be an on-off future action only, but can also be a complex system of regulated cooperation.[357] Just because a regulation is not legally binding does not mean that it is formulated in a more incomplete way. The regulation can be formulated just as complete as a binding international treaty.[358] That soft law is frequently dismissed as *less* law than hard law might lead to the assumption that it is only used for nonrelevant subject matters. That is not true. States or other parties also use soft law in areas that are of great importance to them.[359] The European Union does so as well.[360] Environment is considered as a topic of great importance, and the European Union passed several soft law acts in that area. For example, it passed the *Sixth Environment Action Programme*[361] and the *Environment and Health Action Plan*

---

[353] Shaffer/Pollack, Vol. 94 Minnesota Law Review 706, 721 (2010).

[354] Abbot/Snidal, Vol. 54 International Organization 421, 440 (2000).

[355] Shaffer/Pollack, Vol. 94 Minnesota Law Review 706, 721 (2010).

[356] Guzman, Vol. 16 European Journal of International Law 579, 582 f. (2005).

[357] Hillgenberg, Vol. 10 European Journal of International Law 499, 506 (1999).

[358] Ibid., p. 506 f.

[359] Guzman/Meyer, Vol. 2 Journal of Legal Analysis 171, 183 (2010).

[360] The website http://www.eu-koordination.de/bruessel/rechtsakte-und-soft-laws?start=2 lists several soft law actions of the European Union.

[361] Europa, 2011.

2004–2010.[362] In Germany, soft law is, for example, used in the field of labour law.[363]

Soft law's range and formulation are rather extensive, its content as well. Soft law can "contain rules on, for instance, the relationship of individual obligations to one another, procedures for identifying brach of commitment or for revising the agreement in the light of changes circumstances, and even denunciation."[364] It is thus not precluded from the beginning to *regulate* the requirements for LOD by soft law.

The description of soft law showed that its legitimacy is not completely without doubt. Soft law, however, exists. Several definitions are brought forth to explain soft law. After all, its main feature is that it is not legally binding. That distinguishes it from hard law. Soft law, nevertheless, is *more* than mere commitments. When determining its location in international law, it can be placed in between mere commitments and hard law. Several different instruments are available to actually pass soft law. The content that can possibly be included in soft law is rather broad.

b) Impact of Soft Law

The implication of soft law has not yet been fixed beyond doubt, but is still widely debated.[365] As already mentioned, soft law is not binding in the same way hard law is. That does not mean, however, that it does not have any impact at all. (1) Therefore, the exact effects of soft law and (2) how the courts view it are questionable.

aa) Effects of Soft Law

The following will mainly regard whether soft law has any legal effects, and will also consider other possible effects.

The legal effects of soft law are an issue that is heavily disputed. It is stated that soft law does not have the status of formal law,[366] and it is even explicitly stated that soft law has no legal effects.[367] The opposite view, that soft law has legal effects, exists as well.[368]

---

[362] Europa, 2007.

[363] Däubler, Vol. 28 Neue Zeitschrift für Arbeitsrecht 42, 42 ff. (2011).

[364] Hillgenberg, Vol. 10 European Journal of International Law 499, 507 (1999).

[365] Abbot et al., Vol. 54 International Organization 401, 412 (2000); Cosma/Whish, Vol. 14 European Business Law Review 25, 30 (2003); Stefan, Vol. 14 European Law Journal 753, 755 (2008).

[366] Boyle, Vol. 48 International and Comparative Law Quarterly 901, 902 (1999).

[367] Hillgenberg, Vol. 10 European Journal of International Law 499, 506 (1999); Schachter, Vol. 71 The American Journal of International Law 296, 300 (1977).

[368] Stefan, Vol. 14 European Law Journal 753, 753 (2008).

Apart from the dispute on whether soft law has (real) legal effects or not, soft law regulations are usually complied with anyway. Why is soft law often complied with? General principles of law are mentioned,[369] for example, to achieve legal certainty.[370] Also, legitimate expectations and equality are brought forth to explain compliance with soft law.[371]

Besides the legal effects soft law might or might not have, soft law is considered to have several possible practical effects.[372] It is often stated that soft law creates (legitimate) expectations among individuals.[373] Furthermore, soft law affects behaviour[374] and even shapes the practices of states and other actors.[375] European soft law also often pursues the duty to regulate behavior (without being legally binding).[376] In addition, soft law might even cause the development of hard law.[377]

Even if soft law is considered as *less* than hard law, it can only develop if all parties concerned agree to it. The parties can consequently not claim that the content of the agreement has not been intended.[378] Also, if an action is performed on the basis of the soft law agreement, it is not possible to claim it back as an unjustified enrichment.[379] Thus, soft law is not insignificant and the parties involved cannot just act like no agreement exists.[380] Even though the parties of such an agreement decided against it being binding, it still includes an element of good faith commitment.[381]

As already mentioned, the penalties for a violation of soft law are rather mild,[382] but nonetheless exist. The direct legal sanctions for a violation of hard law do not

---

[369] Ibid., p. 769.

[370] Stefan, Vol. 14 European Law Journal 753, 769 (2008). The European Court of Justice recognises legal certainty as a general principle of the European Union law (von Arnauld, in Riesenhuber/Takayama (ed.) (2006), p. 249).

[371] Stefan, Vol. 14 European Law Journal 753, 769 (2008).

[372] Cosma/Whish, Vol. 14 European Business Law Review 25, 31 (2003); Gersen/Posner, Vol. 61 Stanford Law Review 573, 575 (2008).

[373] Stefan, Vol. 14 European Law Journal 753, 756 (2008).

[374] Gersen/Posner, Vol. 61 Stanford Law Review 573, 573 (2008); cf. Guzman/Meyer, Vol. 2 Journal of Legal Analysis 171, 180 (2010).

[375] Abbot et al., Vol. 54 International Organization 401, 412 (2000); Guzman, Vol. 16 European Journal of International Law 579, 591 (2005); Snyder, Vol. 56 Modern Law Review 19, 33 (1993).

[376] Knauff, Vol. 46 Zeitschrift Europarecht 735, 740 (2011).

[377] Abbot et al., Vol. 54 International Organization 401, 412 (2000).

[378] Baxter, Vol. 29 International and Comparative Law Quarterly 549, 565 (1980); Hillgenberg, Vol. 10 European Journal of International Law 499, 511 (1999); Schachter, Vol. 71 The American Journal of International Law 296, 300 (1977).

[379] Hillgenberg, Vol. 10 European Journal of International Law 499, 511 (1999).

[380] Cf. Cosma/Whish, Vol. 14 European Business Law Review 25, 31 (2003); Hillgenberg, Vol. 10 European Journal of International Law 499, 511 (1999).

[381] Abbot et al., Vol. 54 International Organization 401, 412 (2000); Boyle, Vol. 48 International and Comparative Law Quarterly 901, 902 (1999); cf. Hillgenberg, Vol. 10 European Journal of International Law 499, 510 (1999).

[382] Hillgenberg, Vol. 10 European Journal of International Law 499, 506 (1999).

apply, but diplomatic and moral pressure are common.[383] It is also possible that soft law is enforced in a *soft manner*.[384] That means that control mechanisms are created, which are voluntary for the parties, but are under international and internal political pressure, and can have an impact on the public opinion.[385] In addition, compliance is influenced by the fact that soft law is based on reciprocity. So, one party can counter the other party's violation by also violating the agreement.[386]

The possible (legal) effects of soft law cannot be equated with hard law. Nevertheless, soft law is not useless, but has several (practical) effects.

bb) Consideration in Courts

Courts usually apply hard law to come to a decision.To what extent are courts also allowed to consider soft law? No common opinion exists on this issue.

A judgment by the ECJ approved the consideration of soft law. It refers to recommendations, clarifying that, according to Art. 288 *TFEU*, recommendations are not binding,[387] but that they still have to be considered when national legislation is interpreted.[388] Furthermore, in the *World Wildlife Fund* case, the CFI "held that a Code of Conduct annexed to a decision on transparency may be relied upon by private parties."[389] In a different case, the CFI stated that the European Commission has to comply with its own guidelines.[390]

The European procedural law as such is also receptive to deal with soft law.[391] When looking at the proceedings at the ECJ, soft law might become relevant when bringing (1) an action to annul, Art. 263 *TFEU*, or (2) a preliminary ruling procedure, Art. 267 *TFEU*.[392] (1) According to the wording of Art. 263 *TFEU*, an action to annul is possible only in case of "legal effects", and it explicitly excludes recommendations and opinions.[393] As a consequence, such an action cannot be brought forth against soft law. (2) Art. 267 *TFEU*, which contains the preliminary

---

[383] Chinkin, Vol. 38 International and Comparative Law Quarterly 850, 866 (1989).

[384] Hillgenberg, Vol. 10 European Journal of International Law 499, 511 (1999); Klabbers, Vol. 65 Nordic Journal of International Law 167, 169 (1996).

[385] Hillgenberg, Vol. 10 European Journal of International Law 499, 511 (1999).

[386] Ibid., p. 514.

[387] European Court of Justice [Europäischer Gerichtshof], 13 December, 1989, Case C-322/88, at 7.

[388] Ibid., at 18.

[389] Klabbers, Vol. 67 Nordic Journal of International Law 381, 388 (1998); cf. Judgment of the Court of First Instance (Fourth Chamber, extended composition) of 5 March, 1997, Case T-105/95.

[390] Klabbers, Vol. 67 Nordic Journal of International Law 381, 388 (1998); cf. Judgment of the Court of First Instance (Fifth Chamber, Extended Composition) of 12 December 1996, Case T-380/94.

[391] Knauff, Vol. 46 Zeitschrift Europarecht 735, 736 (2011).

[392] Ibid.

[393] Art. 263 Treaty on the Functioning of the European Union.

ruling procedure, does not contain the restrictions Art. 263 *TFEU* does, so soft law can in principle be the object of Art. 267 *TFEU*.[394] The ECJ already had several preliminary ruling procedures with respect to non-binding legal acts. It stated several times that the circumstances for admission have been fulfilled.[395] In the case *Deutsche Shell AG v Hauptzollamt Hamburg-Harburg*, the ECJ stated, "It is settled case-law that the fact that a measure of Community law has no binding effect does not preclude the Court from ruling on its interpretation in proceedings for a preliminary ruling [. . .]."[396] The ECJ came to the same conclusion in the case *Salvatore Grimaldi v Fonds des maladies professionnelles*.[397] In detail, the question asked to the court in a preliminary ruling procedure must be admissible as such. This is generally possible for soft law.[398] In addition, the ECJ has to decide about the question only "if it considers that a decision on a question is necessary to enable [a court or tribunal of a Member State] to give judgment."[399] Whether this is the case is usually evaluated by the submitting national court.[400] The soft law regulation must have a relevant effect on the outcome of the legal dispute.[401] Against this it can be said that soft law is not legally binding and can therefore not determine a legal decision.[402] The ECJ, in contrast, obligated the courts of the Member States to consider recommendations.[403] That soft law has effects and is not legally unremarkable is also in favour of it being relevant for a legal dispute.[404] As a conclusion, the ECJ generally considers soft law.[405]

The general consideration of soft law instruments in the European Union has increased in the last decade. One reason might be the *White Paper on Governance*, which was published by the European Commission in 2001. It promoted the combination of non-binding legal instruments with hard law. The attitude of courts

[394] Knauff, Vol. 46 Zeitschrift Europarecht 735, 737 (2011).

[395] Ibid.

[396] Judgment of the European Court of Justice (Sixth Chamber) of 21 January, 1993, Case T-188/91.

[397] Judgment of the European Court of Justice (Second Chamber) of 13 December, 1989, Case T-322/88.

[398] Knauff, Vol. 46 Zeitschrift Europarecht 735, 738 (2011).

[399] Art. 267 Treaty on the Functioning of the European Union.

[400] Knauff, Vol. 46 Zeitschrift Europarecht 735, 741 (2011); Wegener, in Calliess/Ruffert (ed.) (2011), Art. 267 at 21 AEUV.

[401] Knauff, Vol. 46 Zeitschrift Europarecht 735, 741 (2011); Thiele (2007), p. 146.

[402] Cf. Baxter, Vol. 29 International and Comparative Law Quarterly 549, 549 (1980); Borgen, Vol. 37 George Washington International Law Rev 573, 642 (2005); Boyle, Vol. 48 International and Comparative Law Quarterly 901, 901 f. (1999); Knauff, Vol. 46 Zeitschrift Europarecht 735, 741 (2011).

[403] European Court of Justice [Europäischer Gerichtshof], 13 December, 1989, Case C-322/88, at 18; Knauff, Vol. 46 Zeitschrift Europarecht 735, 741 (2011).

[404] Cf. Meyer, Vol. 32 Fordham International Law Journal 888, 890 (2009); cf. Snyder, in Daintith (ed.) (1995), p. 64.

[405] Knauff, Vol. 46 Zeitschrift Europarecht 735, 742 (2011).

to consider soft law to a greater extent corresponds to the promotion in the *White Paper on Governance* to make more use of non-legally binding tools.[406]

Another reason for the increasing reference to soft law is the creation of the CFI in 2009.[407] It refers to soft law more often than the ECJ.[408] That soft law instruments have an intensified relevance before courts in the European Union is supported by the fact that Member States increasingly challenge soft law established by the Commission.[409]

It is claimed, however, that an argument based on soft law will not have a big impact in court.[410] Case law shows that courts do not feel bound to consider soft law while rendering a judgment.[411] Many find it undesirable to consider soft law before courts.[412]

Apart from the dispute on whether courts *have to* consider soft law, it is argued that it can constitute a useful point of reference.[413] Also, no prohibition for the courts to consider soft law can be found, making it at least possible for the courts to do so.

Since it is not prohibited for courts to take soft law into consideration, it makes sense to pass soft law, because the possibility that it is considered exists. Whether courts then actually consider soft law cannot be answered in a general manner. It rather depends on the individual court. In addition, it has been worked out that soft law has several effects. It can therefore be assumed that it would not remain without consequences if the European Union would pass soft law for LOD. This possibility will hence be considered in the following.

2) Soft Law for Living Organ Donation

The general findings just made will be applied to the field of LOD. The European Union does not have the competence to establish hard law for LOD. (1) In what ways can LOD be regulated via soft law? (2) Does it make sense at all to regulate LOD by soft law?

---

406 Stefan, Vol. 14 European Law Journal 753, 759 f. (2008).

407 Stefan, Vol. 14 European Law Journal 753, 760 (2008). The Court of First Instance has been renamed by the Treaty of Lisbon and is called General Court now (Horizontal Amendments 2, 2) (n) Treaty of Lisbon, in Vol. 50 Official Journal of the European Union 1, 43 (2007))

408 Stefan, Vol. 14 European Law Journal 753, 760 (2008).

409 Snyder, Vol. 56 Modern Law Review 19, 34 (1993).

410 Klabbers, Vol. 67 Nordic Journal of International Law 381, 382 (1998).

411 Stefan, Vol. 14 European Law Journal 753, 767 (2008).

412 Klabbers, Vol. 65 Nordic Journal of International Law 167, 167 ff. (1996); Klabbers, Vol. 67 Nordic Journal of International Law 381, 381 ff. (1998).

413 European Court of Justice [Europäischer Gerichtshof], 7 March, 2002, Case C-310/99, at 52; Stefan, Vol. 14 European Law Journal 753, 761 (2008).

a) To What Extent Can Living Organ Donation be Regulated via Soft Law?

It is questionable whether (1) LOD can be the content of soft law at all and (2) which soft law instrument(s) in particular would make sense to *regulate* LOD.

aa) Living Organ Donation as Possible Content of Soft Law?

The parties who want to pass an agreement are free in deciding whether it should be binding or not.[414] They can also decide about the content they want to include in the agreement.[415] That also applies to soft law, so that the possible content of soft law is rather broad.[416] Hence, it is possible to regulate the requirements for LOD by soft law. The formulation in such a case can be as complete as the wording of hard law is.[417] Consequently, it would be theoretically possible to pass a complete set of soft law rules with the requirements for LOD.

bb) Possible Soft Law Instrument(s) to Regulate Living Organ Donation

As already mentioned above, several different instruments are used in the European Union to pass soft law.[418] To recapitulate, the following soft law measures have regularly been used by the organs of the European Union: "resolutions; recommendations; declarations; conclusions; green-and white papers; reports; guidelines; action papers; programmes; codes of conduct; communications; memoranda; opinions."[419]

When thinking about regulating LOD, it might make sense to pass a particular recommendation. Recommendations are not binding. The European Union could nevertheless only pass such a recommendation if the principle of conferral is not opposed to it. It is disputed whether this principle applies to recommendations at all. It is stated that this principle does apply,[420] but the opposite that it does not apply is opined as well.[421] The declining opinion argues that the European Commission can pass recommendations whenever it considers them to be necessary without being

[414] Guzman, Vol. 16 European Journal of International Law 579, 580 (2005).

[415] Ibid., p. 582 f.

[416] Cf. Hillgenberg, Vol. 10 European Journal of International Law 499, 506 (1999).

[417] Hillgenberg, Vol. 10 European Journal of International Law 499, 506 f. (1999).

[418] Chinkin, Vol. 38 International and Comparative Law Quarterly 850, 852 (1989); cf. Stefan, Vol. 14 European Law Journal 753, 758 (2008).

[419] Cosma/Whish, Vol. 14 European Business Law Review 25, 46 (2003).

[420] Ruffert, in Callies/Ruffert (ed.) (2011), Art. 288 AEUV at 97; Schroeder, in Streinz (ed.) (2012), Art. 288 at 144; Nettesheim, in Grabitz/Hilf, EU, Art. 249 EGV at 208.

[421] Kotzur, in Geiger/Khan/Kotzur (2010), p. 864.

explicitly granted with the particular competence.[422] Art. 288 *TFEU* contains a catalogue of the legal acts of the European Union. Five different acts are described in four different paragraphs. That leads to the assumption that the different legal acts have to be differentiated. The legal acts enumerated in the first three paragraphs of Art. 288 *TFEU*, namely regulations, directives and decisions are binding. The legal acts listed in paragraph four, namely recommendations and opinions, are, in contrast, not binding. This suggests that recommendations and opinions have to be distinguished from regulations, directives and decisions. It leads to the assumption that the principle of conferral is not applicable in the same way to recommendations and opinions as it is to regulations, directives and decisions. Considering the purpose of the principle of conferral leads to the same result. Its main purpose is to divide the competences between the European Union and its Member States[423] and to limit the competences of the European Union.[424] Considering this, it does not seem necessary to apply the principle of conferral to recommendations, because those legal acts are not binding for the Member States. Therefore, the Member States do not have to comply with them and their competences are not curtailed by a recommendation of the European Union. As worked out, when the European Union wants to pass a recommendation, it does not need to comply with the principle of conferral. Thus, even though the European Union has no all-embracing competence for LOD, it could, nevertheless, pass a recommendation that contains the requirements for LOD.

Besides, incentive measures are also not able to interfere with the responsibility of the Member States to determine their health care policy, the organisation of public health and the medical care.[425] It has been explained above that the European Union usually uses incentive measures to supplement measures of the Member States concerning the content or support national measures. The European Union has no intention of carrying out this sort of function in LOD; rather, it seeks to regulate LOD requirements.

Since the European Union has also already used several other instruments to pass soft law, several instruments are conceivable for the European Union to pass soft law for LOD. The option to pass a recommendation has been introduced. This is especially advisable, because, in comparison to several other potential soft law instruments, recommendations have a legal basis (Art. 288 *TFEU*).[426] To base a regulation on a legal basis provides legal clarity and prevents uncertainty that might arise due to an unknown legal instrument or a legal instrument that has no explicit legal basis.

---

[422] Geiger, in Geiger/Khan/Kotzur (2010), p. 90.

[423] Biervert, in Schwarze (ed.) (2009), Art. 249 at 11.

[424] Cf. Oppermann/Classen/Nettesheim (2011), p. 61.

[425] Cf. Berg, in Schwarze (ed.) (2009), Art. 152 at 35 with respect to the former regulation; cf. Lurger, in Streinz (ed.) (2012), Art. 168 at 38.

[426] Cf. Knauff, Vol. 46 Zeitschrift Europarecht 735, 742 (2011).

b) Would It Make Sense to Regulate Living Organ Donation by Soft Law?

It has to be ascertained whether it would make sense to pass soft law for LOD. Would a soft law regulation influence the practice of LOD? If it has no influence on LOD, it would make no sense for the European Union to pass soft law.

If one recalls the advantages of a unification of laws, it would be preferable to establish soft law in comparison to remaining inactive. Admittedly, this would have fewer implications than passing hard law,[427] but this competence does not exist at the moment. On the contrary, it would have more influence than the absence of any arrangement.

Soft law would make it possible for the European Union to impart its suggestions for LOD regulation to its Member States. And, regardless of the legal effect of soft law, it is usually complied with it.[428]

In addition, it is not seldom for soft law to promote the development of hard law.[429] If the final goal is to create hard law, the soft law regulations might serve as a basis for that. Soft law may already bring about the intended result and is therefore sufficient, making it unnecessary to pass hard law.[430] It could lead the Member States to reassess their rules according to the the European Union's suggestions. The function of soft law — to create coordinated national legislation[431] or to serve as coordinating device[432] — would then be fulfilled.

The option to pass a recommendation has already been introduced and will be analysed further. Since a recommendation "shall have no binding force," passing it would be pointless if it were to have no implications at all. That cannot be expected, though, because recommendations are intended to suggest certain behaviour to the addressee.[433] They actually have political[434] and psychological effects.[435] Recommendations even have legal effects,[436] despite the fact that they are not legally binding. The addressees usually comply with it.[437] In addition, the ECJ has judged that recommendations have to be considered when national legislation is

[427] Cf. Guzman/Meyer, Vol. 2 Journal of Legal Analysis 171, 180 (2010).

[428] P. 260.

[429] Hillgenberg, Vol. 10 European Journal of International Law 499, 501 (1999).

[430] Gersen/Posner, Vol. 61 Stanford Law Review 573, 573 ff. (2008).

[431] Hillgenberg, Vol. 10 European Journal of International Law 499, 501 (1999).

[432] Guzman/Meyer, Vol. 2 Journal of Legal Analysis 171, 188 ff. (2010).

[433] Haratsch/Koenig/Pechstein (2010), p. 179; Streinz (2012), at 518.

[434] Arndt/Fischer/Fetzer (2010), p. 54; Borchardt (2010), at 547; Streinz (2012), at 517.

[435] Borchardt (2010), at 547.

[436] Borchardt (2010), at 547; cf. European Court of Justice [Europäischer Gerichtshof], 13 December, 1989, Case C-322/88, at 18; Knauff, Vol. 46 Zeitschrift Europarecht 735, 737 (2011); Stefan, Vol. 14 European Law Journal 753, 753 (2008) (only with regard to recommendations).

[437] Arndt/Fischer/Fetzer (2010), p. 54.

interpreted.[438] Consequently, the passing of it would have repercussions and would therefore make sense.

In the best case, the Member States would consider the soft law rules as suggestive and would follow them, which would increase the similarity between the national regulations (the advantages have already been clarified). Even if the Member States do not agree with all suggestions made by the European Union, they might at least question (a few of) their own rules and think about how to improve their system of LOD. A recommendation by the European Union will probably at least give ideas to the states on how they can improve their current regulations. A harmonisation of the national laws by passing hard European law would probably be preferable, but if that is impossible right now, a soft law regulation seems desirable.

V. Result

The European Union has no all-embracing competence in the field of LOD.[439] That does not mean, however, that the European Union has no competence at all. Art. 168 *TFEU* is the most relevant provision for the European Union in the field of health protection. It grants the European Union the opportunity to become active in the field of LOD, but this competence does not include the whole area of LOD. The European Union may only pass regulations "setting high standards of quality and safety of organs"[440] and must "not affect national provisions on the donation or medical use of organs."[441] As a consequence, the European Union cannot establish binding requirements for LOD,[442] however, the option to establish soft law remains.[443] The European Union can thus either stay inactive or establish soft law to *regulate* the requirements for LOD. One main characteristic of soft law is that it is not legally binding. It, nevertheless, has effects and addressees often comply with such regulations. It consequently makes sense for the European Union to establish soft law for LOD, for example, a recommendation, in comparison to remaining inactive. The suggested content for such a rule will be worked out in the next chapter, where a best practise proposal for LOD is established.

[438] European Court of Justice [Europäischer Gerichtshof], 13 December, 1989, Case C-322/88, at 18.

[439] Norba (2009), p. 343.

[440] Art. 168, IV (a) Treaty on the Functioning of the European Union.

[441] Ibid., Art. 168.

[442] Gutmann/Schroth, p. 43 with regard to the former Art. 152, 4.

[443] Cf. Cosma/Whish, Vol. 14 European Business Law Review 25, 34 (2003); cf. Klabbers, Vol. 65 Nordic Journal of International Law 167, 171 (1996).

Best Practise Proposal for Living Organ Donation in the European Union

A harmonisation of the national regulations for LOD is desired. In the last chapter, the examination found, however, that the European Union is only competent to pass soft law with respect to the requirements of LOD. The potential content of such a (harmonised) regulation will now be worked out in a best practise proposal for LOD.

I. Introduction: Procedure and Method

Legal experts have shown that LOD regimes create, in part, an *artificial scarcity*.[1] An appropriate legal framework for LOD, one without unjustified legal barriers and misguided rationales, must be established. The purpose of this chapter is twofold: it presents the results of the previous analysis and the policy proposals following from it, in other words, it establishes a *Common Frame of Reference for European Laws on living organ donation (CFR-LOD)*.

A comparison of laws, which is necessary to unify the differing national laws,[2] has been carried out in chapter "Comparative Analysis of European Transplant Laws Regarding Living Organ Donation." The results show that some of the central issues of LOD are regulated homogeneously, while other central issues are regulated differently. The compelling reasons to level these differences and to aim at a common European policy for LOD have already been made explicit. To recapitulate, the Member States of the European Union continually grow together and the level of cooperation between them steadily increases. This development can

[1] Cf. Evans, Vol. 15 Journal of Medical Ethics 17, 19 (1989); Fateh-Moghadam, (2011); cf. Radcliffe Richards, in Weimar/Bos/Busschbach (ed.) (2011), p. 41.

[2] Cf. Beck/Burchhard/Fateh-Moghadam, in Beck/Burchhard/Fateh-Moghadam (ed.) (2011), p. 5; Zweigert/Kötz (1996), p. 16.

L. Lopp, *Regulations Regarding Living Organ Donation in Europe*,

DOI 10.1007/978-3-642-33799-4_7,

also be observed in the field of organ transplantation.[3] European citizens in need of organ transplantation should have equal access to LOD if it is in accordance with justified legal and ethical principles. A harmonisation is, therefore, desirable and possible (the *Common Frame of Reference for European private law*[4] is a fitting example of the harmonisation of European law).

When aiming at harmonisation, one must not forget the diverse cultural backgrounds and legal systems.[5] LOD is embedded in national legal systems with different legal traditions, regulatory techniques, leading principles and case law. A comparison of the general legal principles of consent in some countries, which was considered at the beginning of this thesis, has illustrated this point nicely. Germany and the United Kingdom are fitting examples. In Germany, bodily harm is generally justified if the person injured consents to it.[6] In contrast, actual bodily harm is punishable regardless of the consent of the person concerned in the United Kingdom. The only time bodily harm is not punishable is if a legally recognised exception exists.[7] However, even if the legal foundations in Germany and the United Kingdom differ, the regulatory *telos* — LOD is legitimate under certain circumstances — is equal (the current British regulation on LOD, in certain ways, is even more flexible and liberal than the German one). Hence, a common aim is possible even if different methods, approaches and regulatory techniques are being used.

But not only comparative law gives reasons to expect that there may be different, yet equally functional legal means to serve the end of a justified *CFR-LOD*. The same follows from what *Marmor* calls the "unwritten rules of cross-national policy analysis" in medical care[8] and the problems associated with simple forms of *legal* or *policy transplantation*. In order to avoid gross mistakes in the use of comparative policy analysis, one must not believe in the "'naive' transplantation conception of cross-national learning" which tells us to search "widely for 'best practices'" and to assume, "if found, they can be transplanted without loss from site A to site B. [...] There is plenty of this about in the professional literature, but no social science support for the claim that a practise in one site can be transplanted without adaptation."[9] As there is a *legal* path dependency, there are also formative factors for *policy making* in each country, including the legacies of past policy on

---

[3] Cf. Gutmann/Schroth (2002), p. 41; cf. Prechern-Hauptmann, in Höfling (ed.) (2008), p. 97; Wollenschläger, in Müller-Graff/Schmahl/Skouris (ed.) (2011), p. 447.

[4] See Joint Network on European Private Law (2012).

[5] Cf. von Bar/Zimmermann (2002), p. XXVIII; cf. Meyer (2007), p. 149; Nielsen, in Price/Akveld (ed.) (1997), p. 75; cf. Shinkawa, in Riesenhuber/Takayama (ed.) (2006), p. 389; cf. Zweigert/Kötz (1996), p. 16.

[6] Sec. 228 German Criminal Code.

[7] Fateh-Moghadam (2008), p. 181.

[8] See, also to the following, Marmor, Vol. 7 Health Economics, Policy and Law 19, 19 (2012); see also Klein, Vol. 69 The Milbank Quarterly 275–291 (1991); Marmor/Freeman/Okma, Vol. 7 Journal of Comparative Policy Analysis 331, 333 (2005).

[9] Marmor, Vol. 7 Health Economics, Policy and Law 19, 20 (2012).

institutional features. Nevertheless, there is no reason to believe that different legal and/or policy systems cannot learn from one another.[10]

This thesis so far attempted to define the content of what is on the common policy agenda by referring to medical facts, normative reasoning and to current legal formulations in European countries. It located the basic problems and presented the countries' current solutions to these problems. The Council of the European Union stated: "The importance of organisational aspects of organ procurement and the large differences in practises and performances across Member States show a clear benefit of exchanging best practise between the Member States of the European Union."[11]

The aim of this chapter is therefore to establish a *best practise proposal for LOD in Europe*. Following the Introduction to the *Principles of European Contract Law*, it will be tried "to establish those principles which [are] believed to be best under the existing economic and social conditions in Europe,"[12] while the existing structures in national medical care systems are considered as well.

The chapter "Comparative Analysis of European Transplant Laws Regarding Living Organ Donation" is mainly a collection of the relevant information from the countries considered, while this chapter furthers the data collected.[13] It presents a short summary of the various national regulations on each issue and evaluates the different possibilities available to regulate these issues. The evaluation undertaken here is based on the results concluded in the chapters "Comparative Analysis of European Transplant Laws Regarding Living Organ Donation" and "Analysis of the Normative Arguments That Dominate the Policy Arena About Necessity and Legitimacy of Legal Restrictions in Living Donor Transplantation." In addition, common principles are taken into account, especially those established in the *CFREU*.

II. Establishment of a Best Practise Proposal for Living Organ Donation in the European Union

For good reasons, the general attitude about LOD is positive. Thus, those in favour of restricting LOD must be able to prove that the restriction is legitimate. In addition, since too few donor organs are available, any legal restrictions on LOD

[10] Cf. Meyer (2007), p. 149.

[11] Commission of the European Communities (2008).

[12] Commission on European Contract Law (2006).

[13] Klein clarifies that collecting information means to learn about other countries, while reflecting on that information means to actually learn from their experience (Klein, in Marmor/Freeman/Okma (ed.) (2009), p. 305).

must be justified and any unjustified barriers to LOD should be removed.[14] However, the main priority of a best practise proposal should be to respect the potential donor's informed and voluntary choice and ensure his welfare.[15]

Which of the existing regulations are optimal? (1) I evaluate the different regulations with regard to consent and disclosure and (2) I address the required suitability of the transplantation. (3) The regulations with respect to minors and mentally incapacitated adults are considered. (4) The discussion of the appropriate amount of restriction on the donor-recipient relationship is revived. (5) I illustrate whether it would be preferable to consider LOD as a subsidiary option only. (6) Several procedural issues are examined (7) and last, the optimal social security regulations for donors are worked out.

1) Informed Consent

The transplant laws of the Member States of the European Union are unanimous in their requirement for a donor's informed consent.[16] In Austria, Germany, the Netherlands, Portugal and Switzerland, informed consent is generally sufficient to justify LOD.[17] In England, France, Italy and Spain, in contrast, informed consent is necessary, but not sufficient to justify LOD; rather, an additional legitimising element is required.[18] This difference is rooted in the general national laws of the respective countries. The aim to legitimise LOD can thus be reached differently.

The purpose of informed consent has been made clear and it is unquestionable that the requirement of informed consent has to be part of any transplantation law. However, how the details of this premise must be worked out is unclear. Consent and disclosure will be regarded separately in the following sections. Establishing two separate regulations is preferable to reducing both aspects to a single regulation.[19] This makes the importance of the process of disclosure clear.[20]

---

[14] Fateh-Moghadam (2011), p. 1 ff.; With respect to indirect LOD and unspecified LOD, "the barriers reported most commonly were that these donations were illegal in the country or thought to be unethical and could not be justified" (EULOD WP 2 (2012), DOW: Deliverable 4, p. 12).

[15] Joint Working Party of The British Transplantation Society and The Renal Association (2011), p. 26.

[16] See e.g. Donnelly/Price (ed.) (1997), p. 35. This became apparent in the legal comparison in chapter 3, but is also clarified by the inclusion in the Charter of Fundamental Rights of the European Union (Art. 3 2. (a) of the Charter requires that the free and informed consent of the person concerned must be respected in particular in the fields of medicine and biology).

[17] P. 28 ff.

[18] P. 25 ff.

[19] Cf. Canellopoulou-Bottis, Vol. 7 European Journal of Health Law 427, 433 (2000); cf. Spickhoff, Vol. 45 Zeitschrift für Rechtspolitik 65, 67 (2012).

[20] Canellopoulou-Bottis, Vol. 7 European Journal of Health Law 427, 433 (2000).

a) Consent

All national laws considered require the donor to give valid consent. So, a best practise regulation must explicitly require the donor's consent. The details of the requirement for valid consent partly differ in the countries considered, making individual evaluations necessary.

aa) Capacity to Consent

Consent can only be given by a competent individual.[21] An adequate transplantation law should require the donor's capacity to consent, but need not include a definition of competence, since a definition is typically provided in the national laws.

bb) Voluntariness of Consent

The need for voluntary consent is expressed by all the countries considered. *Gutmann*, for example, considers the lack of donor's voluntariness to be the worst case in LOD (this side of the donor's death).[22] Voluntariness must, consequently, be required in any adequate transplantation law.[23]

As stated in the chapter "Analysis of the Normative Arguments That Dominate the Policy Arena About Necessity and Legitimacy of Legal Restrictions in Living Donor Transplantation," there is no general agreement on the most appropriate definition of voluntariness. None of the national transplant laws contains a definition of voluntariness or criteria to apply in order to prove whether a certain person acts voluntarily or not.[24] A detailed approach to determine how to decide whether the donor's decision was made voluntarily should not be included in the best practise proposal. This would make the best practise proposal too unwieldy and impracticable. It could, however, be mentioned in an explanatory report.

cc) Formal Requirements

The formal requirements for consent differ significantly throughout the countries considered. A best practise proposal should emphasise the necessity of an announcement and the insufficiency of internal consent.[25] In addition, the donor

[21] Cf. Childress/Liverman (2006), p. 270 f.; Kasiske et al., Vol. 7 Journal of the American Society of Nephrology 2288, 2296 (1996).

[22] Information from T. Gutmann.

[23] Cf. Commission of the European Communities (2008).

[24] Cf. Lilie/Krüger, Vol. 81 Der Chirurg 787, 788 (2010); cf. Price (2000), p. 289.

[25] Cf. Schroth, in Schroth et al. (ed.) (2006), p. 93.

should have enough time to reconsider his decision. After consent is given, he should therefore be granted an adequate period of time until the surgery actually takes place.[26]

Are any additional formal requirements recommendable? One potentially important formal requirement is the donor's written consent. Many argue that a written consent is favourable for the doctor and for the potential donor. Firstly, the written consent proves to be valuable if a dispute arises about whether a valid consent was given at all. If the doctor has no written consent, then he has the burden of trying to prove that consent was given. Therefore, many claim that doctors are better prepared for such a dispute if the consent is written down on paper.[27] Secondly, many argue that requiring the written form protects the potential donor, as it underscores the seriousness of his decision. The written consent is supposed to make him cogitate about his decision to donate one more time.[28] In contrast, others argue that the consenting donor is only performing his right of self-determination, and this should not be complicated by any formal requirements.[29] To require the consent to be in writing cannot seriously be considered as making it (more) complicated for the donor to perform his right of self-determination. Also, the majority of the countries demand the consent of the living donor to be in writing. Written consent should, consequently, be required in an adequate transplantation law.

Apart from the formal requirements just listed, a best practise regulation should not contain any additional mandatory formal requirements. Whether consent is to take place "before witnesses, [. . .], before another person other than the physician performing the transplantation, before a public authority, or even with approval by a public authority or a judge" should be decided by the individual countries.[30] Thus, only the minimum requirements, namely that the donor's consent must be announced, that he needs to have an appropriate amount of time to reconsider his decision, and that it must be in writing, should be included in a best practise proposal.

dd) Withdrawal of Consent

Almost all countries allow the potential donor to withdraw his consent. The possibility of withdrawal of consent must be present at all times,[31] and this must be explicitly mentioned in the best practise regulation. The donor must be informed about this option.[32]

---

[26] Wright et al., Vol. 47 Canadian Journal of Surgery 408, 412 (2004).

[27] Bock (1999), p. 153; Schreiber (2004), p. 73.

[28] Schreiber (2004), p. 73.

[29] Bock (1999), p. 152 f.

[30] Nielsen, in Price/Akveld (ed.) (1997), p. 75.

[31] Bakker, in Price/Akveld (ed.) (1997), p. 28.

[32] Hodson, Vol. 26 Journal of Medical Ethics 419, 420 (2000).

ee) Limitation of Consent

An internationally agreed standard is that any LOD which ceases the donor's life, or would be directly life-threatening, is illegal.[33] Several countries additionally declare consent void if it goes against good morals. Defining further *limitations* in the best practise proposal is highly questionable. Apart from those already mentioned, additional limitations should be decided on a case-by-case basis.[34]

ff) Consent of the Recipient

The recipient, due to his right of autonomy, has the right to decline LOD.[35] To make this obvious, the recipient's consent should be demanded by the best practise proposal.

b) Disclosure

All countries considered require the potential donor to become sufficiently informed. This requirement, without doubt, has to be included in any adequate transplantation law. National provisions are not equal in depth. The aim of every disclosure should be to educate the person concerned so that she can come to a self-determined decision.[36] Since the organ removal is a non-therapeutic treatment for the donor,[37] who usually is a medically layperson,[38] the requirements for the amount of information given to the donors should be very high.[39]

aa) Disclosure Matters

The disclosure of LOD's possible consequences, the medical background of the surgery, the protection of the donor and the impact on the recipient are referred to in several national transplant laws. It may seem sufficient to only require the donor to

---

[33] Dickens/Fluss/King, in Chapman/Deierhoi/Wright (ed.) (1997), p. 98; cf. Gutmann/Schroth (2002), p. 49; Pfeiffer (2004), p. 45.

[34] This has been discussed at the 2011 ELPAT-Working Group on Legal Boundaries Conference in Berlin.

[35] Cf. Gutmann/Gerok, in Collins et al. (ed.) (1997), p. 320 f.; cf. Gutmann/Land, Vol. 384 Langenbeck's archives of surgery 515, 516 (1999).

[36] Schroth, in Schroth et al. (ed.) (2006), p. 95.

[37] Fateh-Moghadam et al. in Schroth et al. (ed.) (2006), p. 161.

[38] Bock (1999), p. 144; Schreiber (2004), p. 76.

[39] Bock (1999), p. 145 f.; Edelmann, Vol. 50 Versicherungsrecht 1065, 1067 (1999); Gutmann, Vol. 15 Medizinrecht 147, 151 (1997); Spickhoff, Vol. 59 Neue Juristische Wochenschrift 2075, 2075 (2006).

be informed extensively. However, an actual list of the aspects that have to be included in the donor's disclosure would provide legal clarity and legal certainty.[40] This is, therefore, preferable. The entire (medical) procedure should be explained to the donor.[41] Furthermore, the risk of death must be disclosed[42] as well as any known risks to the donor's health, relating to evidence-based data. Since LOD always involves two *patients*,[43] the donor should be informed about alternative therapies for the recipient.[44] In addition, the recipient's prospect of success and risks should be disclosed to the donor. In a best practise regulation, the donor should be mandatorily informed about any additional, foreseeable circumstances that are relevant to her.[45] Before listing the aspects the disclosure has to contain, the term "especially" should be written.[46] This makes it possible for the jurisprudence to identify and create further, new disclosure matters.[47] One issue open to deliberation is whether the disclosure of the donor should be defined objectively or whether the exact content should be up to the person concerned. In favour of a subjective orientation, many claim that not every patient wants to be burdened with all the details.[48] Others, in contrast, may require an in-depth understanding.[49] An objective approach would provide legal certainty. It could, for example, prevent any conflicts that may arise post-LOD regarding the donor's prior level of knowledge on the topic. It might also be difficult to accurately evaluate exactly what the donor wants to be informed about and what she might rather not want to know. Consequently, a best practise regulation should clearly present what the donor has to be informed about. The listings in Art. 9, Swiss *Transplantation Ordinance* and Sec. 8 II German *Act on the donation, removal and transplantation of organs* can serve as examples.

bb) Comprehension

Only a few national laws explicitly require the donor to understand the information. If the donor does not comprehend the information provided to her, the whole

---

[40] Nickel/Schmidt-Preisigke/Sengler (2001), p. 109.

[41] Land, Vol. 2 Transplant International 168, 170 (1989). See also Principle (23) Directive 2010/53/EU of the European Parliament and of the Council of 7 July 2010 on standards of quality and safety of human organs intended for transplantation.

[42] Joint Working Party of The British Transplantation Society and The Renal Association (2011), p. 33.

[43] Mehrabi et al., Vol. 81 Der Chirurg 794, 800 (2010); Schroth, in Schroth et al. (ed.) (2006), p. 97 f.

[44] Mehrabi et al., Vol. 81 Der Chirurg 794, 800 (2010).

[45] Schroth, in Schroth et al. (ed.) (2006), p. 98.

[46] Spickhoff, Vol. 45 Zeitschrift für Rechtspolitik 65, 68 (2012).

[47] Ibid.

[48] Cf. O'Neill, Vol. 29 Journal of Medical Ethics 4, 6 (2003).

[49] Ibid.

process of disclosure can be thought of as useless. An adequate transplantation law should hence explicitly require the donor to understand the information provided to her.[50]

cc) Formal Requirements

There is no consensus on the disclosure's dependence on formal requirements.

A verbal dialogue between the potential donor and the person who informs her is necessary[51] because it is the only way to determine if the donor understands what the donation means to her.[52] It also allows the potential donor to immediately ask questions. The necessity of a verbal information of the potential donor is also in accord with the jurisdiction of the German Federal Supreme Court.[53] In addition, it is also a good idea for the given information to be presented to the donor in written form as well, which allows the donor a chance to re-read the information whenever she wants to.[54]

The countries that specify who the informant must be require him to be a physician. What kind of physician this has to be, and whether any kind of *third person* has to be present, is not regulated homogeneously. The attendance of a physician who is not part of the transplantation process should be required to ensure that the information is presented in an objective way.[55] Nevertheless, the physician who actually performs the transplantation should be involved in the disclosure as well, because she is familiar with the individual transplantation. The best practise regulation should therefore demand the presence of two physicians: the physician who performs the intended transplantation should inform the donor, and an additional physician who is not part of the transplantation process should be present as well.[56]

---

[50] Cf. Jowsey/Schneekloth, Vol. 22 Transplantation Reviews 192, 195 (2008); cf. Schroth, in Schroth et al. (ed.) (2006), p. 101; cf. Valapour, Vol. 22 Transplantation Reviews 196, 197 (2008); cf. Eyal, in Zalta (ed.) (2011).

[51] Deutsch, in Spickhoff (ed.) (2011), p. 6; Spickhoff, Vol. 59 Neue Juristische Wochenschrift 2075, 2076 (2006); Spickhoff, Vol. 54 World Medical Journal 9, 10 (2008); Spickhoff, Vol. 64 Neue Juristische Wochenschrift 1651, 1654 (2011).

[52] Spickhoff, Vol. 59 Neue Juristische Wochenschrift 2075, 2076 (2006).

[53] Cf. German Federal Supreme Court [Bundesgerichtshof], 7 February, 1984, Case No. VI ZR 174/82, in Vol. 35 Versicherungsrecht Rechtsprechung 465, 467 (1984); cf. German Federal Supreme Court [Bundesgerichtshof], 1 January, 1985, Case No. VI ZR 15/83, in Vol. 36 Versicherungsrecht Rechtsprechung 361, 362 (1985).

[54] Cf. Joint Working Party of The British Transplantation Society and The Renal Association (2011), p. 33.

[55] Esser, in Höfling (ed.) (2003), p. 227; Lilie/Krüger, Vol. 81 Der Chirurg 787, 788 (2010); Weigend/Zielinska, Vol. 14 Medizinrecht 445, 448 (1996).

[56] Cf. Gutmann/Schroth (2002), p. 116.

dd) Disclosure of the Recipient

In the chapter "Comparative Analysis of European Transplant Laws Regarding Living Organ Donation," it has been argued that both the donor and recipient should be informed. A best practise regulation should therefore require the recipient to be informed. The recipient's disclosure should contain information about her health, but also information about the donor's health and possible risks,[57] because recipients usually only accept limited risks for the donor.[58]

2) Suitability of the Transplantation

The transplantation must be medically suitable. This means a benefit for the recipient must be reasonably expected; it also means the potential donor should be suitable to act as living organ donor.[59] To ascertain whether the transplantation is suitable, a medical evaluation of donor and recipient is necessary. The *Directive 2010/53/EU* states: "Member states shall ensure that living donors are selected on the basis of their health and medical history, by suitably qualified or trained and competent professionals."[60] Although these aspects seem self-evident, they should be included in a best practise regulation for clarification purposes.

Many experts recommend requiring a risk-benefit equation in the best practise proposal. Only a few national laws explicitly include a risk-benefit equation. Since LOD always involves one healthy person, the risks should be balanced against the potential benefits.[61] In contrast, many doubt the risk-benefit ratio established by law is sufficient to justify the refusal of certain LOD's. They claim that it is untenable to prevent any LOD that is based on an informed, voluntary and well-considered wish.[62] A best practise regulation should consequently contain the provision of a reasonable risk-benefit equation for the donor, but its requirements should not be too detailed or paternalistic. The decision should be made by the autonomous donor (and the recipient) and transplant team.

---

[57] Cf. Fateh-Moghadam et al., Vol. 22 Medizinrecht 19, 30 (2004); Gutmann (2006), p. 59.

[58] Gutmann (2006), p. 59.

[59] Principle (23) Directive 2010/53/EU of the European Parliament and of the Council of 7 July 2010 on standards of quality and safety of human organs intended for transplantation.

[60] Ibid., Art. 15.

[61] Cf. American Medical Association, Vol. 284 The Journal of the American Medical Association 2919, 2920 (2000); Donnelly/Price (ed.) (1997), p. 52 f.; cf. O'Hara/Bramstedt/Flechner/Goldfarb, Vol. 17 Progress in Transplantation 180, 181 (2007).

[62] Hilhorst et al., Vol. 24 Transplant International 1164, 1165 (2011).

3) Minors and Mentally Incapacitated Adults

All national laws treat minors and mentally incapacitated adults differently than competent adults. The specific regulations for minors and mentally incapacitated adults are rather similar. The *WHO Guiding Principles on Human Cell, Tissue and Organ Transplantation* states: "What is applicable to minors also applies to any legally incompetent person."[63] The majority of countries completely prohibits minors and mentally incapacitated adults from acting as living donors. A few countries, in contrast, allow them to donate an organ while alive if certain circumstances are fulfilled. Determining whether minors and mentally incapacitated adults should be completely prohibited from becoming a living organ donor, or whether they are allowed to do so under specific circumstances, is of great importance. LOD might have severe consequences for the donor. This also applies to the complete prohibition to become a donor, though.

a) Complete Prohibition?

Many believe specific rules for minors and mentally incapacitated adults must be established because they might not be able to evaluate the situation properly.[64] Specific rules are perceived as a needed protection for them.[65] This argument can either be used in favour of a complete prohibition or in favour of allowing them to act as living organ donors under specific circumstances. A complete ban provides legal certainty and maximises the legal protection of minors and mentally incapacitated adults.[66] Also, donations that occur within families, which describes the majority of LOD cases, can be very stressful for the donor, placing him under the pressure of his family. Hence, many assume that mature people are able to handle the pressure better than immature people, who depend on the family to a greater extent.[67]

Opponents of a complete prohibition state that cases might exist in which LOD by a minor is ethically justifiable.[68] They also criticise complete prohibition insofar as it does not take the differences between more or less mature minors into

---

[63] WHO Guiding Principles on Human Cell, Tissue and Organ Transplantation, 2008, Guiding Principle 4.

[64] Leipold (2008), p. 127.

[65] Cf. Cheyette, Vol. 41 Boston College Law Review 465, 468 (2000); cf. Eisenhardt (2007), p. 94. The protection of minors and mentally incapacitated adults from themselves is considered legitimate and should not be equated with hard paternalism (Fateh-Moghadam, in Fateh-Moghadam/Sellmaier/Vossenkuhl (ed.) (2010), p. 26).

[66] Cf. Gutmann/Gerok, in Collins et al. (ed.) (1997), p. 319; cf. Gutmann/Schroth (2002), p. 55.

[67] Swiss Dispatch of 12. September 2001 on a Federal Law on Transplantation of Organs, Tissues and Cells, p. 145.

[68] Cf. Gutmann/Gerok, in Collins et al. (ed.) (1997), p. 319; cf. Gutmann/Schroth (2002), p. 55.

account.[69] According to this view, the most suitable approach to regulate LOD of minors and mentally incapacitated adults is to assess the person's capability individually. In England, the regulation for minors can serve as an example for this potential framework. There, the minor's capability to consent is judged individually.[70]

It is clear that minors and mentally incapacitated adults should receive more protection than competent adults. The most extensive protection would undeniably be to completely ban them from LOD. This, however, would most certainly lead to cases in which a prohibition would be ethically unjustified. Since minors and mentally incapacitated adults can also be protected sufficiently by established, clear legal requirements, under which they may act as living organ donors, a less restrictive approach seems preferable.

b) Suggested Requirements

Minors and mentally incapacitated adults should be allowed to donate a solid organ while alive if specific requirements are fulfilled. The requirements must be very strict, as the protection of minors and mentally incapacitated adults must be of the highest importance. The Joint Working Party of The British Transplantation Society and The Renal Association, for example, states that "[c]hildren should only be considered as living organ donors in exceptionally rare circumstances."[71] All requirements established must therefore be interpreted extremely narrowly. For any issue related to LOD in need of regulation that is not mentioned in the following, the requirements established for adults can be applied.

(1) The donor's informed consent is absolutely essential. The person concerned must be competent to do so. Some mentally incapacitated adults may have the competency to consent. For minors, the capacity to give valid consent can often be answered affirmatively if the minor is older. If the surgeon is unsure of the minor's or the mentally incapacitated adult's capability to consent, he must refuse the LOD. In case a person is incapable of giving consent, this requirement cannot be fulfilled by a parent or a legal guardian. However, it should be required that, in addition to the potential donor's consent, the parents of minors and the custodians of mentally incapacitated adults give their consent as well.
(2) The potential donor and the intended recipient must have a close personal relationship. LOD by a minor most commonly occurs for the benefit of a sick sibling (or even a twin). It is often very difficult for a healthy minor to see his sibling suffering. An ill child has an influence on the whole family life. Furthermore, "transplants between siblings offer the best opportunity for a

---

[69] Cf. Gutmann/Gerok, in Collins et al. (ed.) (1997), p. 319; cf. Gutmann/Schroth (2002), p. 55.

[70] Fateh-Moghadam (2008), p. 204 ff.

[71] Joint Working Party of The British Transplantation Society and The Renal Association (2011), p. 20.

well matched graft [...]."[72] Whether other relationships are sufficient as well is doubtful. It might make sense to allow a minor to donate to benefit a parent, since a parent's death is a traumatic incident for a minor. He should, therefore, be allowed to donate to a sick parent if all other requirements have been fulfilled. Mentally incapacitated adults should also only be allowed to donate for the benefit of a sibling or a parent.

(3) The risks attached to the donations of different organs differ.[73] Living kidney donation contains fewer risks for the living donor than the donation of other organs; it is the most common type of LOD. It consequently makes sense to only allow living kidney donation by minors and mentally incapacitated adults.

(4) Living kidney donation by a minor or a mentally incapacitated adult should be the *ultima ratio*. This should be made clear by law. The Swedish regulation can serve as an example. It only allows LOD by a minor if "medically suitable biological material cannot be taken from anybody else." This means a minor is only permitted to donate if there is no other donor who matches genetically or is equally suitable.

(5) An independent commission should have to approve the intended LOD by a minor or a mentally incapacitated adult.

A best practise regulation would not absolutely prohibit minors and mentally incapacitated adults from becoming organ donors. It would rather contain specific requirements that only allow them to act as living organ donors in very exceptional cases. Sweden can again serve as a model, because no donation of a solid organ by a minor has been approved yet, even though the possibility is granted by law.[74] This indicates that the mere legal possibility of LOD by minors and mentally incapable adults does not always result in the actual realisation of it. This must be kept in mind for the best practise proposal as well.

4) Restrictions on Living Organ Donation by Competent Adults

Should the donor-recipient relationship in LOD be restricted by law? (1) Is a restricted donor-recipient relationship preferable to an unrestricted donor-recipient relationship? (2) How should the different types of LOD that have been introduced in the chapter "Analysis of the Normative Arguments That Dominate the Policy Arena About Necessity and Legitimacy of Legal Restrictions in Living Donor Transplantation" be dealt with?

[72] Ibid., p. 157.

[73] Childress/Liverman (2006), p. 270; Guillod/Perrenoud, in Gutmann et al. (ed.) (2004), p. 158.

[74] Information from A. Lennerling; information from C. Möller.

a) Regulation

(1) The chapter "Analysis of the Normative Arguments That Dominate the Policy Arena About Necessity and Legitimacy of Legal Restrictions in Living Donor Transplantation" considered all arguments about the necessity and legitimacy of legal restrictions of the donor-recipient relationship. It will therefore be used to establish a best practise proposal for this common concern. (2) The extent of the fundamental rights established in the *CFREU* has been described in the chapter "Actions of the European Union So Far." It will be recapitulated in light of the donor-recipient relationship. (3) Finally, a conclusion will be drawn on whether the donor-recipient relationship should be restricted.

aa) Results of the chapter "Analysis of the Normative Arguments That Dominate the Policy Arena About Necessity and Legitimacy of Legal Restrictions in Living Donor Transplantation"

In favour of restricting the donor-recipient relationship, many argue that LOD could be viewed as a violation of the principle of nonmaleficence[75]; it contains risks for the donor. The principle of autonomy is used as a counterargument. This principle implies that every person has a right of self-determination.[76] This includes the right to decide about one's own body.[77] Citing the risks of LOD cannot serve as a valid reason to restrict LOD to direct LOD, because those are equally high for direct LOD as they are for other types of LOD.[78] Opponents claim the risks may be *perceived* differently in direct LOD than in unspecified LOD.[79] Even though one may plausibly consider how risks might be perceived, that does not mean it is a valid reason for prohibiting unspecified LOD.[80] In favour of restricting the donor-recipient relationship, some authors claim that unrelated donors do not benefit as much from LOD as related donors,[81] and that the special relationship between donor and

---

[75] Beauchamp/Childress (2001), p. 113; Bechstein/Moench, Vol. 24 Transplant International 1162, 1163 (2011); Potts/Evans, in Weimar/Bos/Busschbach (ed.) (2008), p. 377.

[76] Dreier, in Dreier (ed.) (2004), before Art. 1 GG at 75, Art. 2 II at 20; Gessmann (ed.) (2009), p. 72; Hügli/Lübcke (ed.) (1997), p. 71; cf. Joint Working Party of The British Transplantation Society and The Renal Association (2011), p. 25; Pöltner (2002), p. 94; Sachs, in Sachs (ed.) (2011), before Art. 1 at 43.

[77] Bundestag printed paper 15/5050 (2005), p. 34; Forkel, Vol. 23 JURA 73, 73 (2001); Gutmann/Schroth (2002), p. 110.

[78] Health Council of the Netherlands (2003); Hilhorst et al., in Weimar/Bos/Busschbach (ed.) (2011), p. 380; Matas et al., in Gutmann et al. (ed.) (2004), p. 195; Patel/Chadha/Papalois, Vol. 3 Experimental and Clinical Transplantation 181, 183 (2011); Veatch (2000), p. 189.

[79] Cf. Ross, Vol. 30 The Journal of Law, Medicine and Ethics 440, 442 (2002).

[80] Ibid.

[81] Cf. Patel/Chadha/Papalois, Vol. 3 Experimental and Clinical Transplantation 181, 181 (2011); cf. Persson et al., in Weimar/Bos/Busschbach (ed.) (2008), p. 271; cf. Roff, Vol. 33 Journal of Medical Ethics 437, 439 (2007).

recipient *exclusively* justifies LOD.[82] Both statements may be true, but neither is convincing enough to prohibit unspecified LOD. The most relevant feature of any LOD must be the donor's concern for the needs of others. This applies to any LOD, regardless of the relationship between the donor and the recipient.[83]

The second relevant aspect of the donor-recipient relationship is safeguarding the donor's voluntariness. Many believe that unspecified LOD is connected to involuntariness. The German Bundestag, for example, claims that only familial relationships or other close relationships guarantee voluntariness in LOD.[84] Social research, in contrast, demonstrates that organ donors with a competent, informed and voluntary decision to act altruistically towards strangers can be found in great numbers.[85] Involuntary decisions could occur in any type of LOD,[86] while every type of LOD also contains cases that are the result of a voluntary decision. The best approach, therefore, is to judge each case individually and to avoid the view that only certain donor-recipient relationships should be allowed.[87]

Third, some believe that the general legalisation of LOD would increase the pressure on living individuals to become organ donors. This belief has led many to demand that the donor-recipient relationship should be restricted.[88] These reservations can be easily dispelled: giving people the *chance* to choose to help, or even save a person, cannot be equated with imposing a *duty* on people to actually become an unspecified living organ donor.

Fourth, unspecified LOD[89] and cross-over LOD[90] are associated with organ trade. The possibility that unspecified LOD might cause commercialisation and organ trade can be nipped in the bud, though, if the donation occurs within a pool of organs that is handled by a neutral agency, which coordinates the organ

---

[82] Bundestag printed paper 15/5050 (2005), p. 74; Council of Europe Committee of Ministers, Resolution CM/Res(2008)4 on adult-to-adult living donor liver transplantation, 4b; cf. den Hartogh (2008), p. 83; World Health Organization, WHO Guiding Principles on Human Organ Transplantation, Guiding Principle 3.

[83] Hilhorst et al., Vol. 24 Transplant International 1164, 1167 (2011).

[84] Bundestag printed paper 13/4355 (1996), p. 20; see also German Federal Constitutional Court [Bundesverfassungsgericht], 11 August, 1999, Case No. 1 BvR 218/98, at 3b.

[85] Cf. German Federal Constitutional Court [Bundesverfassungsgericht], 11 August, 1999, Case No. 1 BvR 218/98, at 25; Hilhorst, Vol. 8 Ethical Theory and Moral Practice 197, 200 (2005); Landolt et al., Vol. 76 Transplantation 1437, 1441 (2003); cf. Massey et al., Vol. 10 American Journal of Transplantation 1445, 1445 (2010); cf. Spital, Vol. 71 Transplantation 1061, 1063 (2001).

[86] Gutmann (2006), p. 28 ff.

[87] Cf. Choudhry, Vol. 29 Journal of Medical Ethics 169, 169 (2003); Gutmann/Schroth (2002), p. 63; Price (2000), p. 328; cf. Sanner/Dew/Busschbach, in Weimar/Bos/Busschbach (ed.) (2008), p. 194 f.

[88] Bundestag printed paper 15/5050 (2005), p. 40, 42 f., 48, 73 f.; cf. Schroth, in Schroth et al. (ed.) (2006), p. 82, who immediately after presenting this argument invalidates it.

[89] Cf. Evans, Vol. 15 Journal of Medical Ethics 17, 18, (1989); German Federal Constitutional Court [Bundesverfassungsgericht], 11 August, 1999, Case No. 1 BvR 218/98, at 3b.

[90] Bundestag printed paper 15/5050 (2005), p. 42, 73 f.

transplantation, organ allocation and maintains the anonymity of the donor.[91] Furthermore, past experiences of LOD between persons who were not closely related to each other demonstrate that no uncontrollable danger of commercialisation exists.[92] The concern with respect to cross-over LOD is dispelled by referencing the donor's motives: to help a relative.[93]

None of the arguments provided in favour of a restriction of the donor-recipient relationship seems tenable. There is no class of donors and no kind of donor-recipient relationship that should be generally dismissed by law.

bb) *Charter of Fundamental Rights of the European Union*

The chapter "Possibilities of Harmonisation" demonstrated how a complete prohibition of LOD, as well as unjustified legal barriers and misguided rationales, would be in conflict with Art. 2 (1), Art. 3 (1) and Art. 35 *CFREU*. The chapter "Analysis of the Normative Arguments That Dominate the Policy Arena About Necessity and Legitimacy of Legal Restrictions in Living Donor Transplantation" concluded that no understandable reasons exist to generally require certain donor-recipient relationships by law. Such a restriction can thus be considered as an unjustified legal barrier. The *CFREU* is hence in accord with the findings made in chapter "Analysis of the Normative Arguments That Dominate the Policy Arena About Necessity and Legitimacy of Legal Restrictions in Living Donor Transplantation"; both are against only allowing LODs with specific donor-recipient relationships.

Since potential donors who are not included in the allowed circle of donors in a concrete case cannot donate, this restriction is thus essentially a complete prohibition for them. A complete prohibition of LOD is without doubt in conflict with the *CFREU*, meaning that then a restriction of the donor-recipient relationship is also in conflict with the *CFREU*.

cc) Interim Result

Chapter "Analysis of the Normative Arguments That Dominate the Policy Arena About Necessity and Legitimacy of Legal Restrictions in Living Donor Transplantation" and the *CFREU* both argue against a restriction of the donor-recipient relationship for LOD. It seems preferable to evaluate concrete risks, rather than to distinguish between relationships.[94] As a consequence, a best practise regulation

---

[91] Gutmann/Schroth (2002), p. 16, p. 121 f.; cf. Schreiber, in Rittner/Paul (ed.) (2005), p. 64.

[92] Danovitch, in Weimar/Bos/Busschbach (ed.) (2011), p. 392; cf. Gutmann, Vol. 15 Medizinrecht 147, 150 (1997); cf. Swiss Dispatch of 12. September 2001 on a Federal Law on Transplantation of Organs, Tissues and Cells, p. 98.

[93] de Klerk (2010), p. 14; Gutmann/Schroth (2002), p. 118; Veatch (2000), p. 187.

[94] Cf. Hilhorst et al., Vol. 24 Transplant International 1164, 1165 (2011).

should not contain a restriction of the donor-recipient relationship, but should require case-by-case decisions.[95]

b) Special Cases

Chapter "Analysis of the Normative Arguments That Dominate the Policy Arena About Necessity and Legitimacy of Legal Restrictions in Living Donor Transplantation" showed that several types of LOD are medically possible. The main argument in favour of allowing many types of LOD is that they are often able to overcome any incompatibility between the willing donor and the intended recipient, making it possible to increase the amount of donor organs. Critics fear the potential donor might be inflicted with increasing pressure. Overall, I concluded that cross-over LOD, living paired cascade exchange, unspecified LOD and pool donation seem to be legitimate. Unbalanced living paired exchanged, directed altruistic LOD, unspecified non-directed donation catalysing cascade exchanges and list-paired exchanges, in contrast, seem to be rather problematic, although there is hardly a reason for *legally* banning them.

A best practise regulation cannot mention every type of LOD. This would cause an impracticable extension of the law. Also, since the medical possibilities increase steadily, the law would be obsolete very quickly and, thus, would need to be revised continuously. A best practise regulation should, therefore, not determine which types of LOD are to be permitted or prohibited. Rather, one should be able to interpret the best practise proposal's particular regulations in light of the different types of LOD, and then judge whether or not the LOD is legitimate. This approach is in the style of the *Principles of European Contract Law*. Those rules are also intended to be very flexible, making it possible to consider future developments in the legal thinking of contract law. The *Principles* do not regulate every eventuality, because that would be too detailed and concrete, constricting a further development of European contract law.[96]

Most types of LOD can only exist if a sufficient national scheme has been established. Whether the countries or national transplantation systems do so is up to them, although there are good reasons for such policies. The Netherlands[97] and the United Kingdom[98] have, e.g., established such a scheme for cross-over LOD.

---

[95] Cf. Gutmann et al., in Terasaki (ed.) (1995), p. 356; cf. Gutmann (2011), p. 5; Land, Vol. 2 Transplant International 168, 173 (1989) with regard to spouse-to-spouse donation.

[96] von Bar/Zimmermann (2002), p. XXIX.

[97] P. 104 f.

[98] Joint Working Party of The British Transplantation Society and The Renal Association (2011), p. 161.

5) Subsidiarity

When deciding whether LOD should be subsidiary, two aspects need to be considered. (1) One must evaluate whether LOD should be subordinate to post-mortem organ donation. (2) The relation between LOD and alternative therapies must be clarified.[99]

a) Relationship Between Living Organ Donation and Post-Mortem Organ Donation

Chapter "Comparative Analysis of European Transplant Laws Regarding Living Organ Donation" showed that there is no consensus between the countries on whether LOD should be subsidiary to post-mortem organ donation or not.

Most argue that LOD should be subsidiary because they believe this is the only way to ensure the protection of potential living organ donors, since LOD requires a healthy person to become part of a surgery that contains risk.[100] This is sometimes considered as a violation of the principle of nonmaleficence.[101]

In contrast, one strong argument against the principle of subsidiarity is that the donor has a right of self-determination.[102] This even applies to self-harming behaviour.[103] The potential organ recipient has a right of self-determination as well. So, if she has a choice between receiving a LOD or a post-mortem organ donation, she should be able to choose which organ she wants to receive.[104] Living kidney donation leads to better short-term and long-term results than post-mortem kidney donation.[105] As a consequence, the principle of subsidiarity might lead to cases in which the potential recipient is forced by law to receive the worse treatment.[106] Another convincing argument against the principle of subsidiarity is that too few donor organs are available. It is, therefore, highly questionable for a patient, who could receive a donor organ from a living person, to be forced to receive an organ donated by a deceased person. Furthermore, this also entails that

---

[99] Gutmann/Schroth (2002), p. 76.

[100] Childress/Liverman (2006), p. 270; Guillod/Perrenoud, in Gutmann et al. (ed.) (2004), p. 158.

[101] Cf. Gutmann/Land, Vol. 384 Langenbeck's archives of surgery 515, 516 (1999); Schutzeichel (2002), p. 36.

[102] Cf. Beauchamp/Childress (2001), p. 63; cf. Gutmann/Schroth, in Oduncu/Schroth/Vossenkuhl (ed.) (2003), p. 282.

[103] Gutmann, in Schroth et al. (ed.) (2006), p. 220.

[104] Edelmann, Vol. 50 Versicherungsrecht 1065, 1068 (1999).

[105] Esser, in Höfling (ed.) (2003), p. 201; Gutmann, in Schroth et al. (ed.) (2005), Sec. 8 at 22; cf. Land/Gutmann, Vol. 35 Transplantation Proceedings 926, 928 (2003); Lilie/Krüger, Vol. 81 Der Chirurg 787, 787 (2010); Norba (2009), p. 60; Schreiber (2004), p. 102; Teubner (2006), p. 5; Ugowski (1998), p. 22.

[106] Edelmann, Vol. 50 Versicherungsrecht 1065, 1068 (1999); Gutmann, Vol. 15 Medizinrecht 147, 152 (1997); Gutmann (2011), p. 9.

the next person on the waiting list will not receive that particular organ donated post-mortem.[107]

When weighing the conflicting arguments, the right of self-determination that results from the principle of autonomy must be rated rather highly. This right is protected by the *CFREU*, which confirms that all countries of the European Union consider this a fundamental right. Although this right is not absolute, the arguments mentioned in favour of the principle of subsidiarity are still not convincing. Potential living organ donors should without doubt be protected, but this is also possible without restricting their right of self-determination; for example, safety by procedure[108] would be a less drastic and therefore preferable action.

To summarise, LOD should not be subsidiary to post-mortem donation.[109] Nevertheless, post-mortem organ donation should never be regarded as secondary to LOD, as it is a very valuable means to alleviate the organ shortage.[110]

b) Relationship Between Living Organ Donation and Alternative Therapies

Several laws contain a provision stating that performing LOD is only legitimate when other methods of therapy are less effective or do not provide comparable efficiency. Art. 12 d Swiss *Transplantation Act* states, "Organs, tissues and cells may be removed from a living person if: [...] the recipient cannot be treated with any other therapeutic method with comparable benefit." This regulation clarifies that LOD is the *ultima ratio*, but does not cause a patient suffering from kidney failure to depend on the worse option of dialysis if LOD is possible.[111] Hence, a regulation like the one in Switzerland is recommendable.

6) Procedural Issues

As this best practise proposal suggests eliminating legal restrictions on the donor-recipient relationship, it seems important to establish procedural safeguards in addition to the material aspects just presented ("Safety by procedure"[112]). In liberal societies, people generally are allowed to behave unwisely, to harm or cause disadvantages for themselves without the law protecting them by stipulating a certain procedure (e.g., people go to casinos to waste their money without serious

[107] Gutmann (2006), p. 74; Gutmann, in Middel et al. (ed.) (2010), p. 33.

[108] Cf. Gutmann et al., in Terasaki (ed.) (1995), p. 356 ff.; cf. Gutmann (2011), p. 6; Gutmann (2011), p. 9.

[109] Gutmann/Schroth (2002), p. 82; Gutmann (2006), p. 84.

[110] Cf. Schreiber, in Rittner/Paul (ed.) (2005), p. 63.

[111] Gutmann (2006), p. 78.

[112] Cf. Gutmann et al., in Terasaki (ed.) (1995), p. 356 ff.; cf. Gutmann (2011), p. 6.

restrictions, and all kinds of unhealthy behaviour, such as smoking, are not forbidden by law). However, it may be specific to LOD that it does not only concern the potential donor, but also the surgeon and the transplantation team. This distinguishes LOD from self-harming behaviour that only concerns the acting person and gives rise to the idea of procedural safeguards, especially to ensure the autonomy of the donors' decisions. It will therefore be examined next what procedural regulations should be part of a best practise regulation.

a) Commission for Living Organ Transplantation

Most countries have established some kind of independent authority as a safeguard for LOD.[113] Both the type of the commission and its composition differ, though.

First, the donor's consent and autonomy play a major role in LOD.[114] Adequate procedures are necessary to ensure the donor acts voluntarily.[115] According to some experts, hospitals should "institutionalize a careful and comprehensible screening procedure" to evaluate the donor's motivation.[116] The regulation in the German *Act on the donation, removal and transplantation of organs* could serve as a model for an adequate screening procedure (although Germany, having severe limitations concerning the classes of living donors, does not consistently rely on a safety-by-procedure-system).[117] According to the said act, a commission consisting of a physician, a lawyer and a psychologist must investigate and confirm that there is no substantiated reason to assume that the donor's consent is not being given freely and that the organ is not the object of prohibited organ trade.[118] The commission's decisions are not binding for the surgeon,[119] but still have a considerable impact.[120] A best practise regulation should clarify whether decisions by the commission in charge are binding or mere suggestions.[121] In favour of making decisions binding, some assert that the regulation is a safeguard for the donors.[122] In contrast, it might interfere with the donor's right to self-determination or with the recipient's fundamental right to life and health.[123] In addition, some question if the surgeon should

[113] Cf. Manyalich et al., Vol. 41 Transplantation Proceedings 2021, 2022 (2009).

[114] Gutmann (2006), p. 49 f.

[115] Gutmann (2006), p. 49 f.; Wagner/Fateh-Moghadam, Vol. 56 Soziale Welt 73, 79 (2005).

[116] Cf. Gutmann et al., in Terasaki (ed.) (1995), p. 356.

[117] Sec. 8 III 2 Act on the donation, removal and transplantation of organs.

[118] Ibid.

[119] Fateh-Moghadam, Vol. 21 Medizinrecht 245, 246 (2003); Nickel/Schmidt-Preisigke/Sengler (2001), p. 114.

[120] Fateh-Moghadam, Vol. 21 Medizinrecht 245, 254 (2003); Fateh-Moghadam et al., Vol. 22 Medizinrecht 82, 90 (2003); Wagner/Fateh-Moghadam, Vol. 56 Soziale Welt 73, 88 (2005).

[121] Lilie, Enquete Kommission (2004), p. 9.

[122] Cf. Gutmann (2006), p. 55 f.; cf. Fateh-Moghadam, Vol. 21 Medizinrecht 245, 245 (2003).

[123] Cf. Gutmann (2006), p. 55 f.

(or can) be released from having the final responsibility to decide whether the LOD will be executed or denied.[124] Nevertheless, because of the great importance of voluntariness, the commission's decisions should be binding. The disadvantages such a provision might have should be considered in light of the commission's composition and the decision-making process.[125] Apart from the aspects just mentioned, the acting of the commission can be classified as *weak paternalism*.[126] Even if the commission's decisions were not binding, they would still have relevance for the realisation of donors' and recipients' fundamental rights. This, however, is not considered problematic as long as the commission's decision is based on the aspect of voluntariness and interprets this in an appropriate way.[127] The details of the German Commissions' decision-making criteria, which are stipulated in regulations from 16 different Länder-legislation, however, are far from being well-defined. This causes arbitrariness and uncertainty. A best practise regulation should not adopt this approach.[128] Rather, even federalist countries should provide a regulation that applies to the whole country uniformly.[129]

Seeking a uniform procedure, it seems useful to require the approval of an independent commission for every LOD. This is usually required in countries that do not consider informed consent as sufficient to justify LOD. The British Human Tissue Authority may provide guidance here. It was established "to oversee and control the working of the [Human Tissue] Act."[130] The United Kingdom does not restrict the donor-recipient relationship, but every LOD needs to be approved by the Human Tissue Authority through an independent assessment process.[131] The Human Tissue Authority is a state authority with a wide range of people from different professions. It turns down offers of inappropriate LODs. It is suggested that this model works well.[132] Only a few rejections have taken place, numerically eight applications to perform a LOD have been turned down since 2006.[133] Clinicians as well as most patients like the approach, inter alia, because it is very flexible and considers each case individually.[134]

[124] Cf. Fateh-Moghadam, Vol. 21 Medizinrecht 245, 253 (2003); cf. Gutmann/Schroth (2002), p. 35.

[125] Cf. Gutmann (2006), p. 56.

[126] See Fateh-Moghadam, in Fateh-Moghadam/Sellmaier/Vossenkuhl (ed.) (2010), p. 26–48.

[127] Information from B. Fateh-Moghadam.

[128] Cf. Gutmann (2006), p. 62 f.; cf. Rixen, in Höfling (ed.) (2008), p. 76 f.

[129] Cf. Rixen, in Höfling (ed.) (2008), p. 76 f.

[130] Joint Working Party of The British Transplantation Society and The Renal Association (2011), p. 16.

[131] Sec. 13 ff. Human Tissue Act; information from D. Price; Joint Working Party of The British Transplantation Society and The Renal Association (2011), p. 16.

[132] Information from C. Rudge.

[133] Information from K. Rigg.

[134] Information from C. Rudge.

Third, a few countries demand every LOD to be disclosed. This makes the collection of data of LOD and the traceability of all organs donated by living persons possible. The duty to disclose every LOD to a central commission should therefore be written down in a best practise regulation.[135] This could be modelled on the Bulgarian regulation. The Bulgarian *Law of the Transplantation of Organs, Tissues and Cells* stipulates that each medical establishment that removes organs from living donors has to inform the Executive Agency for Transplantation thereof at least 7 days in advance.[136] After the organ removal, the medical establishment has to register the procedure within 7 days with the Executive Agency for Transplantation.[137] This makes it possible to keep track of LOD.

The procedural approach that should be preferred in a best practise proposal is still undecided, especially with respect to establishing a commission for LOD.

(1) Not all countries stipulate certain procedures by law at all. Denmark, Latvia, the Netherlands and Switzerland, for example, neither restrict the donor-recipient relationship nor require any particular procedures. Countries with no established legal procedures for LOD often establish procedures at the hospitals or transplant centres themselves. The Netherlands, Switzerland and the USA can serve as examples, among others. None of these countries stipulates any procedures or commissions by law. In the Netherlands and Switzerland, the hospital/transplant centre responsible for the donor evaluates him carefully and follows certain guidelines. The guidelines are not prescribed by law, but are made by the hospitals/transplant centres themselves.[138] They are generally based on their own research and take into account international consensus documents like the *Consensus Statement of the Amsterdam Forum on the Care of the Live Kidney Donor* (2004) or the *Ethics Statement of the Vancouver Forum on the Live Lung, Liver, Pancreas, and Intestine Donor* (2006).[139] In the USA, Congress passed the *National Organ Transplant Act* in 1984.[140] It does not contain rules about procedures, but the hospitals nevertheless work according to specific procedures and professional standards. The Organ Procurement and Transplantation Network (OPTN) that has been created under the *National Organ Transplant Act*[141] established several guidelines, for example, *Guidance for the Medical Evaluation of Potential Living Liver Donors* and *Guidance for the Informed Consent of Living Donors*.[142] In addition, the OPTN/UNOS Living Donor Committee is involved in each case. It "considers issues relating to the donation and transplantation of organs

---

[135] National data protection regulations must be complied with.

[136] Art. 29 (1) Law of the Transplantation of Organs, Tissues and Cells.

[137] Ibid., Art. 29 (2).

[138] Information from M. Bos.

[139] See e.g. The Ethics Committee of the Transplantation Society, Vol. 78 Transplantation 491–492 (2004); Pruett, et al., Vol. 81 Transplantation 1386–87 (2006).

[140] Transplant Living (2012).

[141] Organ Procurement and Transplantation Network (United Network for Organ Sharing (2012)).

[142] U.S. Department of Health & Human Services (2009).

from living donors to recipients. The committee makes recommendations to improve the process of living donation and transplantation."[143] Such guidelines are generally more flexible than legal rules, and they allow for innovation as a result of pluralistic approaches, because the guidelines are easily adapted to changing situations.[144] One negative aspect of this approach is that not one consistent procedure exists nationwide. This might cause legal uncertainty and unequal treatment of patients or make it difficult to supervise the hospital's compliance to those instructions and professional standards or even to sanction non-compliance. To sum up, this approach is good and seems sufficient for countries with hospitals that are highly developed and take an active part in the ethical and legal discussions about LOD. In these cases, legal rules on LOD procedures might be superfluous. (2) Another idea would be to stipulate the legal rules for decentralised procedures that take place at hospitals. Thus, legal rules would exist, but would not call for a state commission. This approach might only make sense for general provisions which are open for individual development. It might be possible to combine the advantages of legal rules with the advantages of hospital established procedures. A minimal version of this model can be found in Switzerland. Art. 10 *Transplantation Ordinance* (not the Swiss *Federal Act on the Transplantation of Organs, Tissues and Cells* itself) which stipulates that an independent person, one who is experienced with LOD, must ensure that the LOD is voluntary and free of charge. The person in charge has to keep records of his meeting with the potential donor, and he must keep this record separate from the patient's records. If the potential donor is rejected, he may obtain a second opinion. (3) A further approach would be a system of multiple, but state-run commissions or commissions defined by public law, respectively, based on a common legal basis. Germany, being the main example, established incomplete rules with respect to decentralized commissions. Only parts of the procedures are regulated by federal law. This causes legal uncertainty, because the details of the working methods of the commission differ in the federal states. To stipulate procedures by law incompletely is inconsistent. This approach is therefore not worth following. In Germany, the federal lawmaker himself should establish the essential rules for how the commission should proceed, he should lay down that donor and recipient have to be heard personally by the commission and he should determine the composition of the commission. In addition, the commission should follow a standardised approach with respect to procedures, but also to material decision criteria.[145] (4) The United Kingdom, in contrast, created a central state commission by law, the Human Tissue Authority. This has the positive effect that the same working methods and criteria are applied nationwide, guaranteeing legal certainty and equal treatment of comparable cases. This also makes supervision possible.

---

[143] Transplant Living (2012).

[144] Information from M. Bos.

[145] Gutmann (2011), p. 6 f.

The functions of the regulations in the United Kingdom and Germany are comparable after all, but the approach in the United Kingdom is more consistent. It creates complete procedures by law that are uniform nationwide. The countries that choose to establish procedures by law should hence follow the model of the United Kingdom. Also, when again considering the donor-recipient relationship, the United Kingdom is more flexible than Germany. The law as such does not make any general restrictions, but every LOD is evaluated by the Human Tissue Authority. The German regulation, in contrast, restricts LOD from the beginning on and involves a commission in addition.[146] The United Kingdom, consequently, is better suited to cope with the recommendation to provide "safety by procedure."[147]

All in all, it can be seen as advantageous for a hospital to create its own procedures, but a central regulation has advantages as well. Since the countries are very different, it is impossible to determine one regulation to be best for all countries considered. It can be concluded, though, that a commission should consist of independent[148] experts from different professions to examine and ensure the voluntariness of the donor (and the recipient). It should apply consistent criteria and apply them in an equal manner. A commission must usually authorise each LOD individually in countries that do not regard a donor's consent as sufficient to justify LOD.

b) Transplanting Person and Location of Transplantation

It is an agreed upon standard that a physician has to carry out the organ removal and the transplantation, but further requirements differ. In addition, several laws in the countries considered contain rules about the permissible locations. LOD contains a surgery. That it needs to be carried out by a physician in an appropriate hospital or transplantation centre is self-evident.[149] A best practise regulation does not need to include that requirement explicitly.

c) Post-Care

Only a few national laws mandatorily require care to be continued after the transplantation. The main aim of post-care is to guarantee ideal medical and psychological care for the donor and recipient, ensuring the success of the transplantation in the long run.[150] Appropriate post-care for donor and recipient should

---

[146] Ibid., p. 3.

[147] Cf. Gutmann et al., in Terasaki (ed.) (1995), p. 356 ff.

[148] Cf. Gutmann (2011), p. 6 f.

[149] Cf. Joint Working Party of The British Transplantation Society and The Renal Association (2011), p. 143.

[150] Esser, in Höfling (ed.) (2003), p. 231.

be stipulated by any adequate transplantation law.[151] This is in accordance with the *Directive 2010/53/EU*. It requests that "[m]ember states shall endeavour to carry out the follow-up of living donors [. . .]."[152] The current change in Dutch legislation may serve as a model for this requirement. It encourages living organ donors to undergo regular check-ups after the donation. The *Ministerial Decree on Basic Health Insurance*, which came into force on 1 January, 2012, declares that the arising expenses will be reimbursed by the basic health insurance. *Bos*, a member of the Health Council of the Netherlands, explained that the purpose of this decree is "to make the cost of yearly follow-up health check-up of living donors exempt from the mandatory own risk deductible (leading to loss of no-claim). This means that for living donors the cost of the yearly health check-ups is not considered as being part of the own risk deductible/no claim (every insured Dutch citizen has to pay the first Euro 220,- of their health care cost from their own pocket). This measure was taken to avoid that healthy living donors who have no other health care costs, would skip their yearly check-up because it would cost them money."[153]

d) Donor Registry

That donors are rarely registered in a centralised database is unacceptable.[154] The key justification of LOD is the donor's informed consent.[155] Legitimate interest in the facts for a medical disclosure makes it necessary to establish a registry that records perioperative complications and impairments.[156] It is, in addition, an effective instrument to ensure a constant post-care of the donor.[157]

The *Swiss Organ Living Donor Health Registry* might serve as a model.[158] It was established in 1993 and was the first living donor registry worldwide.[159] One aim of the registry is to include and analyse post-LOD complications and problems in the long term and on a regularly basis.[160] In addition, the registry should ensure that

[151] Cf. Joint Working Party of The British Transplantation Society and The Renal Association (2011), p. 184.

[152] Art. 15 4. Directive 2010/53/EU of the European Parliament and of the Council of 7 July 2010 on standards of quality and safety of human organs intended for transplantation.

[153] Information from M. Bos.

[154] Cf. Art. 15 3. Directive 2010/53/EU of the European Parliament and of the Council of 7 July 2010 on standards of quality and safety of human organs intended for transplantation; cf. Eigler, Vol. 122 Deutsche Medizinische Wochenschrift 1398, 1401 (1997); cf. Commission of the European Communities (2008); cf. Gutmann (2006), p. 85.

[155] Gutmann (2006), p. 85.

[156] Eigler, Vol. 122 Deutsche Medizinische Wochenschrift 1398, 1401 (1997).

[157] Gutmann (2006), p. 85.

[158] Cf. Gutmann (2006), p. 85, who mentions this with regard to the German Act on the donation, removal and transplantation of organs.

[159] Bundestag printed paper 15/5050, p. 56; Gutmann (2006), p. 85.

[160] Bundestag printed paper 15/5050, p. 54 f.; Gutmann (2006), p. 86.

post-care is executed by the physician and that the donor's medical care for the LOD is not neglected. The registry is also intended to provide a fast reaction to any adverse effects.[161] Art. 16 e *Transplantation Ordinance* contains details on how to ensure quality and the donor's post-transplant follow-up. According to the *Ordinance*, medical and psychological data must be collected prior to the surgery, and afterwards with the donor's consent. The donor must be offered the chance to receive lifelong check-ups. The results of his check-ups should be evaluated on a regular basis and made available to all transplant centres. The donor must also be informed of the results of his check-up, and should be advised if any additional measures are necessary. The results relevant for health will be incorporated in the disclosure of future donors.

A best practise regulation should demand that "[m]andatory organ-specific registries of transplant recipients and donors should be assembled to enable annual review of data."[162] The constant collection of data must be obligatory,[163] and a legal foundation is necessary.[164] When establishing a donor registry, (national) data protection regulations must be complied with.[165]

7) Social Security Regulations for Donors

With respect to the social security regulations for donors, no consensus exists. LOD is very expensive and might cause financial or economic disadvantages for the donor.[166] The aim should be to protect her in the best possible way[167]; she should be completely carefree.[168] This should be regulated by law.[169] The following will concentrate on (1) the insurance for the donor and (2) her reimbursement.

[161] Bundestag printed paper 15/5050, p. 54 f.

[162] Delmonico et al., Vol. 378 The Lancet 1414, 1414 (2011).

[163] Gutmann (2006), p. 86.

[164] Ibid.

[165] Cf. Art. 15 3. Directive 2010/53/EU of the European Parliament and of the Council of 7 July 2010 on standards of quality and safety of human organs intended for transplantation.

[166] Gutmann (2006), p. 89; Neft, Vol. 20 Neue Zeitschrift für Sozialrecht 566, 566 (2011); cf. Sterner et al., Vol. 21 Pediatric Nephrology 1357, 1357 f. (2006).

[167] Cf. EULOD WP 2 (2012), DOW: Deliverable 4, p. 14; Gutmann (2006), p. 98; cf. Interessengemeinschaft Nierenlebendspende (2012).

[168] Lilie/Krüger, Vol. 81 Der Chirurg 787, 792 (2010).

[169] Gutmann (2006), p. 98; cf. Interessengemeinschaft Nierenlebendspende (2012).

a) Insurance

With respect to the donor's insurance, Switzerland can serve as a "gold standard": It demands the surgeon who removes an organ from a living person to ensure that the person concerned is insured against any other potential serious consequences connected to the organ removal.[170] This rule holds the insurer liable. The insurer must, in any case, compensate the recipient's costs for her illness if no LOD is available.[171] Additional details have been specified by the Federal Council in Art. 11 *Transplantation Ordinance*.[172]

The new German regulation can be seen as the second best social security solution. The German Federal Health Minister *Bahr* announced that every living organ donor should be entitled to the recipient's health insurance with respect to health treatment, pre- and post-care, rehabilitation and sick pay.[173] The Federal Council furthermore demands social accident insurance cover all complications that might result from the LOD.[174]

b) Reimbursement

A best practise regulation should explicitly clarify that all the donor's possible expenses resulting from the LOD have to be reimbursed.[175] The results of a recent survey, which included 113 kidney transplant units from 25 European Union Member States and 15 Non-member States (i.e. 40 countries in Europe) and 39 liver transplant units from 18 European Union Member States and 6 Non-member States (i.e. 24 countries in Europe), shows that the current situation is insufficient, because 54 % of the kidney donors, and even about 70 % of the liver donors, were not reimbursed.[176]

In Switzerland, the recipient's insurance company is responsible for all costs and for an "appropriate compensation for loss of earning or other expenses incurred by the donor in connection with removal."[177] In addition, Sec. 12 *Transplantation Ordinance* addresses the issue of expense allowance.[178] The donor is compensated for all expenses she has in connection to the donation. This includes travel expenses, the costs she had with respect to the clarification of her being suitable as living organ donor, lifelong check-ups and any costs the donor may incur if she

[170] Art. 14, 1 Federal Act on the Transplantation of Organs, Tissues and Cells.

[171] Ibid., Art. 14, 2.

[172] This has been presented in chapter "Comparative Analysis of European Transplant Laws Regarding Living Organ Donation".

[173] Ibid.

[174] Ibid.

[175] Cf. Bundestag printed paper 15/5050 (2005), p. 76 f.; cf. Gutmann (2006), p. 99 f.

[176] EULOD WP 2 (2012), DOW: Deliverable 4, p. 9 ff.

[177] Art. 14, 2 Federal Act on the Transplantation of Organs, Tissues and Cells.

[178] Cf. ibid., Art. 14, 4.

requires assistance, either personal assistance at home for herself or for others if she is their primary caretaker. The duties just stated also exist if the removal or donation cannot be performed. Swiss regulation considers cases where the recipient's insurer is unknown. If the recipient's insurer is unknown, the Confederation has to bear the costs.[179] Since this regulation is extensive and protects the donor adequately, yet practicable and clear, a best practise proposal should follow the Swiss approach.[180]

III. Result

This best practise proposal for LOD has been established from the following sources: conclusions drawn in chapters "Comparative Analysis of European Transplant Laws Regarding Living Organ Donation" and "Analysis of the Normative Arguments That Dominate the Policy Arena About Necessity and Legitimacy of Legal Restrictions in Living Donor Transplantation," through a consideration of the common principles laid down in the *CFREU* and an examination of the relevant general national laws. The best practise proposal considered all issues of LOD in need of regulation and achieved extensive unification.

[179] Ibid., Art. 14, 3.

[180] Cf. Gutmann (2011), p. 8.

Summary

All countries considered have too few donor organs and do not completely exhaust the potential of organ transplantation. Effort should therefore be made to increase the amount of donor organs — post-morgen organ donation as well as LOD. The reasons for the existing organ shortage are, of course, not *only* of legal nature. With respect to LOD, there are, however, *also* legal reasons for this shortage. As legal experts have shown, LOD regimes create, in part, an *artificial scarcity*.[1] The final aim of this dissertation was therefore to work out which rules would be best for the issues of LOD in need of regulation.

1. The first step in the process of harmonisation was to compare the national laws for LOD. This comparison showed that several issues of LOD in need of regulation are regulated similarly in the countries involved, while substantial differences exist as well.

   a) The countries commonly hold that no citizen has a duty to donate an organ. They all require consent and disclosure to justify LOD. In some countries, however, consent plays a striking part, but is not sufficient to justify LOD. Others, in contrast, consider consent sufficient to justify medical operations, hence ranking the patient's right of self-determination highly.

   aa) Not all national laws explicitly state that the donor must give consent. However, consenting prior to the organ removal is considered necessary to justify LOD, regardless of whether it is explicitly required or not. All countries also agree on requiring the capacity to consent and the consent to be voluntarily, but only some countries pledge a commission to ensure the voluntariness. Examining the formal requirements imposed on consent shows that a significant amount of countries demand a written consent. A small amount of national laws states additional requirements. It is generally very important to ensure the donor's autonomy; he can

[1] Cf. Evans, Vol. 15 Journal of Medical Ethics 17, 19 (1989); Fateh-Moghadam, (2011); cf. Radcliffe Richards, in Weimar/Bos/Busschbach (ed.) (2011), p. 41.

DOI 10.1007/978-3-642-33799-4_8,

withdraw his consent at any time before the actual organ removal. The countries also agree upon limiting the scope of the consent. Any LOD that ceases the donor's life or that would be directly life-threatening is prohibited. LODs against good morals are also typically prohibited. In order to ensure the recipient's autonomy, he must consent as well.

bb) Since the donor is usually a medical layperson, he should be informed of the procedure prior to his consent. The content of the donor's disclosure differs from country to country. Possible consequences, medical background, the protection of the donor and the impact on the recipient are uniquely considered. Some countries explicitly require the donor to understand the explanation provided to him. The issue of possible formal requirements arises as well with regard to the disclosure of the donor. A significant amount of countries not only require the donor's consent to be in writing, but also require the disclosure of the donor. Sometimes they demand a certain person — usually a physician — to inform the donor. As stated above, the recipient must also consent and, therefore, must be informed about the procedure as well.

b) LOD aims at healing the sick recipient, so the intended living organ transplantation has to be suitable. A medical observation is necessary to evaluate this question in any case, but only a small amount of national transplant laws explicitly require this. Regarding the suitability of the transplantation, the recipient's anticipated benefit is usually required. It seems self-evident that the donor must be suitable to act as such; however, several national laws explicitly mention this requirement. A potential donor may not take unlimited risks. So, the prohibition of any LOD that leads to the donor's death or that could be directly life-threatening for him is common. Beyond this consensus, the national laws differ with regard to the ceiling on permissible risks. In addition to considering donor and recipient separately, some countries require a risk-benefit equation.

c) Minors and mentally incapacitated adults have to be considered separately since they are usually especially in need of protection. The majority of countries either completely prohibit minors from becoming living organ donors, or only allow them to act as a living organ donor under several specific requirements. Several countries have the same rules for mentally incapacitated adults as they have for minors. However, not all countries apply the same regulations. Similar to minors, a significant number of countries completely prohibit mentally incapacitated adults from becoming living organ donors. In contrast, there are also several countries that do not exclude mentally incapacitated adults as living organ donors a priori, but stipulate special requirements that must be fulfilled.

d) One central issue of LOD by mentally capable adults is the restriction of the donor-recipient relationship. None of the countries demand a genetic relation between the donor and recipient. It differs, however, whether any kind of relationship is necessary or whether donor and recipient do not have to be related at all. Only a small amount of countries require a specific donor-recipient relationship without providing an open clause. Several countries restrict the donor-recipient relationship insofar as they list the relationships that are specifically allowed, but also contain an open clause. However, there are also countries that do not restrict the donor-recipient relationship at all.

e) The subsidiarity of LOD contains two aspects: subsidiarity of LOD to post-mortem organ donation and subsidiarity of LOD to alternative therapies. With regard to both aspects, there is no consensus in the national laws.

f) A significant amount of countries also address numerous procedural issues connected to LOD. Some national transplant laws explicitly declare that only doctors are allowed to perform LODs and that the necessary medical operations have to take place at suitable locations. A few countries consider the period after the LOD. They stipulate by law that donor or recipient or both have to take part in a post-care procedure. The involvement of an independent authority in the process of LOD is very common. However, differences exist with regard to its composition and the assigned tasks. Several countries require an authority's approval to justify every LOD or just in selected cases. Commissions and procedural safeguards in general are usually established by state law, but sometimes the details are stipulated by federal law. Some countries do not have a state commission, instead the hospitals/transplant centres themselves have established certain procedures. A few countries demand surveillance of the LOD by the appointed authority. Other countries instruct that every LOD has to be communicated to an authority. Legal protection is also considered by several countries. Usually, legal protection provided by the national law can be used for matters of LOD as well. In addition, several countries established particular rules to either regulate the supervision over compliance or that contain specific legal actions available to appeal a refused LOD.

g) The donor incurs substantial costs while being involved in LOD. Consequently, social security regulations for the donor must be implemented. Since LOD is not a therapeutic treatment for the donor, it is often the case that the donor does not have to pay for her expenses. The institution responsible for covering the donor's costs differs. Liability lies mostly with the designated state or with the recipient's health insurance.

Almost all of the countries considered established specific legal rules for LOD. However, only some regulations are similar, substantial differences exist as well.

2. The arguments with respect to the necessity and legitimacy of a restriction of LOD have been analysed afterwards. Several arguments in favour of restricting LOD are brought forth, but those are also subject to severe criticism.

   a) LOD could be seen as a violation of the principle of nonmaleficence and the donor's risks are mentioned in favour of restricting the donor-recipient relationship. Those against restricting the donor-recipient relationship mention the principle of autonomy. If this principle is applied, every person has a right of self-determination. We must consider behaviour that might not be comprehensible for others, e. g., unspecified LOD. The right of self-determination is ensured by the patient's informed consent, necessary for LOD in all countries. Several countries even consider informed consent as sufficient to justify LOD, but not all. Even if informed consent is provided, its scope is limited. LOD is not legal in case it would end the donor's life or would be directly life-threatening. Several countries additionally declare consent that would be against 'good morals' void. Nevertheless, a general prohibition of unspecified LOD would be in conflict with the autonomy of the persons who want to donate. Arguing that only direct LOD should be permitted because the risks involved with LOD are too high is also incorrect since the risks for all types of LOD are equal. One could say that the risks may be *perceived* differently in direct and unspecified LOD. Even though it is not implausible to consider how others might perceive the risks, it does not serve as a valid reason for actually prohibiting unspecified LOD. In favour of restricting the donor-recipient relationship, some claim that unrelated donors do not benefit as much from LOD as related donors, and also that the special relationship between donor and recipient *exclusively* justifies LOD. It can be agreed with the assumption that prohibiting direct LOD creates a heavier burden for the donor and the recipient. However, this is not a good enough reason to prohibit unspecified LOD. Even though special responsibilities do exist in close personal relationships, this does not mean that these special responsibilities should be the *only* normative foundation of LOD. In conclusion, the most relevant aspect of any LOD is that the donor cares for the needs of others. This applies to all (noncommercial) LOD, regardless of the relatedness of donor and recipient.

b) There is no common opinion on the connection between related LOD and (in) voluntariness, or unrelated LOD and (in)voluntariness. The holders of the second approach are consequently in favour of prohibiting unrelated LOD. The German Bundestag, e. g., claims that only a family relationship or another close relationship guarantees voluntariness in LOD. Social research, however, demonstrates that organ donors with a competent, informed and voluntary decision to act altruistically towards strangers can be found in great numbers. On the one hand, with respect to any type of LOD, involuntary decisions might occur. On the other hand, every type of LOD might contain cases that are without doubt the result of a voluntary decision. The best approach, therefore, is to judge each case individually and not simply prohibit certain donor-recipient relationships. Besides, several reasons for why more women act as living organ donors and more men act as recipients have been listed. As long as every LOD is justified in itself, the gender bias is irrelevant and a gender imbalance cannot be considered as *malum in se*.

c) Some are concerned that a general legalisation of LOD would increase the pressure on living individuals. The rationale of this — that it is more important for the state to protect its citizens from having to face hard choices than to give them the chance to choose whether to help or even save a person in need — is unsound. In addition, permitting unspecified LOD does not inevitably raise the pressure on all citizens, because it cannot be equated with the duty to actually become an unspecified living organ donor. Critics allege that intended and performed LOD negatively influence the donor-recipient relationship. This, however, is not predominantly problematic, since positive effects occur as well.

d) Unspecified LOD and cross-over LOD are associated with organ trade. The possibility that unspecified LOD might cause organ trade can be nipped in the bud, though, if the donation occurs within a pool of organs that is handled by a neutral agency, which coordinates the organ transplantation, organ allocation and maintains the anonymity of the donor. The numerous experiences with different types of LOD between persons who are not closely related to each other, including experiences of unspecified LOD, demonstrate that no uncontrollable danger of commercialisation exists with respect to them. The concerns with respect to cross-over LOD are dispelled with reference to the donor's motives, namely to help a relative. Organ trade is thus not a convincing argument to restrict the donor-recipient relationship.

e) Is LOD just and equal? Some question the fairness of LOD, since it is not available to every patient in need. The solution, to prohibit LOD in general, would not improve the situation of any of the patients. It would only have a negative impact on the patients who already have a living organ donor. And, even if not every person suffering from organ failure knows a living person who will donate, each LOD shortens the waiting list, thus benefiting all candidates. This approach can also be used to promote the reduction of LOD restrictions, since it would most likely enlarge the amount of donor organs. The existing gender inequality is problematic only if it is based upon an exploitation of women. With respect to the fact that people from minority group's act less often as living organ donors and also profit less from LOD, an effort should be made to increase the amount of organ donations from people that belong to minorities.

f) Minors and mentally incapacitated adults have to be regarded individually. Several arguments in favour of completely prohibiting them from acting as living organ donors exist. However, there are also valid arguments against such a complete prohibition. The arguments with respect to both approaches mainly address direct LOD. Unspecified LOD is not even considered.

g) Several different types of LOD that can either be classified as indirect LOD or as unspecified LOD are (medically) possible. All types of LOD have the positive effect of overcoming incompatibility between the willing donor and the intended recipient, increasing the amount of donor organs. The various types are also criticised, though. Most fear that the pressure inflicted upon the potential donor might increase. Overall, with respect to cross-over LOD, living paired cascade exchange, unspecified LOD and pool donation, the advantages seem to outweigh the disadvantages. Unbalanced living paired exchanged, directed altruistic LOD, unspecified non-directed donation catalysing cascade exchanges and list-paired exchanges, in contrast, seem to be rather problematic, although there is hardly a reason for *legally* banning them.

h) Procedures that restrict LOD were considered as well. Several countries do not restrict the donor-recipient relationship, but do have specific procedures that must be followed.

Thus, with the exception of specific arguments concerning special models of LOD exchanges, none of the arguments brought forth for the necessity and the legitimacy of most of the legal restrictions proposed for LOD seems cogent.[2] There is no class of donors and no category of LOD that should be generally dismissed by law. Rather, the solution seems to be a procedural one. Careful case-by-case decisions should be made.[3]

3. Since the Member States of the European Union are continually growing more and more together, it makes sense to level the existing differences. The Europen Union has already become active. It released the *Communication from the Commission to the European Parliament and the Council – Organ Donation and Transplantation: Policy Actions at EU Level* in 2007. This *Communication* did not remain without consequences; the *Action plan on Organ Donation and Transplantation* and the *Directive 2010/53/EU* have been passed. The *Action plan* involves three important challenges for organ donation, namely to increase organ availability, to improve the transplantation systems with regard to efficiency and accessibility and to raise the standards of quality and safety. It, however, is not binding to the Member States. The *Directive 2010/53/EU*, in contrast, is binding with respect to its aim. It addresses LOD, but does not regulate the requirements for LOD. It rather concentrates on aspects of quality and safety of LOD. The effect the *Directive 2010/53/EU* has right now is therefore rather mild. However, it might produce a further dynamism with regard to a Europeanizing of the transplantation laws in future.

4. The European Union can make decisions that are binding for its Member States. In particular, Art. 168 *TFEU* grants competences to the European Union to make law in the field of public health. This also includes LOD. According to Art. 168 IV (a) *TFEU*, the European Union can pass rules ensuring high standards of quality and the safety of organs. The European Union has no all-embracing competence to regulate LOD, and it is especially not competent to harmonise the requirements for LOD. The *CRFEU* does not produce a divergent result; it does not grant competences to the European Union, and it is only effective within the competences of the European Union. However, the option to *regulate* the requirements for LOD with soft law remains since soft law is not (legally)

[2] Cf. Daar et al., Vol. 11 Transplantation Review 95, 99 (1997); cf. Evans, Vol. 15 Journal of Medical Ethics 17, 20 (1989); Gutmann/Schroth (2002), p. 61; Gutmann (2006), p. 47; cf. Hilhorst et al., Vol. 24 Transplant International 1164, 1168 (2011).

[3] Cf. Gutmann et al., Terasaki (ed.) (1995), p. 356; Land, Vol. 2 Transplant International 168, 173 (1989) with regard to spouse-to-spouse donation.

binding. It nevertheless has effects, and addressees often comply with it. Hence, to create soft law for the requirements of LOD would make sense.

5. The best practise proposal for LOD, i.e. a *Common Frame of Reference for European Laws on LOD (CFR-LOD),* was established from the conclusions drawn in the comparison of laws and in the analysis of the necessity and legitimacy of legal restrictions in LOD. The common principles laid down in the *CFREU* and the relevant general national laws have been considered as well.

   a) The donor's *informed consent* is absolutely necessary, and in a few countries even sufficient to justify LOD. A best practise regulation must consequently explicitly require the donor's consent. He must have the competence to consent and his consent must be given voluntarily. The minimum formal requirements include: an appropriate amount of time for the donor to reconsider his decision and a written consent. A best practise regulation should not contain any additional mandatory formal requirements. Even after the donor has given his consent, he may withdraw it at any time. This must be explicitly mentioned in the best practise regulation and the donor must be informed about this option. The donor's informed consent can under no circumstances justify LODs that cease his life or would be directly life-threatening. No further limit on the legal ability of donors to give consent should be stipulated by law. The recipient has to give consent as well.
   A valid informed consent requires the disclosure of the donor. A best practise regulation should clearly present what the donor has to be informed about. The whole (medical) procedure and potential risks should be explained to him. The donor should be informed about alternative therapies for the recipient, about the prospect of success and the risks for the recipient. The donor should be mandatorily informed about any additional, foreseeable circumstances that are relevant to him. The donor needs to understand the information provided to him. With respect to the person providing the disclosure: the physician who performs the intended transplantation has to inform the donor, and another physician with no part in the transplantation process should be present as well. The recipient should also be informed, namely about his health and about the donor's health and risks, because recipients usually only accept limited risks for the donor.
   b) The transplantation must be suitable. The best practise proposal should contain a risk-benefit equation without providing very detailed or paternalistic requirements, leaving the decision to the autonomous donor (and recipient) and transplant team.
   c) A best practise regulation would not absolutely prohibit minors and mentally incapacitated adults from becoming living organ donors, but would contain specific requirements that only allow them to donate a kidney in very exceptional cases. The minors or mentally incapacitated adults informed consent is necessary. The person concerned must have the capacity to give informed consent. Additionally, the parents or the custodian of persons of full age must give their consent. Donor and recipient must have a close personal

relationship. Living kidney donation by a minor or a mentally incapacitated adult must remain the *ultima ratio* and, furthermore, an independent commission must give its approval.

d) A possible restriction of the donor-recipient relationship has been rejected after impartially reviewing the key arguments in this debate and considering the *CFREU*. It is advisable to evaluate concrete risks, rather than to distinguish between relationships. A best practise regulation should, consequently, not contain a restriction of the donor-recipient relationship, but should require case-by-case decisions. With respect to different models of LOD, such as cross-over LOD and unspecified LOD, there is hardly a reason for *legally* banning them. They, however, can only exist if a sufficient national scheme has been established. Whether the countries or national transplantation systems want to do so is up to them, although there are good reasons for such policies.
e) LOD should not be subsidiary to post-mortem donation; post-mortem donation should, however, never be regarded as secondary to LOD. With respect to the relationship between LOD and alternative therapies, LOD should be the *ultima ratio*, but a patient suffering from kidney failure should not have to depend on the worse option of dialysis, even though LOD is possible. National transplant laws should therefore also not hinder preemptive kidney transplantations. The Swiss regulation that states "[o]rgans, tissues and cells may be removed from a living person if: [...] the recipient cannot be treated with any other therapeutic method with comparable benefit" should be adapted.
f) Several procedural regulations are recommendable as safeguards for LOD ("Safety by procedure"). Most countries have established some kind of procedure and an independent commission as a safeguard for LOD. Four different categories with respect to procedures can be classified among the countries considered. (1) Some countries have not established legal procedures for LOD. Those countries often established procedures at the hospitals/transplant centres themselves, though. This is recommendable for countries with hospitals/transplant centres that are professionally well-developed and take an active part in the ethical and legal discussions related to LOD. (2) Another idea would be to stipulate rules by law for decentralised procedures that take place at the hospitals. This approach might make it possible to combine the advantages of a legal rule with the advantages of innovative and pluralistic procedures established by the hospitals themselves. (3) Another approach is a system of multiple, but state-run commissions or commissions defined by public law. The procedures are only partly regulated by federal law. This could lead to legal uncertainty, because the details differ in the federal states. (4) Other countries created a central state commission by law. This has the advantage that the same working methods and criteria are applied nationwide, guaranteeing legal certainty and equal treatment of comparable cases. This is, in comparison to the third approach, more consistent and is more likely to provide "safety by procedure". Due to the pre-existing

diversity among the countries considered, it seems impossible to unify the national approaches just introduced. In principle, all these procedural models may be adequate. Apart from that, a commission should consist of independent experts from different professions to examine and, as far as possible, ensure the voluntariness of the donor (and the recipient). It should apply consistent criteria in an equal manner. In the countries that do not consider the donor's consent as sufficient to justify LOD, the commission additionally has to authorise every LOD.

Appropriate post-care for donor and recipient should be explicitly stipulated in a best practise regulation. The *Directive 2010/53/EU* requests that "[m]ember states shall endeavour to carry out the follow-up of living donors [. . .]." The Netherlands can take on a role as a trailblazer for that. Its legislation now encourages living organ donors to undergo regular check-ups after the LOD.[4] Also, a donor registry that records perioperative complications and impairments is necessary to ensure continual post-care and to increase the amount of relevant information contained in future medical disclosures.

g) Since LOD is very expensive and is no therapeutic treatment for the donor, social security regulations should be very extensive, eliminating the donor's financial responsibilities. The donor should be protected in the best possible way; he should be completely carefree. The Swiss regulation demands the surgeon who removes an organ from a living person to ensure that the person concerned is insured against any potential serious consequences connected to the organ removal.[5] It holds the insurer liable who would have to compensate the recipient's costs for his illness if no LOD were available.[6] Additional details have been specified by the Federal Council in Art. 11 *Transplantation Ordinance*. Since the Swiss rule is very extensive, it can be used as a model for social security regulations. With respect to the reimbursement of living organ donors, a recent survey revealed that the practical situation is very insufficient, numerically 54 % of the included kidney donors, and even about 70 % of the considered liver donors, were not reimbursed.[7] It hence has to be regulated explicitly that all the donor's possible expenses resulting from the LOD have to be reimbursed.[8] The Swiss regulation can again serve as a model. There, the recipient's insurance company is explicitly responsible for all costs and for an "appropriate compensation for loss of earning or other expenses incurred by the donor in connection with removal."[9]

---

[4] Information from M. Bos.

[5] Art. 14, 1 Federal Act on the Transplantation of Organs, Tissues and Cells.

[6] Ibid., Art. 14, 2.

[7] EULOD WP 2, Living Organ Donation Practices in Europe, p. 9 ff.

[8] Cf. Bundestag printed paper 15/5050 (2005), p. 76 f.; cf. Gutmann (2006), p. 99 f.

[9] Art. 14, 2 Federal Act on the Transplantation of Organs, Tissues and Cells.

All in all, this best practise proposal considered all issues of LOD in need of regulation. It shows how it is possible to achieve extensive unification of the transplant laws, which differ considerably at the moment. However, the European Union can only establish such unification with soft law.

References

I. Bibliography

Abbot, Kenneth/Keohane, Robert/Moravcsik, Andrew/Slaughter, Anne-Marie/Snidal, Duncan, The Concept of Legalization, Vol. 54 International Organization 401–419 (2000) (Abbot et al., Vol. 54 International Organization 401, p. (2000))

Abbot, Kenneth/Snidal, Duncan, Hard and Soft Law in International Governance, Vol. 54 International Organization 421–456 (2000) (Abbot/Snidal, Vol. 54 International Organization 421, p. (2000))

Ackerman, Paul/Thistlethwaite, Richard/Ross, Lainie Friedman, Attitudes of Minority Patients with End-Stage Renal Disease Regarding ABO-Incompatible List-Paired Exchanges, Vol. 6 American Journal of Transplantation 83–88 (2006) (Ackerman/Thistlethwaite/Ross, Vol. 6 American Journal of Transplantation 83, p. (2006))

Ach, Johann/Anderheiten, Michael/Quante, Michael, Ethik der Organtransplantation, Erlangen 2000(Ach/Anderheiten/Quante (2000), p.)

Adams, Patricia/Cohen, David/Danovitch, Gabriel/Edington, Mark/Gaston, Robert/Jacobs, Cheryl/Luskin, Richard/Metzger, Robert/Peters, Thomas/Siminoff, Laura/Veatch, Robert/ Rothberg-Wegman, Lynn/Bartlett, Stephen/Brigham, Lori/Burdick, James/Gunderson, Susan/ Harmon, William/Matas, Arthur/Thistlethwaite, Richard/Delmonico, Francis, The nondirected live-kidney donor: ethical considerations and practice guidelines, Vol. 74 Transplantation 582–590 (2002) (Adams et al., Vol. 74 Transplantation 582, p. (2002))

Ahrens, Hans-Jürgen/von Bar, Christian/Fischer, Gerfried/Spickhoff, Andreas/Taupitz, Jochen (ed.), Festschrift für Erwin Deutsch, Köln 1999 (Author, in Ahrens et al. (ed.) (1999), p.)

Aigner, Gerhard, An overview of legal aspects in organ transplantation – what are the family rights?, Vol. 9 Annals of Transplantation 11–14 (2004) (Aigner, Vol. 9 Annals of Transplantation 11, p. (2004))

Akkina, Sanjeev/Muster, Heather/Steffens, Eugenia/Kim, Joseph/Kasiske, Bertram/Israni, Ajay, Donor Exchange Programs in Kidney Transplantation: Rationale and Operational Details From the North Central Donor Exchange Cooperative, Vol. 57 American Journal of Kidney Diseases 152–158 (2011) (Akkina et al., Vol. 57 American Journal of Kidney Diseases 152, p. (2011))

Alfani, Dario/Pretagostini, Renzo/Rossi, Massimo/Poli, L./Bruzzone, Paolo/Colonnello, M./De Simone, P./Berloco, P./Persijn, G./Smits, J./Cortesini, Raffaello, Living Unrelated Kidney Transplantation: A 12-Year Single Center Experience, Vol. 29 Transplantation Proceedings 191–194 (1997) (Alfani et al., Vol. 29 Transplantation Proceedings 191, p. (1997))

Amelung, Knut/Beulke, Werner/Lilie, Hans (ed.), Strafrecht – Biorecht - Rechtsphilosophie: Festschrift für Hans Ludwig Schreiber zum 70. Geburtstag, Heidelberg 2003 (Author, in Amelung/Beulke/Lilie (ed.) (2003), p.)

L. Lopp, *Regulations Regarding Living Organ Donation in Europe*,
DOI 10.1007/978-3-642-33799-4, © Springer-Verlag Berlin Heidelberg 2013

American Medical Association, Consensus Statement on the Live Organ Donor, Vol. 284 The Journal of the American Medical Association 2919–2925 (2000) (American Medical Association, Vol. 284 The Journal of the American Medical Association 2919, p. (2000))

Ankeny, Rachel, The Moral Status of Preferences for Directed Donation: Who Should Decide Who Gets Transplantable Organs?, Vol. 10 Cambridge Quarterly of Healthcare Ethics 387–398 (2001) (Ankeny, Vol. 10 Cambridge Quarterly of Healthcare Ethics 387, p. (2001))

AOK Bundesverband, Stellungnahme des AOK Bundesverbandes zum Fragenkatalog zur öffentlichen Anhörung „Organlebendspende" der Enquete-Kommission „Ethik und Recht der modernen Medizin" am 1.März 2004 (AOK Bundesverband (2004), p.)

Arnason, Wayne, Directed Donation, The Relevance of Race, Vol. 21 Hastings Center Report 13–19 (1991) (Arnason, Vol. 21 Hastings Center Report 13, p. (1991))

(von) Arnauld, Andreas, Strukturelle Fragen des allgemeinen Persönlichkeitsrechts, Vol. 40 Zeitschrift für Urheber- und Medienrecht 286–292 (1996) (von Arnauld, Vol. 40 Zeitschrift für Urheber- und Medienrecht 286, p. (1996))

(von) Arnauld, Andreas/Musil, Andreas (ed.), Strukturfragen des Sozialverfassungsrechts, Tübingen 2009 (Author, in von Arnauld/Musil (ed.) (2009), p.)

(von) Arnauld, Andreas/Hufeld, Ulrich (ed.), Systematischer Kommentar zu den Lissabon-Begleitgesetzen, Baden-Baden 2011 (Author, in von Arnauld/Hufeld (ed.) (2011), p.)

Arndt, Hans-Wolfgang/Fischer, Kristian/Fezer, Thomas, Europarecht, Heidelberg 2010 (Arndt/Ficher/Fezer (2010), p.)

Ärztezeitung, Versicherung muss Folgeschäden eine Organspende zahlen, 2012, retrieved 20 June, 2012 from http://www.aerztezeitung.de/politik_gesellschaft/organspende/?sid=813461 (Ärztezeitung (2012))

Ayanian, Joan/Cleary, Paul/Weissman, Joel/Epstein Arnold, The effect of patients' preferences on racial differences in access to renal transplantation, Vol. 341 The New England Journal of Medicine 1661–1669 (2011) (Ayanian et al., Vol. 341 The New England Journal of Medicine 1661, p. (2011))

(von) Bar, Christian/Zimmermann, Reinhard, Grundregeln des Europäischen Vertragsrechts, München 2002 (von Bar/Zimmermann (2002), p.)

(von) Bar, Christian, A Common Frame of Reference for European Private Law – Academic Efforts and Political Realities, Vol. 12 Electronic Journal of Comparative Law 1–10 (2008) (von Bar (2008), p.)

(von) Bar, Christian/Clive, Eric/Schulte-Nölke, Hans/Beale, Hugh/Herre, Johnny/Huet, Jérôme, Storme, Matthias/Swann, Stephen/Varul, Paul/Veneziano, Anna/Zoll, Fryderyk (ed.), Principles, Definitions and Model Rules of European Private Law, Draft Common Frame of Reference (DCFR), Outline Edition, Prepared by the Study Group on a European Civil Code and the Research Group on EC Private Law (Acquis Group), Based in part on a revised version of the Principles of European Contract Law, München 2009 (von Bar et al. (ed.) (2009), Draft Common Frame of Reference, p.)

Barendt, Eric, An introduction to Constitutional Law, Oxford 1998 (Barendt (1998), p.)

Barta, Heinz/Weber, Karl (ed.), Rechtsfragen der Transplantationsmedizin in Europa, Wien 2001 (Author, in Barta/Weber (ed.) (2001), p.)

Barta, Heinz, Zivilrecht – Grundriss und Einführung in das Rechtsdenken, Innsbruck 2004, retrieved 13 January, 2012 from http://www.uibk.ac.at/zivilrecht/buch/kap4_0.xml?Sec.=1; Sec.-view=true (Barta (2004))

Baxter, R., International law in "her infinite variety", Vol. 29 International and Comparative Law Quarterly 549–567 (1980) (Baxter, Vol. 29 International and Comparative Law Quarterly 549, p. (1980))

Beauchamp, Tom/Childress, James, Principles of Biomedical Ethics, 5. edition, Oxford 2001 (Beauchamp/Childress (2001), p.)

Bechstein, Wolf/Moench, Christian, The right to refuse, Vol. 24 Transplant International 1162–1163 (2011) (Bechstein/Moench, Vol. 24 Transplant International 1162, p. (2011))

Beck, Susanne/Burchhard, Christoph/Fateh-Moghadam, Bijan (ed.), Strafrechtsvergleichung als Problem und Lösung, Baden-Baden 2011 (Author, in Beck/Burchhard/Fateh-Moghadam (ed.) (2011), p.)

(di) Bella, Marco, Entwurf eines Gesetzes zur Änderung des Transplantationsgesetzes, Vol. 9 Rechtsdepesche für das Gesundheitswesen 38–39 (2012) (di Bella, Vol. 9 Rechtsdepesche für das Gesundheitswesen 38, p. (2012))

Ben-Haim, Menahem/Emre, Sukru/Fishbein, Thomas/Sheiner, Patricia/Bodian, Carol/Kim-Schluger, Leona/Schwartz, Myron/Miller, Charles, Critical graft size in adult-to-adult living donor liver transplantation: Impact of the recipient's disease, Vol. 7 Liver Transplantation 948–953 (2001) (Ben-Haim et al., Vol. 7 Liver Transplantation 948, p. (2001))

Benke, Nikolaus/Meissel, Franz-Stefan/Luggauer, Karl, Juristenlatein, Wien 1997 (Benke/Meissel/Luggauer (1997), p.)

Benn, Stanley, The Principle of Respect for Persons, in: Benn, A Theory of Freedom, Cambridge 1988, p. 103-121 (Benn (1988))

Berka, Walter, Die Grundrechte, Grundfreiheiten und Menschenrechte in Österreich, Wien 1999 (Berka (1999), p.)

Biller-Andorno, Nikola, Gender imbalance in LOD, Vol. 5 Medicine, Health Care and Philosophy 199–204 (2002) (Biller-Andorno, Vol. 5 Medicine, Health Care and Philosophy 199, p. (2002))

Bloembergen, Wendy/Port, Friedrich/Mauger, Elizabeth/Briggs, Josephine/Leichtman, Alan, Gender Discrepancies in Living Related Renal Transplant Donors and Recipients, Vol. 7 Journal of the American Society of Nephrology 1139–1144 (1996) (Bloembergen et al., Vol. 7 Journal of the American Society of Nephrology 1139, p. (1996))

Bock, Nadine, Rechtliche Voraussetzungen der Organentnahme von Lebenden und Verstorbenen, Frankfurt on the Main 1999 (Bock (1999), p.)

Bonde, Jens Peter, 2003, Incentive measures, retrieved 08 April, 2011 from http://en.euabc.com/word/562 (Bonde (2003))

Borchardt, Klaus-Dieter, Die rechtlichen Grundlagen der Europäischen Union, 4. Edition, Wien 2010 (Borchardt (2010), at)

Borgen, Christopher, Resolving treaty conflicts, Vol. 37 George Washington International Law Rev 573–648 (2005) (Borgen, Vol. 37 George Washington International Law Rev 573, p. (2005))

Bowdish, Michael/Pessotto, Renzo/Barbers, Richard/Schenkel, Felicia/Starnes, Vaughn/Barr, Mark, Long-term Pulmonary Function After Living-Donor Lobar Lung Transplantation in Adults, Vol. 79 The Annals of Thoracic Surgery 418–425 (2005) (Bowdish et al., Vol. 79 The Annals of Thoracic Surgery 418, p. (2005))

Boyle, Alan, Some reflections on the relationship of treaties and soft law, Vol. 48 International and Comparative Law Quarterly 901–913 (1999) (Boyle, Vol. 48 International and Comparative Law Quarterly 901–913 (1999))

Brandstetter, Wolfgang/Kopetzki, Christian (ed.), Organtransplantationen, Medizinische und rechtliche Aspekte der Verwendung menschlicher Organe zu Heilzwecken, Wien 1987 (Author, in Brandstetter/Kopetzki (ed.) (1987), p.)

Broelsch, Christoph (ed.), Nordrhein Westfälische Akademie der Wissenschaften, Organlebendspende, Diskurs zu ethischen, rechtlichen, theologischen und praktischen Aspekten, Symposium der Nordrhein-Westfälischen Akademie der Wissenschaften 2005, Paderborn 2006 (Author, in Broelsch (ed.) (2006), p.)

Broering, Dieter/Walter, Jessica/Braun, Felix/Rogiers, Xavier, Current Status of Hepatic Transplantation, Vol. 45 Current Problems in Surgery 587–661 (2008) (Broering et al., Vol. 45 Current Problems in Surgery 587, p. (2008))

Brown, Robert/Russo, Mark/Lai, Michelle/Shiffman, Michtell/Richardson, Michael/Everhart, James/Hoofnagle, Jay, A Survey of Liver Transplantation from Living Adult Donors in the United States, Vol. 348 The New England Journal of Medicine 818–825 (2003) (Brown et al., Vol. 348 The New England Journal of Medicine 818, p. (2003))

Bundesrat, Bundesrat billigt Reform der Organspende, 2012, retrieved June 20, 2012 from http://www.bundesrat.de/cln_235/nn_6898/DE/service/thema-aktuell/12/20120615-Organspende.html?__nnn=true (Bundesrat (2012)

Bundestag, Bundestag stimmt über neue Organspende-Regelung ab, 2012, retrieved June 20, 2012 from http://www.bundestag.de/dokumente/textarchiv/2012/39014904_kw21_sp_organspende/index.html (Bundestag (2012)

Bundesverband der Organtransplantierten, EU-Richtlinie zur Qualität und Sicherheit von Organen verabschiedet, 2012, retrieved 24 February, 2012 from http://www.bdo-ev.de/front_content.php?idcat=643&idart=11113 (Bundesverband der Organtransplantierten (2012))

Bundeszentrale für gesundheitliche Aufklärung, Repräsentativbefragung: Wissen, Einstellung und Verhalten der Allgemeinbevölkerung zur Organspende, 2010, retrieved 5 May, 2011 from http://www.dso.de/swf/bzga2010_1.swf (Bundeszentrale für gesundheitliche Aufklärung (2010))

Burgtaller, Manfred/Schütz, Hannes, Wiener Kommentar StGB, 2. edition, Wien 2010 (Burgstaller/Schütz (2010), Sec. at)

Calliess, Christian, Die neue Europäische Union nach dem Vertrag von Lissabon, Tübingen 2010 (Calliess (2010), p.)

Calliess, Christian/Ruffert, Matthias (ed.), EUV, AEUV Kommentar, 4. edition, München 2011 (Author, in Calliess/Ruffert (ed.) (2011), Art. at)

Canellopoulou-Bottis, Maria, A New Law on Organ Donation in Greece: One More Effort to Advance Transplants, Vol. 7 European Journal of Health Law 427–439 (2000) (Canellopoulou-Bottis, Vol. 7 European Journal of Health Law 427, p. (2000))

Chapman, Jeremy/Deierhoi, Mark/Wright, Celia (ed.), Organ and tissue donation for transplantation, London 1997 (Author, in Chapman/Deierhoi/Wright (ed.) (1997), p.)

Charité, Lebertransplantation, Lebendspende-Lebertransplantation, 2010, retrieved 24 March, 2011 from http://www.charite.de/avt/medizin/lebergallengangbauchspeicheldruese/ldltx/ldltx.htm (Charité (2010))

Cheyette, Cara, Organ harvests from the legally incompetent: An argument against compelled altruism, Vol. 41 Boston College Law Review 465–515 (2000) (Cheyette, Vol. 41 Boston College Law Review 465, p. (2000))

Childress, James/Liverman, Catharyn, Organ donation, Opportunities for action, Washington D. C. 2006 (Childress/Liverman (2006), p.)

Chinkin, Christine, The challenge of soft law: Development and change in international law, Vol. 38 International and Comparative Law Quarterly 850–866 (1989) (Chinkin, Vol. 38 International and Comparative Law Quarterly 850, p. (1989))

Choudhry, Sujit/Daar, Abdallah /Radcliffe Richards, Janet/Guttmann, Ronald/Hoffenberg, Raymond/Lock, Margaret/Sells, Robert/Tilney, Nicholas, Unrelated living organ donation: ULTRA needs to go, Vol. 29 Journal of Medical Ethics 169–170 (2003) (Choudhry et al., Vol. 29 Journal of Medical Ethics 169, p. (2003))

Christians, Allison, Hard Law & Soft Law in International Taxation, Vol. 25 Wisconsin International Law Journal 1–10 (2007) (Christians, Vol. 25 Wisconsin International Law Journal 1, p. (2007))

Christman, John, "Autonomy in Moral and Political Philosophy", The Stanford Encyclopedia of Philosophy (Spring 2011 Edition), Edward N. Zalta (ed.), retrieved 10 May, 2012, from http://plato.stanford.edu/archives/spr2011/entries/autonomy-moral/ (Christman (2011))

Cini, Michelle, From Soft Law to Hard Law? Discretion and Rule-making in the Commission's State Aid Regime, EUI Working Papers, 2000(Cini (2000), p.)

Collins/Dubernard/Persijn/Land, Procurement and Preservation of Vascularized Organs, Dordrecht 1997 (Author, in Collins et al. (ed.) (1997), p.)

Commission of the European Communities, Communication from the Commission to the European Parliament and the Council, Organ Donation and Transplantation: Policy Actions at EU Level, Brussels 2007 (Commission of the European Communities (2007), p.)

Commission, of the European Communities, Communication from the Commission, Action plan on Organ Donation and Transplantation (2009–2015): Strengthened Cooperation between Member States, Brussels 2008 (Commission of the European Communities (2008), p.)
Commission on European Contract Law, Introduction to the Principles of European Contract Law, 2006, retrieved 09 March, 2012 from http://frontpage.cbs.dk/law/commission_on_european_-contract_law/survey_pecl.htm (Commission on European Contract Law (2006))
Committee on the Environment, Public Health and Food Safety, Explanatory Statement, Report 1 April 2008 on organ donation and transplantation: Policy actions at EU-level, 2008, retrieved 04 July, 2011 from http://www.europarl.europa.eu/sides/getDoc.do?type=REPORT&reference=A6-2008-0090&language=EN (Committee on the Environment (2008))
Constantinesco, Léontin-Jean, Rechtsvergleichung Vol. II, Die rechtsvergleichende Methode, Cologne 1972 (Constantinesco (1972), p.)
Coppen, Remco, Organ donation, policy and legislation, With a special reference to the Dutch Organ Donation Act, Utrecht 2010 (Coppen (2010), p.)
Cosma, Hakon/Whish, Richard, Soft law in the Field of EU Competition Policy, Vol. 14 European Business Law Review 25–56 (2003) (Cosma/Whish, Vol. 14 European Business Law Review 25, p. (2003))
Cotler, Scott/Cotler, Sheldon/Gambera, Michele/Benedetti, Enrico/Jensen, Donald/Testa, Guiliano, Adult living donor liver transplantation: Perspectives from 100 liver transplant surgeons, Vol. 9 Liver Transplantation 637–644 (2003) (Cotler et al., Vol. 9 Liver Transplantation 637, p. (2003))
Council of Europe, Who we are, retrieved 22 March, 2011 from http://www.coe.int/aboutCoe/index.asp?page=quisommesnous&l=en (Council of Europe)
Council of the European Union, Commission Staff Working Document Accompanying document to the Communication from the Commission to the European Parliament and the Council, Organ Donation and Transplantation: Policy Actions at EU Level, Impact Assessment, Brussels 2007 (Council of the European Union (2007))
Council of the European Union, Commission Staff Working Document accompanying the Proposal for a Directive of the European Parliament and of the Council on standards of quality and safety of human organs intended for transplantation and the Communication from the Commission Action plan on Organ Donation and Transplantation (2009–2015): Strengthened Cooperation between Member States, Brussels 2008 (Council of the European Union (2008), p.)
Council of the European Union, 2010, Council improves safety of transplantation and facilitates the exchange of human organs in the EU (Council of the European Union (2010))
Council of Europe, Council of Europe in brief, retrieved 12 April, 3011 from http://www.coe.int/aboutCoe/Default.asp?l=en (Council of Europe)
Creifelds, Carl/Weber, Klaus, Rechtswörterbuch, 20. edition, München 2011 (Creifelds/Weber (ed.) (2011), p.)
Cremer, Wolfram, Lissabon-Vertrag und Grundgesetz, Vol. 32 JURA 296–307 (2010) (Cremer, Vol. 32 JURA 296, p. (2010))
Cronin, Antonia, Is it unethical for doctors to encourage healthy adults to donate a kidney to a stranger? No, Vol. 343 British Medical Journal (2011) (Cronin, Vol. 343 British Medical Journal (2011))
Crowley-Matoka, Megan/Switzer, Galen, Nondirected Living Donation: A Survey of Current Trends and Practices, Vol. 79 Transplantation 515–519 (2005) (Crowley-Matoka/Switzer, Vol. 79 Transplantation 515, p. (2005))
Daar, Abdallah/Land, Walter/Yahya, Taher/Schneewind, Klaus/Gutmann, Thomas/Jakobsen, Arnt, Living-Donor Renal Transplantation: Evidence-based Justification For An Ethical Option, Vol. 11 Transplantation Reviews 95–109 (1997) (Daar et al., Vol. 11 Transplantation Reviews 95, p. (1997))
Daintith, Terence (ed.), Implementing EC Law in the United Kingdom: Structures for Indirect Rule, *Ort 1995* (Author, in Daintith (ed.) (1995), p. 64)

Dahlke, Marc/Popp, Felix/Eggert, Nadine/Hoy, Ludwig/Tanaka, Hideaki/Sasaki, Katsunori/Piso, Pompiliu/Schlitt, Hans, Differences in Attitude Toward Living and Postmortal Liver Donation in the United States, Germany, and Japan, Vol. 46 Psychosomatics 58–64 (2005) (Dahlke et al., Vol. 46 Psychosomatics 58, p. (2005))

Däubler, Wolfgang, Soft law und kollektives Arbeitsrecht, Wittenberg-Prozess – Verhaltensrichtlinien von Unternehmen – Ehrenkodex für Betriebsräte, Vol. 28 Neue Zeitschrift für Arbeitsrecht 42–47 (2011) (Däubler, Vol. 28 Neue Zeitschrift für Arbeitsrecht 42, p. (2011))

Dawson, Mark, Soft Law and the Rule of Law in the European Union: Revision or Redundancy?, Italy 2009 (Dawson (2009), p.)

Delmonico, Francis, Exchanging Kidneys – Advances in Living-Donor Transplantation, Vol. 18 The New England Journal of Medicine 1812–1814 (2004) (Delmonico, Vol. 18 The New England Journal of Medicine 1812, p. (2004))

Delmonico, Francis/Surman, Owen, Is the live-organ donor your patient?, Vol. 76 Transplantation 1257–1260 (2003) (Delmonico/Surman, Vol. 76 Transplantation 1257, p. (2003))

Delmonico, Francis/Dominguez-Gil, Beatriz/Matesanz, Rafael/Noel, Luc, A call for government accountability to achieve national self-sufficiency in organ donation and transplantation, Vol. 378 The Lancet 1414–1418 (2011) (Delmonico et al., Vol. 378 The Lancet 1414, p. (2011))

Deutsch, Erwin/Spickhoff, Andreas, Medizinrecht, Arztrecht, Arzneimittelrecht, Medizinprodukterecht und Transfusionsrecht, 6. edition, Berlin 2006 (Deutsch/Spickhoff (2006), p.)

Deutsche Stiftung Organtransplantation, Nierentransplantation, 2010, retrieved 24 March, 2011 from http://www.dso.de/transplantation/organe/niere_idx.html (Deutsche Stiftung Organtransplantation (2010))

Deutsche Stiftung Organtransplantation, Lebertransplantation, 2010, retrieved 24 March, 2011 fromhttp://www.dso.de/transplantation/organe/leber_idx.html (Deutsche Stiftung Organtransplantation (2010))

Donnelly, Peter/Price, David (ed.), EUROTOLD, Project Management Group, Questioning Attitudes to Living Donor Transplantation, Leicester 1997 (Donnelly/Price (ed.) (1997), p.)

Dor, Frank/Massey, Emma/Frunza, Mihaela/Johnson, Rachel/Lennerling, Annette/Lovén, Charlotte/Mamode, Nizan/Pascalev, Assya/Sterckx, Sigrid/Van Assche, Kristof/Zuidema, Willij/Weimar, Willem, New Classification of ELPAT For Living Organ Donation, 2011 Transplantation (Dor et al., 2011 Transplantation, p.)

Dreier, Horst (ed.), Grundgesetz Kommentar, 2. edition, Tübingen 2004 (Author, in Dreier (ed.), Art. at)

Dreßke, Bettina/Schulze, Maren/Braun, Felix/Walter, Jessica/Kohl, Martina/Schulz-Jürgensen, Sebastian/Krause, Martin/Schrappe, Martin/Burdelski, Martin/Bröring, Dieter, Lebendspende-Lebertransplantation beim Kind, Vol 81 Der Chirurg 813–819 (2010) (Dreßke et. al., Vol 81 Der Chirurg 813, p. (2010))

Dworkin, Gerald, The law relating to organ transplantation in England, Vol. 33 The Modern Law Review 353–377 (1979) (Dworkin, Vol. 33 The Modern Law Review 353, p. (1970))

Dworkin, Gerald, The Theory and Practice of Autonomy, New York 1988 (Dworkin (1988))

Dworkin, Ronald, Life's Dominion, London 1993 (Dworkin (1993))

Ebert, Kurt Hanns, Rechtsvergleichung, Einführung in die Grundlagen, Bern 1978 (Ebert (1978), p.)

Edelmann, Hervé, Ausgewählte Probleme bei der Organspende unter Lebenden, Vol. 50 Versicherungsrecht, 1065–1069 (1999) (Edelmann, Vol. 50 Versicherungsrecht 1065, p. (1999))

Edelstein, Ludwig, From the Hippocratic Oath: Text, Translation and Interpretation, 1943, retrieved 22 March, 2011 from http://guides.library.jhu.edu/content.php?pid=23699&sid=190555 (Edelstein (1943))

Ehlers, Dirk, Die Grundfreiheiten des europäischen Gemeinschaftsrechts (Teil I), Vol. 23 JURA 266–275 (2001) (Ehlers, Vol. 23 JURA 266, p. (2001))

Ehlers, Dirk (ed.), European Fundamental Rights and Freedoms, Berlin 2007 (Author, in Ehlers (ed.) (2007), p.)

Ehlers, Dirk (ed.), Europäische Grundrechte und Grundfreiheiten, 3. edition, Berlin 2009 (Author, in Ehlers (ed.) (2009), p.)

Ehlers, Dirk, Der Vorrang des Europäischen Unionsrechts, Vol. 33 JURA 187–193 (2011) (Ehlers, Vol. 33 JURA 187, p. (2011))

Ehrenzeller, Bernhard/Mastronardi, Philippe/Schweizer, Rainer/Vallender, Klaus (ed.), Die schweizerische Bundesverfassung, Kommentar, Zurich 2002 (Author, in Ehrenzeller et al. (ed.) (2002), p.)

Eigler, Friedrich-Wilhem, Das Problem der Organspende vom Lebenden, Vol. 122 Deutsche Medizinische Wochenschrift 1398–1401 (1997) (Eigler, Vol. 122 Deutsche Medizinische Wochenschrift 1398, p. (1997))

Eisenhardt, Ulrich, Einführung in das Bürgerliche Recht, 5. edition, Heidelberg 2007 (Eisenhardt (2007), p.)

Elliot, Carl, Doing harm: living organ donors, clinical research and The Tenth Man, Vol. 21 Journal of Medical Ethics 91–96 (1995) (Elliot, Vol. 21 Journal of Medical Ethics 91, p. (1995))

Elsässer, Antonellus, Die Transplantation lebend gespendeter Organe aus heutiger christlicher Sicht, Vol. 12 Transplantationsmedizin 184–188 (2000) (Elsässer, Vol. 12 Transplantationsmedizin 184, p. (2000))

Epstein, Miran/Danovitch, Gabriel, Is altruistic-directed living unrelated organ donation a legal fiction?, Vol. 24 Nephrology Dialysis Transplantation 357–360 (2009) (Epstein/Danovitch, Vol. 24 Nephrology Dialysis Transplantation 357, p. (2009))

Erim, Yesim/Malagó, Massimo/Valentin-Gamazo, Camino/Senf, Wolfgang/Broelsch, Christoph, Guidelines for the psychosomatic evaluation of living liver donors: analysis of donor exclusion, Vol. 35 Transplantation Proceedings 909–910 (2003) (Erim/Malagó/Valentin-Gamazo/Senf/Broelsch, Vol. 35 Transplantation Proceedings 909, p. (2003))

Erim, Yesim/Beckmann, Mingo/Gerken, Guido/Paul, David/Senf, Wolfgang/Beckebaum, Susanne, Psychosomatic aspects of living donor liver transplantation/Psychosomatische Aspekte der Leberlebendspende, Vol. 81 Der Chirurg 820–825 (2010) (Erim et. al., Vol. 81 Der Chirurg 820, p. (2010))

Ermecke, Gustav, Katholische Moraltheologie/3: Die spezielle Moral, 10th Edition, Muenster 1961 (Ermecke (1961), p.)

Eser, Albin, Medizin und Strafrecht: Eine schutzgutorientierte Problemübersicht, Vol. 97 Zeitschrift für die gesamte Strafrechtswissenschaft 1–46 (1985) (Eser, Vol. 97 Zeitschrift für die gesamte Strafrechtswissenschaft 1, p. (1985))

Esser, Dirk, Verfassungsrechtliche Aspekte der Lebendspende von Organen zu Transplantationszwecken, Dusseldorf 2000 (Esser (2000), p.)

(The) Ethics Committee of the Transplantation Society, The Consensus Statement of the Amsterdam Forum on the Care of the Live Kidney Donor, Vol. 78 Transplantation 491–492 (2004) (The Ethics Committee of the Transplantation Society, Vol. 78 Transplantation (2004), p.)

EULOD, WP 2, DOW: Deliverable 4, Living Organ Donation Practices in Europe, 2012 (EULOD WP 2 (2012), DOW: Deliverable 4, p.)

Eurofound, Soft Law, 2011, retrieved 10 May, 2012, from http://www.eurofound.europa.eu/areas/industrialrelations/dictionary/definitions/softlaw.htm (Eurofound (2011))

Europa, Zusammenfassungen der EU Gesetzgebungen, Aktionsplan Umwelt und Gesundheit 2004–2010, 2007 retrieved 19 June, 2012 from http://europa.eu/legislation_summaries/public_health/health_determinants_environment/l28145_de.htm (Europa (2007))

Europa, Zusammenfassungen der EU Gesetzgebungen, Sechstes Aktionsprogramm für die Umwelt, 2011 retrieved 19 June, 2012 from http://europa.eu/legislation_summaries/agriculture/environment/l28027_de.htm (Europa (2011))

Europa, Zusammenfassungen der EU Gesetzgebungen, Qualitäts- und Sicherheitsstandards für zur Transplantation bestimmte menschliche Organ, 2012 retrieved 25 April, 2012 from http://

europa.eu/legislation_summaries/public_health/threats_to_health/sp0008_de.htm (Europa (2012))

European Commission, Human Organ Transplantation in Europe: An Overview, 2003 (European Commission (2003))

European Commission, Organ Donation and Transplantation, Policy Options at EU Level, Consultation Document, 2006, p. 6, retrieved 22 March, 2011 from http://ec.europa.eu/health/ph_threats/human_substance/oc_organs/consultation_paper.pdf (European Commission (2006))

(The) European Data Protection Supervisor, Opinion of the European Data Protection Supervisor on the Proposal for a Directive of the European Parliament and of the Council on standards of quality and safety of human organs intended for transplantation, Vol. 52 Official Journal of the European Union 1–14 (2009) (The European Data Protection Supervisor, Vol. 52 Official Journal of the European Union 1, p. (2009))

European Parliament, Position of the European Parliament adopted at first reading on 19 May 2010 with a view to the adoption of Directive 2010/. . ./EU of the European Parliament and of the Council on standards of quality and safety of human organs intended for transplantation, 2010, retrieved April 14, 2011 from http://www.europarl.europa.eu/sides/getDoc.do?type=TA&reference=P7-TA-2010-0181&language=EN (European Parliament (2010))

European Parliament, Information Office Germany, Kürzere Wartezeiten bei Organspenden, 2010, retrieved 22 March, 2011, from http://www.europarl.de/view/de/Aktuell (European Parliament, Information Office Germany (2010))

European Union, Vol. 53 Official Journal of the European Union 1–403 (2010) (European Union, Vol. 53 Official Journal of the European Union 1, p. (2010))

European Urology, 2005, retrieved 24 March, 2011 from http://www.europeanurology.com/uploads/eur_articles/S0302-2838%2807%2901253-5/assets/gr1.jpg (European Urology (2005))

Eurotransplant, The EU Directive on standards of quality and safety, 2012 http://www.eurotransplant.org/cms/index.php?page=efretos_directive (Eurotransplant (2012))

Evans, Martyn, Organ donations should not be restricted to relatives, Vol. 15 Journal of Medical Ethics 17–20 (1989) (Evans, Vol. 15 Journal of Medical Ethics 17, p. (1989))

Faden, Ruth/Beauchamp, Tom, A History and Theory of Informed Consent, New York 1986 (Faden/Beauchamp (1986), p.)

Fateh-Moghadam, Bijan, Zwischen Beratung und Entscheidung – Einrichtung, Funktion und Legitimation der Verfahren vor den Lebendspendekommissionen gemäß § 8 Abs. 3 S. 2 TPG im bundesweiten Vergleich, Vol. 21 Medizinrecht 245–257 (2003) (Fateh-Moghadam, Vol. 21 Medizinrecht 245, p. (2003))

Fateh-Moghadam, Bijan/Schroth, Ulrich/Gross, Christiane/Gutmann, Thomas, Die Praxis der Lebendspendekommissionen – Eine empirische Untersuchung zur Implementierung prozeduraler Modelle der Absicherung von Autonomiebedingungen im Transplantationswesen, Teil 1: Freiwilligkeit, Vol. 22 Medizinrecht 19–35 (2004) (Fateh-Moghadam et al., Vol. 22 Medizinrecht 19, p. (2004))

Fateh-Moghadam, Bijan/Schroth, Ulrich/Gross, Christiane/Gutmann, Thomas, Die Praxis der Lebendspendekommissionen – Eine empirische Untersuchung zur Implementierung prozeduraler Modelle der Absicherung von Autonomiebedingungen im Transplantationswesen, Teil 2: Spender-Empfänger-Beziehung, Organhandel, Verfahren, Vol. 22 Medizinrecht 82–90 (2004) (Fateh-Moghadam et al., Vol. 22 Medizinrecht 82, p. (2004))

Fateh-Moghadam, Bijan, Die Einwilligung in die Lebendorganspende. Die Entfaltung des Paternalismusproblems im Horizont differenter Rechtsordnungen am Beispiel Deutschlands und Englands, München 2008 (Fateh-Moghadam (2008), p.)

Fateh-Moghadam, Bijan/Sellmaier, Stephan/Vossenkuhl, Wilhelm (ed.), Grenzen des Paternalismus, Stuttgart 2010 (Author, in Fateh-Moghadam/Sellmaier/Vossenkuhl (ed.) (2010), p.)

Fateh-Moghadam, Bijan, Künstliche Organverknappung: Rechtliche Schranken der Lebendorganspende als ethisches Problem, 2011, unpublished manuscript (Fateh-Moghadam (2011), p.)

Feinberg, Joel, Harm to Self. The Moral Limits of the Criminal Law, Vol. III, New York 1986 (Feinberg (1986))

Feinberg, Joel, Autonomy, in: John Christman (ed.), The Inner Citadel: Essays on Individual Autonomy, New York 1989, p. 27-53 (Feinberg, in Christman (ed.) (1989))

Ferrari, Paolo/de Klerk, Marry, Paired kidney donations to expand the living donor pool, Vol. 22 Journal of Nephrology 699–707 (2009) (Ferrari/de Klerk, Vol. 22 Journal of Nephrology 699, p. (2009))

Finnish Ministry of Health and Social Affairs, Emphasis on the donor's opinion in organ donations, 2010, retrieved 24 March, 2011 from http://www.stm.fi/en/pressreleases/pressrelease/view/1516653#ens (Finnish Ministry of Health and Social Affairs (2010))

First, Roy, Controversies in Organ Donation: Minority Donation and Living Unrelated Renal Donors, Vol. 29 Transplantation Proceedings 67–69 (1997) (First, Vol. 29 Transplantation Proceedings 67, p. (1997))

Fischer, Gerfried/Lilie, Hans (ed.), Ärztliche Verantwortung im europäischen Rechtsvergleich, Köln 1999 (Author, in Fischer/Lilie (ed.) (1999), p.)

Fischer, Klemens, Der Vertrag von Lissabon, 2. Edition, Baden-Baden 2010 (Fischer (2010), p.)

Fischer, Thomas (ed.), Strafgesetzbuch und Nebengesetze, 58. edition, München 2011 (Fischer (ed.) (2011), Sec. at.)

Folke Schuppert, Gunnar/Neidhardt, Friedhelm, Gemeinwohl – Auf der Suche nach Substanz, Berlin 2002 (Folke Schuppert/Neidhardt (2002), p.)

Forkel, Hans, Das Persönlichkeitsrecht am Körper, gesehen besonders im Lichte des Transplantationsgesetzes, Vol. 23 JURA 73–79 (2001) (Forkel, Vol. 23 JURA 73, p. (2001))

Forsberg, Anna/Nilsson, Madeleine/Krantz, Marie/Olausson, Michael, The essence of living parental liver donation – donors' lived experiences of donation to their children, Vol. 8 Pediatric Transplantation 372–380 (2004) (Forsberg et al., Vol. 8 Pediatric Transplantation 372, p. (2004))

Fox, Renee/Swazey, Judith, Spare Parts: Organ Replacement in American Society, Oxford 1999 (Fox/Swazey (1999), p.)

Frade/Fonseca/Dias/Henriques/Martins/Santos/Sarmento/Lopes, Impact Assessment in Living Kidney Donation: Psychosocial Aspects in the Donor, Vol. 40 Transplantation Proceedings 677–681 (2008) (Frade et al., Vol. 40 Transplantation Proceedings 677, p. (2008))

Frenz, Walter/Götzkes, Vera, Europäische Gesundheitspolitik nach Lissabon, Vol. 28 Medizinrecht 613–618 (2010) (Frenz/*Götzkes*, Vol. 28 Medizinrecht 613, p. (2010))

Frunza, Mihaela/Frunza, Sandu/Bobb, Catalin Vasile/Grad, Ovidiu, Altruistic living unrelated organ donation at the crossroads of ethics and religion. A case study, Vol. 9 Journal for the Study of Religions and Ideologies 3–24 (2010) (Frunza et al., Vol. 9 Journal for the Study of Religions and Ideologies 3, p. (2010))

Fuggle, Susan/Allen, Joanne/Johnson, Rachel/Collett, Dave/Mason, Philip/Dudley, Christopher/Rudge, Christopher/Bradley, Andrew/Watson, Christopher, Factors Affecting Graft and Patient Survival After Live Donor Kidney Transplantation in the UK, Vol. 89 Transplantation 694–701 (2010) (Fuggle et al., Vol. 89 Transplantation 694, p. (2010))

Geiger, Rudolf/Kahn, Daniel-Erasmus/Kotzur, Markus, EUV, AEUV, München 2010 (Author, in Geiger/Kahn/Kotzur (2010), p.)

Gentry, Sommer/Segev, Dorry/Montgomery, Robert, A Comparison of Populations Served by Kidney Paired Donation and List Paired Donation, Vol. 5 American Journal of Transplantation 1914–1921 (2005) (Sommer/Segev/Montgomery, Vol. 5 American Journal of Transplantation 1914, p. (2005))

German National Ethics Council, Increasing the number of organ donations: A pressing issue for transplant medicine in Germany, Berlin 2007 (German National Ethics Council (2007), p.)

Gersen, Jacob/Posner, Eric, Soft Law: Lessons from Congressional Practice, Vol. 61 Stanford Law Review 573-628 (Gersen/Posner, Vol. 61 Stanford Law Review 573, p. (2008))

Gessmann, Martin (ed.), Philosophisches Wörterbuch, 23. edition, Stuttgart 2009 (Gessmann (ed.) (2009), p.)
Gesundheit Österreich GmbH, 2009, Transplant-Annual Report 2009, p. 77, retrieved 24 March, 2011 from http://www.goeg.at/cxdata/media/download/berichte/TX_JB_2009.pdf (Gesundheit Österreich GmbH (2009))
Ghods, Ahad, Ethical Issues and Living Unrelated Donor Kidney Transplantation, Vol. 3 Iranian Journal of Kidney Diseases 183–191 (2009) (Ghods, Vol. 3 Iranian Journal of Kidney Diseases 183, p. (2009))
Gilligan, Carol, Die andere Stimme, München 1982 (Gilligan (1982), p.)
Gillon, Raanan, Transplantation: A Framework for Analysis of the Ethical Issues, Vol. 22 Transplantation Proceedings 902–903 (1990) (Gillon, Vol. 22 Transplantation Proceedings 902, p. (1990))
Glannon, Walter/Ross, Lainie Friedman, Do Genetic Relationships Create Moral Obligations in Organ Transplantation?, Vol. 11 Cambridge Quarterly of Healthcare Ethics 153–159 (2002) (Glannon/Ross, Vol. 11 Cambridge Quarterly of Healthcare Ethics 153, p. (2002))
Glannon, Walter, Is it unethical for doctors to encourage healthy adults to donate a kidney to a stranger? Yes, Vol. 343 British Medical Journal (2011) (Glannon, Vol. 343 British Medical Journal (2011))
Godlee, Fiona, Kidney wanted, Vol. 343 British Medical Journal (2011) (Godlee, Vol. 343 British Medical Journal (2011))
Gohh, Reginald/Morrissey, Paul/Madras, Peter/Monaco, Anthony, Controversies in organ donation: the altruistic living donor, Vol. 16 Nephrology Dialysis Transplantation 619–621 (2001) (Gohh et al., Vol. 16 Nephrology Dialysis Transplantation 619, p. (2001))
Grabenwarter, Christoph, Europäische Menschenrechtskonvention, 4. edition, München 2009 (Grabenwarter (2009), p.)
Grabitz, Eberhard/Hilf, Meinhard/Nettesheim, Martin (ed.), Das Recht der Europäischen Union, Kommentar, München 2011 (Author, in Grabitz/Hilf/Nettesheim (ed.) (2011), Art. at)
Gregor, Mary (ed.), The Cambridge Edition of the Works of Immanuel Kant, Practical Philosophie, Cambridge 1996 (Kant, in Gregor (ed.) (1996), p.)
Großfeld, Bernhard/Yamauchi, Koresuke/Ehlers, Dirk/Ishikawa, Toshiyuki (ed.), Probleme des deutschen, europäischen und japanischen Rechts, Berlin 2006 (Author, in Großfeld et al. (ed.) (2006), p.)
Gruessner, Rainer/Benedetti, Enrico (ed.), Living Donor Organ Transplantation, USA 2008 (Author, in Gruessner/Benedetti (ed.) (2008), p.)
Gründel,Johannes, Ethische Probleme bei der Lebendspende von Organen II., Vol. 5 Zeitschrift für Transplantationsmedizin 70–74 (1993) (Gründel, Vol. 5 Zeitschrift für Transplantationsmedizin 70, p. (1993))
Gubernatis, Gundolf, Ein Markt für Organe, 2002, retrieved 14 July, 2011from http://www.sueddeutsche.de/wissen/transplantation-ein-markt-fuer-organe-1.628346 (Gubernatis (2002))
Gutmann, Thomas, Lebendspende von Organen – Nur unter Verwandten?, Vol. 27 Zeitschrift für Rechtspolitik 111–114 (1994) (Gutmann, Vol. 27 Zeitschrift für Rechtspolitik 111, p. (1994))
Gutmann, Thomas, Probleme einer gesetzlichen Regelung der Lebendspende von Organen, Vol. 15 Medizinrecht 147–155 (1997) (Gutmann, Vol. 15 Medizinrecht 147, p. (1997))
Gutmann, Thomas, Gesetzgeberischer Paternalismus ohne Grenzen? Zum Beschluss des Bundesverfassungsgerichts zur Lebendspende von Organen, Vol. 52 Neue Juristische Wochenschrift 3387–3389 (1999) (Gutmann, Vol. 52 Neue Juristische Wochenschrift 3387, p. (1999))
Gutmann, Thomas/Land, Walter, Ethics regarding living-donor organ transplantation, Vol. 384 Langenbecks archives of surgery, München 1999 (Gutmann/Land, Vol. 384 Langenbeck's archives of surgery p. (1999))
Gutmann, Thomas, Freiwilligkeit als Rechtsbegriff, München 2000 (Gutmann (2000), p.)
Gutmann, Thomas/Schroth, Ulrich, Organlebendspende in Europa, Berlin 2002 (Gutmann/Schroth (2002), p.)

Gutmann, Thomas/Schneewind, Klaus/Schroth, Ulrich/Schmidt, Volker/Elsässer, Antonellus/ Land, Walter/Hillebrand, Günther (ed.), unter Mitarbeit von Bijan Fateh-Moghadam/Eva Wunderer, Grundlagen einer gerechten Organverteilung, Medizin Psychologie Recht Ethik Soziologie, Berlin 2003 (Author, in Gutmann et al. (ed.) (2003), p.)

Gutmann, Thomas, Stellungnahme zur öffentlichen Anhörung „Organlebendspende" der Enquete-Kommission Ethik und Recht der modernen Medizin am 1. März 2004 (Gutmann (2004), p.)

Gutmann, Thomas/Daar, Abdallah/Sells, Robert/Land, Walter (ed.), Ethical, Legal and Social Issues in Organ Transplantation, Lengerich 2004 (Author, in Gutmann et al. (ed.) (2004), p.)

Gutmann, Thomas, Für ein neues Transplantationsgesetz. Eine Bestandsaufnahme des Novellierungsbedarfs im Recht der Transplantationsmedizin, Berlin 2006 (Gutmann (2006), p.)

Gutmann, Thomas, Stellungnahme als Einzelsachverständiger in der öffentlichen Anhörung des Ausschusses für Gesundheit des Deutschen Bundestages am 29. Juni 2011, Ethische und rechtliche Aspekte von Organspenden. Zur Notwendigkeit einer Novellierung des Transplantationsgesetzes (Gutmann (2011), p.)

Guzman, Andrew, The Design of International Agreements, Vol. 16 European Journal of International Law 579–612 (2005) (Guzman, Vol. 16 European Journal of International Law 579, p. (2005))

Guzman, Andrew/Meyer, Timothy, International Soft Law, Vol. 2 Journal of Legal Analysis 171–225 (2010) (Guzman/Meyer, Vol. 2 Journal of Legal Analysis 171, p. (2010))

Häckel, Horst, Organtransplantation am Fallbeispiel Österreich, 1997, retrieved 22 March, 2011 from http://www.lpb-bw.de/publikationen/organ/organ6.htm (Häckel (1997))

Hamza, Amir/Loertzer, Hagen/Wicht, Andreas/Rettkowski, Olaf/Koch, Eileen/Fornara, Paolo, Umfrage zur Überkreuznierentransplantation in Deutschland, Vol. 45 Der Urologe 60–66 (2006) (Hamza et al., Vol. 45 Der Urologe 60, p. (2006))

Haratsch, Andreas/König, Christian/Pechstein, Matthias, Europarecht, 7. edition, Tuebingen 2010 (Haratsch/König/Pechstein (2010), p.)

Hasenöhrl, Ute, Zivilgesellschaft, Gemeinwohl und Kollektivgüter, 2005, retrieved 22 July, 2011 from http://bibliothek.wzb.eu/pdf/2005/iv05-401.pdf (Hasenöhrl (2005))

Hauptmann, Paul/O´Connor, Kevin, Procurement and Allocation of Solid Organs for Transplantation, Vol. 336 The New England Journal of Medicine 422–431 (1997) (Hauptmann/O´ Connor, Vol. 336 The New England Journal of Medicine 422, p. (1997))

(den) Hartogh, Govert, Farewell to non-commitment, Decision systems for organ donation from an ethical viewpoint, The Hague, 2008 (den Hartogh (2008), p.)

Health Council of the Netherlands, New Options for Organ Donation, 2003, retrieved 21 June, 2011 from http://www.gezondheidsraad.nl/sites/default/files/summary%20orgaandonatie.pdf (Health Council of the Netherlands (2003))

Health Council of the Netherlands, Living Donor List Exchange: an addition to the Dutch living kidney donor programme, 2007, retrieved 21 June, 2011 from http://www.gezondheidsraad.nl/en/publications/living-donor-list-exchange-addition-dutch-living-kidney-donor-programme (Health Council of the Netherlands (2007))

Henderson, Antonia/Landolt, Monica/McDonald, Michael/Barrable, William/Soos, John/ Gourlay, William/Allison, Colleen/Landsberg, David, The Living Anonymous Kidney Donor: Lunatic or Saint, Vol. 3 American Journal of Transplantation 203–213 (2003) (Henderson et al., Vol. 3 American Journal of Transplantation 203, p. (2003))

Herdegen, Matthias, Europarecht, 13. Edition, München 2011 (Herdegen (2011), p.)

Heuer, Stefanie/Conrads, Christoph, Aktueller Stand der Transplantationsgesetzgebung 1997, Vol. 15 Medizinrecht 195–202 (1997) (Heuer/Conrads, Vol. 15 Medizinrecht 195, p. (1997))

Hilhorst, Medard, Directed Altruistic Living Organ Donation: Partial but Not Unfair, Vol. 8 Ethical Theory and Moral Practice 197–215 (2005) (Hilhorst, Vol. 8 Ethical Theory and Moral Practice 197, p. (2005))

Hilhorst, Medard/Wijsbek, Henri/Erdman, Ruud/Metselaar, Herold/van Dijk, Gert/Zuidema, Willij/Weimar, Willem, Can we turn down autonomous wishes to donate anonymously?,

Vol. 24 Transplant International 1164–1169 (2011) (Hilhorst et al., Vol. 24 Transplant International 1164, p. (2011))

Hillgenberg, Hartmut, A Fresh Look at Soft Law, Vol. 10 European Journal of International Law 499–515 (1999) (Hillgenberg, Vol. 10 European Journal of International Law 499, p. (1999))

Hippen, Benjamin, In Defense of a Regulated Market in Kidney from Living Vendors, Vol. 30 Journal of Medicine and Philosophy 593–626 (2005) (Hippen, Vol. 30 Journal of Medicine and Philosophy 593, p. (2005))

Hobe, Stephan, Europarecht, 6. edition, München 2010 (Hobe (2011), p.)

Hodson, Margaret, Transplantation using lung lobes from living donors, Vol. 26 Journal of Medical Ethics 419–421 (2000) (Hodson, Vol. 26 Journal of Medical Ethics 419, p. (2000))

Höfling, Wolfram (ed.), Transplantationsgesetz, Berlin 2003 (Author, in Höfling (ed.) (2003), p.)

Höfling, Wolfram (ed.), Die Regulierung der Transplantationsmedizin in Deutschland: Eine kritische Bestandsaufnahme nach 10 Jahren Transplantationsgesetz, Tübingen 2008 (Author, in Höfling (ed.) (2008), p.)

Hofer, Pascal, Das Recht der Transplantationsmedizin in der Schweiz, Berlin 2006 (Hofer (2006), p.)

Hohmann, Elmar Sebastian, Das Transplantationswesen in Deutschland, Österreich und der Schweiz, unter Einbeziehung ethischer und rechtspolitischer Aspekte, Frankfurt on the Main 2003 (Hohmann (2003), p.)

Hoyer, Jochem, A nondirected kidney donation and its consequences: personal experience of a transplant surgeon, Vol. 76 Transplantation 1264–1265 (2003) (Hoyer, Vol. 76 Transplantation 1264, p. (2003))

Hoyer, Peter, Commercial living non-related organ transplantation: a viewpoint from a developed country, Vol. 21 Pediatric Nephrology 1364–1368 (2006) (Hoyer, Vol. 21 Pediatric Nephrology 1364, p. (2006))

Hügli, Anton/Lübcke, Poul (ed.), Philosophielexikon, Hamburg 1997 (Hügli/Lübcke (ed.) (1997), p.)

Human Tissue Authority, Organ donation FAQ's, What is an Independent Assessor and what is their role?, 2010, retrieved 21 October, 2011 from http://www.hta.gov.UK/donations.cfm (Human Tissue Authority (2010))

Iliopoulos-Strangas, Julia (ed.), Soziale Grundrechte in Europa nach Lissabon, Eine rechtsvergleichende Untersuchung der nationalen Rechtsordnungen und des europäischen Rechts, Baden-Baden 2010 (Author, in Iliopoulos-Strangas (ed.) (2010), p.)

Interessengemeinschaft Nierenlebendspende, 2012, retrieved 07 June, 2012 from http://www.nierenlebendspende.com/ (Interessengemeinschaft Nierenlebendspende (2012))

Ismail, Sohal/Massey, Emma/Luchtenburg, Annemarie/Claassens, Lily/Zuidema, Willij/Busschbach, Jan/Weimar, Willem, Religious attitudes towards living kidney donation among Dutch renal patients, Vol. 14 Medicine, Health Care and Philosophy (2011) (Ismail et al., Vol. 14 Medicine, Health Care and Philosophy (2011))

Jackson, Mary, Directive 2010/53/EU (to be transposed into Irish law by 27 August, 2012), retrieved 25 April, 2012 from http://www.dohc.ie/issues/organ_directive_201053EU/mary_jackson.pdf?direct=1 (Jackson (2012))

Jarass, Hans, Das allgemeine Persönlichkeitsrecht im Grundgesetz, Vol. 42 Neue Juristische Wochenschrift 857–862 (1989) (Jarass, Vol. 42 Neue Juristische Wochenschrift 857, p. (1989))

Jarass, Hans (ed.), Charta der Grundrechte der Europäischen Union, Kommentar, München 2010 (Jarass (ed.) (2010), Art. at)

Jarass, Hans/Pieroth, Bodo (ed.), Grundgesetz für die Bundesrepublik Deutschland, 11. edition, München 2011 (Jarass/Pieroth (ed.) (2011), Art. at)

Jendrisak, Martin/Hong, Barry/Shenoy, Surendra/Lowell, Jeffrey/Desai, Niraj/Chapman,Will/Vijayan, Anitha/Wetzel,Richard/Smith, Merry/Wagner, Jessica/Brennan, Susann/Brockmeier, Diane/Kappel, Dean, Altruistic Living Donors: Evaluation for Nondirected Kidney or Liver Donation, Vol. 6 American Journal of Transplantation 115–120 (2006) (Jendrisak et al., Vol. 6 American Journal of Transplantation 115, p. (2006))

Joint Council of Europe/United Nations Study, Trafficking in organs, tissues and cells and trafficking in human beings for the purpose of removal of organs, 2009, retrieved 25 July, 2011 from http://www.coe.int/t/dghl/monitoring/trafficking/docs/news/OrganTrafficking_study.pdf (Joint Council of Europe/United Nations Study (2009))
Joint Network on European Private Law, Joint Network on European Private Law — EU Sixth Framework Programme "Network of Excellence", 2012, retrieved 15 May, 2012 from www.copecl.org (Joint Network on European Private Law (2012))
Joint Working Party of The British Transplantation Society and The Renal Association, United Kingdom Guidelines for Living Donor Kidney Transplantation, 3. edition, 2011 (Joint Working Party of The British Transplantation Society and The Renal Association (2011), p.)
Jowsey, Sheila/Schneekloth, Terry, Psychosocial factors in living organ donation: clinical and ethical challenges, Vol. 22 Transplantation Reviews 192–195 (2008) (Jowsey/Schneekloth, Vol. 22 Transplantation Reviews 192, p. (2008))
Jung, Andrea, Die französische Rechtslage auf dem Gebiet der Transplantationsmedizin, Vol. 14 Medizinrecht 355–365 (1996) (Jung, Vol. 14 Medizinrecht 355, p. (1996))
Jus@Publicum, BSG: Zum Unfall-Begriff nach Lebendnierenspende, 2012, retrieved 20 June, 2012 from https://jusatpublicum.wordpress.com/2012/05/10/bsg-zum-unfall-begriff-nach-lebendnierenspende/ (Jus@Publicum (2012))
Kalchschmid, Gertrud, Die Organtransplantation, Überlegungen de lege lata und de lege ferenda, Wien 1997 (Kalchschmid (1997), p.)
Kant, Immanuel, Metaphysik der Sitten, Werkausgabe Vol. VIII, W. Weischedel (ed.), Frankfurt on the Main (1968 [1797]) (Kant (1968 [1797]), p.)
Kasiske, Bertram/Ravenscraft, Mark/Ramos, Eleanor/Gaston, Robert/Bia, Margaret/Danovitch, Gabriel, The Evaluation of Living Renal Transplant Donors: Clinical Practice Guidelines, Vol. 7 Journal of the American Society of Nephrology 2288–2313 (1996) (Kasiske et al., Vol. 7 Journal of the American Society of Nephrology 2288, p. (1996))
Kaufmann, Arthur/Bemmann, Günter/Kraus, Detlef/Volk, Klaus (ed.), FS für Paul Bockelmann zum 70. Geburtstag, München 1979 (Author, in Kaufmann et al. (ed.) (1979), p.)
Kayler, Liise/Meier-Kriesche, Herwig-Ulf/Punch, Jeffrey/Campbell, Darrell/Leichtman, Alan/ Magee, John/Rudich, Steven/Arenas, Juan/Merion, Robert, Gender imbalance in living donor renal transplantation, Vol. 73 Transplantation 248–252 (2002) (Kayler et al., Vol. 73 Transplantation 248, p. (2002))
Keller, Kelly, Bed of Life: A Discussion of Organ Donation, Its Legal and Scientific History, and a Recommended Opt-Out Solution to Organ Scarcity, The Comment, Vol. XXXII Stetson Law Review 855–896 (2003) (Keller, Vol. XXXII Stetson Law Review 855, p. (2003))
Kienapfel, Diethelm, Grundriß des österreichischen Strafrechts, Besonderer Teil, Vol. I Delikte gegen Personenwerte, 3. edition, Wien 1990 (Kienapfel (1990), p. at)
Kindhäuser, Urs (ed.), Strafgesetzbuch, Lehr- und Praxiskommentar, 4. edition, Baden-Baden 2010 (Kindhäuser (ed.) (2010) Sec. at.)
Kingreen, Andreas, Zur Inanspruchnahme von Gesundheitsleistungen im europäischen Binnenmarkt, Vol. 54 Neue Juristische Wochenschrift 3382–3385 (2001) (Kingreen, Vol. 54 Neue Juristische Wochenschrift 3382, p. (2001))
Kirste, Günter, Stellungnahme zum Fragenkatalog der öffentlichen Anhörung „*Organlebendspende*“ der Enquete-Kommission “Ethik und Recht der modernen Medizin” des Deutschen Bundestages am 1. März 2004, Protokoll 15/14, Freiburg 2004 (Kirste (2004), p.)
Kirste, Günter, Organspendemangel als Begründung für die Lebendspende?, Vol. 81 Der Chirurg 778–786 (2010) (Kirste, Vol. 81 Der Chirurg 778, p. (2010))
Klabbers, Jan, The Redundancy of Soft Law, Vol. 65 Nordic Journal of International Law 167–182 (1996) (Klabbers, Vol. 65 Nordic Journal of International Law 167, p. (1996))
Klabbers, Jan, The Undesirability of Soft Law, Vol. 67 Nordic Journal of International Law 381–391 (1998) (Klabbers, Vol. 67 Nordic Journal of International Law 381, p. (1998))

Klein, Rudolf, Risks and Benefits of Comparative Studies: Notes from Another Shore, Vol. 69 The Milbank Quarterly 275–291 (1991) (Klein, Vol. 69 The Milbank Quarterly 275, p. (1991))
(de) Klerk, Marry, The Dutch Living Donor Exchange Program, Barendrecht 2010 (De Klerk (2010), p.)
Kluge, Eike-Henner, Designated Organ Donation: Private Choice in Social Context, Vol. 19 Hastings Center Report 10–16 (1989) (Kluge, Vol. 19 Hastings Center Report 10, p. (1989))
Knauff, Matthias, Europäisches Soft Law als Gegenstand des Vorabentscheidungsverfahrens, Vol. 46 Zeitschrift Europarecht 735–745 (2011) (Knauff, Vol. 46 Zeitschrift Europarecht 735, p. (2011))
König, Peter, Biomedizinkonvention des Europarats, EU und deutsches Organhandelsverbot, Vol. 23 Medizinrecht 22–26 (2005) (König, Vol. 23 Medizinrecht 22, p. (2005))
Kopetzki, Christian, Organgewinnung zu Zwecken der Transplantation, Wien 1988 (Kopetzki (1988), p.)
Korff, Wilhelm (ed.), Lexikon der Bioethik, Guetersloh 2000 (Author, in Korff (ed.) (2000), p.)
Kranenburg, Leonieke/Zuidema, Willij/Erdmann, Ruud/Weimar, Willem/Passchier, Jan/ Busschbach, Jan, The psychological evaluation of Samaritan kidney donors: a systematic review, Vol. 38 Psychological Medicine 177–185 (2008) (Kranenburg et al., Vol. 38 Psychological Medicine 177, p. (2008))
Kruse, Andreas, Strukturen des Erlebens und Verhaltens bei chronischer Erkrankung im Alter, Bonn 1986 (Kruse (1986), p.)
Kühn, Hermann Christoph, Die Motivationslösung, Neue Wege im Recht der Organtransplantation, Berlin 1998 (Kühn (1998), p.)
Land, Walter, The problem of living organ donation: Facts, thoughts, and reflections, Vol. 2 Transplant International 168–179 (1989) (Land, Vol. 2 Transplant International 168, p. (1989))
Land, Walter/Gutmann, Thomas, Breaking barriers in living donor organ transplantation: a European perspective, Vol. 35 Transplantation Proceedings 926–929 (2003) (Land/Gutmann, Vol. 35 Transplantation Proceedings 926, p. (2003))
Landolt, Monica/Henderson, Antonia/Gourlay, William/McDonald, Michael/Soos, John/ Barrable, William/Landsberg, David, They talk the talk: Surveying attitudes and judging behavior about living anonymous kidney donation, Vol. 76 Transplantation 1437–1444 (2003) (Landolt et al., Vol. 76 Transplantation 1437, p. (2003))
Laufhütte, Heinrich/Rissing-van Saan, Ruth/Tiedemann, Klaus, Leipziger Kommentar StGB, 12. edition, Vol. 2, Sec. 32–55, Berlin 2006 (Author, in Laufhütte/Rissing-van Saan/Tiedemann (ed.) (2006), Sec. at)
Law Commission, Consultation Paper No 139, Criminal Law, Consent in the Criminal Law, A Consultation Paper, 2012, retrieved 24 January, 2012 from http://www.lawcom.gov.uk/docs/ cp195_web.pdf (Law Commission (2012), p.)
Lawin, Peter/Lenhart, Franz-Peter/Peter, Klaus (ed.), Organtransplantation heute, Anästhesie – Chirurgie – Intensivmedizin, Stuttgart 1990 (Author, in Lawin/Lenhart/Peter (ed.) (1990), p.)
Leible, Stefan, Werbung für Lebensmittel und Kosmetika. Möglichkeiten und Grenzen hoheitlicher Regulierung, Vol. 3 Zeitschrift für das gesamte Lebensmittelrecht 285–315 (2010) (Leible, Vol. 3 Zeitschrift für das gesamte Lebensmittelrecht 285, p. (2010))
Lenz, Carl Otto (ed.), EG-Handbuch, Recht im Binnenmarkt, 2. edition, Herne 1994 (Author, in Lenz (ed.) (1994), p.)
Leipold, Dieter, BGB I Einführung und Allgemeiner Teil, 5. edition, Tübingen 2008 (Leipold (2008), p.)
Lennerling, Annette/Forsberg, Anna/Nyberg, Gudrun, Becoming a living kidney donor, Vol. 76 Transplantation 1243–1247 (2003) (Lennerling/Forsberg/Nyberg, Vol. 76 Transplantation 1243, 1244 (2003))
Lennerling, Annette/Forsberg, Anna/Meyer, Käthe/Nyberg, Gudrun, Motives for becoming a living kidney donor, Vol. 19 Nephrology Dialysis Transplantation 1600–1605 (2004 (Lennerling et al., Vol. 19 Nephrology Dialysis Transplantation 1600, p. (2004))

Levinsky, Norman, Organ Donation by Unrelated Donors, Vol. 343 The New England Journal of Medicine 430–436 (2000) (Levinsky, Vol. 343 The New England Journal of Medicine 430, p. (2000))

Lieberwirth, Rolf, Latein im Recht, 5. edition, Berlin 2007 (Lieberwirth (2007), p.)

Lilie, Hans/Krüger, Matthias, Ethische und rechtliche Voraussetzungen der Lebendspende, Vol. 81 Der Chirurg 787–793 (2010) (Lilie/Krüger, Vol. 81 Der Chirurg 787, p. (2010))

Ludwigs, Markus, Rechtsangleichung nach Art. 94, 95 EG-Vertrag, Baden-Baden 2004 (Ludwigs (2004), p.)

Lüke, Gerhard/Ress, Georg/Will, Michael (ed.), Rechtsvergleichung, Europarecht und Staatenintegration, Gedächnisschrift für Léontin-Jean Constatinesco, Cologne 1983 (Author, in Lüke/Ress/Will (ed.) (1983), p.)

(von) Mangoldt, Hermann/Klein, Friedrich/Starck, Christian (ed.), Kommentar zum Grundgesetz, 6. edition, München 2010 (Author, in von Mangoldt/Klein/Starck (ed.) (2010), Art. at)

Manyalich, Marti/Ricart, Assumpta/Martínez, Irene/Balleste, Chloë /Paredes, David/Vilardell, Jordi/Avsec, Danica/Leonidio Dias, Leonidio/Fehrman-Eckholm, Ingela/Hiesse, Christian/Kyriakides, George/Line, Pal-Dag/Maxwell, Andy/Costa, Alessandro Nanni/Paez,Gladys/Turcu, Rosana/Walaszewski, Janusz, EULID Project: European Living Donation and Public Health, Vol. 41 Transplantation Proceedings 2021–2024 (2009) (Manyalich et al., Vol. 41 Transplantation Proceedings 2021, p. (2009))

Marmor, Theodore/Freeman, Richard/Okma, Kieke, Comparative Perspectives and Policy Learning in the World of Health Care, Vol. 7 Journal of Comparative Policy Analysis 331–248 (2005) (Marmor/Freeman/Okma, Vol. 7 Journal of Comparative Policy Analysis 331, p. (2005))

Marmor, Theodore/Freeman, Richard/Okma, Kieke (ed.), Comparative Studies and the Politics of Modern Medical Care, New Haven 2009 (Author, in Marmor/Freeman/Okma (ed.) (2009), p.)

Marmor, Theodore, The unwritten rules of cross-national policy analysis, Vol. 7 Health Economics, Policy and Law 19–20 (2012) (Marmor, Vol. 7 Health Economics, Policy and Law 19, p. (2012))

Massey, Emma/Kranenburg, Leonieke/Zuidema, Willij/Hak, Gert/Erdmann, Ruud/ Hilhorst, Medard/IJzermans, Jan/Busschbach, Jan/Weimar, Willem, Encouraging Psychological Outcomes After Altruistic Donation to a Stranger, Vol. 10 American Journal of Transplantation 1445–1452 (2010) (Massey et al., Vol. 10 American Journal of Transplantation 1442, p. (2010))

Matesanz, Rafael (ed.), Council of Europe, International Figures on Donation and Transplantation-2010, Newsletter Transplant, Vol. 16 No. 1 September 2011 (Matesanz (ed.), Council of Europe (2011), p.)

Matas, Arthur/Garvey,Catherine /Jacobs, Cheryl/Kahn, Jeffrey, Nondirected donation of kidneys from living donors, Vol. 343 The New England Journal of Medicine 433–436 (2000) (Matas et al., Vol. 343 The New England Journal of Medicine 433, p. (2000))

Matas, Arthur/Bartlett, Stephen/Leichtman, Alan/Delmonico, Francis, Morbidity and Mortality After Living Kidney Donation, 1999–2001: Survey of United States Transplant Centers, Vol. 3 American Journal of Transplantation 830–834 (2003) (Matas et al., Vol. 3 American Journal of Transplantation 830, p. (2003))

Matesanz, Rafael (ed.), Council of Europe, International Figures on Donation and Transplantation – 2010, Newsletter Transplant, Vol. 16 No. 1 September 2011 (Matesanz (ed.), Council of Europe (2011), p.)

Maunz, Theodor/Dürig, Günter (ed.), Grundgesetz Kommentar, München 2011 (Author, in Maunz/Dürig (ed.) (2011), Art. at)

McCormick, Thomas, Principles of Bioethics, 1998, retrieved 24 March, 2011 from http://depts.washington.edu/bioethx/tools/princpl.html#prin2 (McCormick (1998), http://depts.washington.edu/bioethx/tools/princpl.html#prin2)

MedicineNet, Definition of Histocompatibility, 2003, retrieved 24 March, 2011 from http://www.medterms.com/script/main/art.asp?articlekey=23075 (MedicineNet (2003))

Mehrabi, Arianeb/Fonouni, Hamidreza/Golriz, Mohammad/Schmied, Bernd/Tahmasbirad, Morva/Weitz, Jürgen/Büchler, Markus/Zeier, Martin/Schmidt, Jan, Lebendspende-Nierentransplantation, Vol. 81 Der Chirurg 794–803 (2010) (Mehrabi et al., Vol. 81 Der Chirurg 794, p. (2010))

Menikoff, Jerry, Organ swapping, Vol. 29 The Hastings Centre Report 28–33 (1999) (Menikoff, Vol. 29 The Hastings Centre Report 28, p. (1999))

Meyer, Olaf, Principles of Contract Law und nationales Vertragsrecht: Chancen und Wege für eine Internationalisierung der Rechtsanwendung, Baden-Baden 2007 (Meyer (2007), p.)

Meyer, Timothy, Soft law as delegation, Vol. 32 Fordham International Law Journal 888–942 (2009) (Meyer, Vol. 32 Fordham International Law Journal 888, p. (2009))

Meyer, Jürgen (ed.), Charta der Grundrechte der Europäischen Union, 3. edition, Baden-Baden 2011 (Author, in Meyer (ed.) (2011), Art. at)

Michaels, Ralf, The Functional Method of Comparative Law, Duke Law School Faculty Scholarship Series, Paper 26, 2005 (Michaels (2005), p.)

Michielsen, Paul, Criteria for selecting organ donors and recipients, Vol. 16 Health Policy 117–120 (1990) (Michielsen, Vol. 16 Health Policy 117, p. (1990))

Middel/Pühler/Lilie/Vilmar (ed.), Novellierungsbedarf des Transplantationsrechts. Bestandsaufnahme und Bewertung, München 2010 (Author, in Middel et al. (ed.) (2010), p.)

Montgomery, Robert/Gentry, Sommer/Marks, William/Warren, Daniel/Hiller, Janet/Houp, Julie/ Zachary, Andrea/Melancon, Keith/Maley, Warren/Rabb, Hamid/Simpkins, Christopher/Segev, Dorry, Domino paired kidney donation: a strategy to make best use of live non-directed donation, Vol. 368 The Lancet 419–421 (2006) (Montgomery et al., Vol. 368 The Lancet 419, p. (2006))

Morrissey, Paul, In Support of List Paired Exchange, Vol. 6 American Journal of Transplantation 434–435 (2006) (Morrissey, Vol. 6 American Journal of Transplantation 434, p. (2006))

Nicholas, Moussis, Access to the European Union, 2011, retrieved April 05, 2011 from http://europedia.moussis.eu/books/Book_2/2/3/2/index.tkl?all=1&pos=28 (Moussis (2011))

Mueller, Paul/Case, Ellen/Hook, Christopher, Responding to offers of altruistic living unrelated kidney donation by group associations: an ethical analysis, Vol. 22 Transplantation Review 200–205 (2008) (Mueller/Case/Hook, Vol. 22 Transplantation Review 200, p. (2008))

Mühlbacher, Ferdinand, Presseinformation für Tree of Life – Förderung der Lebendspende, Medizinische Universität Wien, Wien 2009 (Mühlbacher (2009), p.)

Müller-Graff, Peter-Christian/Schmahl, Stefanie/Skouris, Vassilios, Europäisches Recht zwischen Bewährung und Wandel, Festschrift für Dieter H. Scheuing, Baden-Baden 2011 (Author, in Müller-Graff/Schmahl/Skouris (ed.) (2011), p.)

(von) Münch, Ingo/Kunig, Philip (ed.), Grundgesetz Kommentar, 6. edition, München 2012 (Author, in von Münch/Kunig (ed.) (2012), Art. at)

Münkler, Herfried/Fischer, Karsten (ed.), Gemeinwohl und Gemeinsinn im Recht, Berlin 2002 (Author, in Münkler/Fischer (ed.) (2002), p.)

Muthny, Fritz/Suwelack, Barbara/Wolters, Heiner/Michalke, Denise, Lebendnierenspende: Spender-Empfänger-Konstellationen und medizinpsychologische Abklärung bei Lebendnierenspende in Münster, 2010 Transplantationsmedizin 106-110 (Muthny et al., 2010 Transplantationsmedizin 106, p.)

Najarian, John/Chavers, Blanche/McHugh, Lois/Matas, Arthur, 20 years or more of follow-up of living kidney donors, Vol. 340 The Lancet 807–810 (1992) (Najarian et al., Vol. 340 The Lancet 807, p. (1992))

National Cancer Institute, Dictionary of Cancer Terms, cold ischemia time, 2012, retrieved 24 March, 2011 from http://www.cancer.gov/dictionary/?CdrID=630927 (National Cancer Institute (2012))

Neft, Hans, Die Überkreuz-Lebendspende im Lichte der Restriktionen des Transplantationsgesetzes, Vol. 13 Neue Zeitschrift für Sozialrecht 519–522 (2004) (Neft, Vol. 13 Neue Zeitschrift für Sozialrecht 519, p. (2004))

Neft, Hans, Die versicherungsrechtliche Absicherung des Organlebendspenders – Bestandsaufnahme und Reformvorschläge, Vol. 20 Neue Zeitschrift für Sozialrecht 566–571 (2011) (Neft, Vol. 20 Neue Zeitschrift für Sozialrecht 566, p. (2011))

Neuberger, James/Price, David, Role of living liver donation in the United Kingdom, Vol. 327 British Medical Journal 676–679 (2003) (Neuberger/Price, Vol. 327 British Medical Journal 676, p. (2003))

Neuberger, James, Making an offer you can't refuse? A challenge of altruistic donation, Vol. 24 Transplant International 1159–1161 (2011) (Neuberger, Vol. 24 Transplant International 1159, p. (2011))

Neumann, Ulf/Neuhaus, Peter/Schmeding, Maximilian, Lebendspende-Lebertransplantation beim Erwachsenen, Living donor liver transplantation in adults, Vol. 81 Der Chirurg 804–812 (2010) (Neumann/Neuhaus/Schmeding, Vol. 81 Der Chirurg 804, p. (2010))

Neyer, Christian/Scheuring, Stefanie, Ethik und Organtransplantation, 2004 Österreichische Pflegezeitschrift 12-15 (Neyer/Scheuring, 2004 Österreichische Pflegezeitschrift p.)

NHS Blood and Transplant, Transplant Activity in the United Kingdom, 2008–2009, retrieved 24 March, 2011 from http://www.organdonation.nhs.uk/ukt/statistics/transplant_activity_report/archive_activity_reports/pdf/ukt/transplant_activity_uk_2008-2009.pdf (NHS Blood and Transplant (2008–2009))

NHS Blood and Transplant, Transplant Activity in the United Kingdom, 2009–2010, retrieved 24 March, 2011 from http://www.uktransplant.org.uk/ukt/statistics/transplant_activity_report/archive_activity_reports/pdf/ukt/activity_report_2009_10.pdf (NHS Blood and Transplant (2010))

Nickel, Lars/Schmidt-Preisigke, Angelika/Sengler, Helmut, Transplantationsgesetz, Stuttgart 2001 (Nickel/Schmidt-Preisigke/Sengler (2001), p.)

Nickel, Lars/Preisigke, Angelika, Zulässigkeit einer Überkreuz-Lebendspende nach dem Transplantationsgesetz, Vol. 22 Medizinrecht 307–310 (2004) (Nickel/Preisigke, Vol. 22 Medizinrecht 307, p. (2004))

Norba, Daniela, Rechtsfragen der Transplantationsmedizin aus deutscher und europäischer Sicht, Berlin 2009 (Norba (2009), p.)

Nowak, Carsten, Europarecht nach Lissabon, Baden-Baden 2011(Nowak (2011), p.)

Nys, Herman, The Belgian Law on retrieval and transplantation of organs, Power Point Presentation (Nys, The Belgian Law on retrieval and transplantation of organs, p. 5 (Power Point Presentation))

Nys, Herman, International Encyclopeadia of Laws, Medical Law, Supplement 59 Belgium, Great Britain 2009 (Nys (2009), p.)

Oduncu, Fuat/Schroth, Ulrich/Vossenkuh, Wilhelm (ed.), Transplantation, Organgewinnung- und –allokation, Vol. 2, Göttingen 2003 (Author, in Oduncu/Schroth/Vossenkuhl (ed.) (2003), p.)

O'Hara Jr, Jerome/Bramstedt, Katrina/Flechner, Stewart/Goldfarb, David, Ethical issues surrounding high-risk kidney recipients: implications for the living donor, Vol. 17 Progress in Transplantation 180–182 (2007) (O'Hara et al., Vol. 17 Progress in Transplantation 180, p. (2007))

O'Neill, Onora, Some limits of informed consent, Vol. 29 Journal of Medical Ethics 4–7 (2003) (O'Neill, Vol. 29 Journal of Medical Ethics 4, p. (2003))

Oniscu, Gabriel/Forsythe, John, An Overview of Transplantation in Culturally Diverse Regions, Vol. 38 Annals Academy of Medicine 365–369 (2009) (Oniscu/Forsythe, Vol. 38 Annals Academy of Medicine 365, p. (2009))

Oppermann, Thomas/Classen, Claus Dieter/Nettesheim, Martin, Europarecht, 5. edition, München 2011 (Oppermann/Classen/Nettesheim (2011), p.)

Pancevski, Bojan, Bulgarian hospital admits role in illegal transplants, Vol. 367 The Lancet 461 (2006) (Pancevski, Vol. 367 The Lancet 461, p. (2006))

Park, Kiil/Moon, Jang/Kim, Soon Il/Kim, Yu Seun, Exchange donor program in kidney transplantation, Vol. 67 Transplantation 336–338 (1999) (Park et al., Vol. 67 Transplantation 336, p. (1999))

Parzeller, Markus/Henze, Claudia/Bratzke, Hansjürgen, Gewebe- und Organtransplantation – Verfehlte und praxisferne Regelungen im Transplantationsgesetz, Vol. 87 Kritische Vierteljahresschrift 371–196 (2004) (Parzeller/Henze/Bratzke, 87 Kritische Vierteljahresschrift 371, p. (2004))

Papachristou, Christina/Walter, Marc/Dietrich, Kerstin/Danzer, Gerhard/Klupp, Jochen/Klapp, Burghard/Frommer, Jörg, Motivation for Living-Donor Liver Transplantation from the Donor's Perspective: An In-Depth Qualitative Research Study, Vol. 78 Transplantation 1506–1514 (2004) (Papachristou et al., Vol. 78 Transplantation 1506, p. (2004))

Patel, Shaneel Rajendra/Chadha, Priyanka/Papalois, Vassilios, Expanding the Live Kidney Donor Pool: Ethical Considerations Regarding Altruistic Donors, Paired and Pooled Programs, Vol. 3 Experimental and Clinical Transplantation 181–186 (2011) (Patel/Chadha/Papalois, Vol. 3 Experimental and Clinical Transplantation 181, p. (2011))

Pereira, David, Vol. 18 European Journal of Health Law 1 (2011) (Pereira, Vol. 18 European Journal of Health Law 1, p. (2011))

Pfeiffer, Alexandra, Die Regelung der Lebendorganspende im Transplantationsgesetz, Frankfurt on the Main, 2004 (Pfeiffer (2004), p.)

Pöltner, Günther, Grundkurs Medizin-Ethik, Wien 2002 (Pöltner (2002), p.)

Porth, Carol Mattson, Essentials of Pathophysiology, China 2010 (Porth (2010), p.)

Posener, Paul, Rechtslexikon, Berlin 1909 (Posener (1909, p.)

Prechtl, Peter/Burkhard, Franz Peter (ed.), Metzler Lexikon Philosophie, 3. edition, Stuttgart 2008 (Prechtl/Burkhard (ed.) (2008), p.)

President's Council on Bioethics, Session 2: Living Organ Donation: Outcomes and Ethics, 2004, retrieved 11 July, 2011 from http://bioethics.georgetown.edu/pcbe/transcripts/sept06/session2.html (Speaker (2011))

(The) President of the European Economic and Social Committee, Opinion of the European Economic and Social Committee on the Proposal for a Directive of the European Parliament and of the Council on standards of quality and safety of human organs intended for transplantation, 2009 Official Journal of the European Union 64-67 (The President of the European Economic and Social Committee, 2009 Official Journal of the European Union, 64, p.)

Price, David/Akveld, Hans (ed.), Living Organ Donation in the Nineties: European Medico-Legal Perspectives, EUROTOLD Project, Leicester 1997 (Author, in Price/Akveld (ed.) (1997), p.)

Price, David, Legal and Ethical Aspects of Organ Transplantation, Cambridge 2000 (Price (2000), p.)

Price, David, Human Tissue in Transplantation and Research, A Model Legal and Ethical Donation Framework, Cambridge 2010 (Price (2010), p.)

Pruett, Timothy L/Tibell, Annika/Alabdulkareem, Abdulmajeed/Bhandari, Mahendra/Cronin, David/Dew, Mary/Dib-Kuri, Arturo/Gutmann, Thomas/Matas, Arthur/McMurdo, Lisa/Rahmel, Axel/Rizvi, Adibul Hasan/Wright, Linda/Delmonico, Francis, The Ethics Statement *of* the Vancouver Forum on the Live Lung, Liver, Pancreas, and Intestine Donor, Vol. 81 Transplantation (2006), p. 1386-87 (Pruett, et al., Vol. 81 Transplantation (2006), p.)

Pschyrembel, Klinisches Wörterbuch, unter der Leitung von Helmut Hildebrandt, 258. edition, Berlin 1998 (Pschyrembel (1998), p.)

Pühler, Wiebke/Middel, Claus-Dieter/Schirmer, Horst Dieter/Hübner, Marlis, Zur Mitteilung der EU-Kommission zu Organspende und –transplantation (KOM (2007) 275 endg.), Vol. 25 Medizinrecht 584–589 (2007) (Pühler et al., Vol. 25 Medizinrecht 584, p. (2007))

Ratner, Lloyd/Rana, Abbas/Ratner, Emily/Ernst, Victoria/Kelly, Joan/Kornfeld, Donald/Cohen, David/Wiener, Ilona, The Altruistic Unbalanced Paired Kidney Exchange: Proof of Concept and Survey of Potential Donor and Recipient Attitudes, Vol. 89 Transplantation 15–22 (2010) (Ratner et al., Vol. 89 Transplantation 15, p. (2010))

Rawls, John, Eine Theorie der Gerechtigkeit, Frankfurt on the Main 1975 (Rawls (1975), p.)

Rawls, John, Political Liberalism. New York 1993 (Rawls (1993))

Reference.Md, cold ischemia, 2007, retrieved 24 March, 2011 from http://www.reference.md/files/D050/mD050377.html (Reference.Md (2007))

Riesenhuber, Karl/Takayama (ed.), Rechtsangleichung: Grundlagen, Methoden und Inhalte, Berlin 2006 (Author, in Riesenhuber/Takayama (ed.) (2006), p.)

Ritter, Joachim (ed.), Historisches Wörterbuch der Philosophie, Basel 1971 (Ritter (ed.) (1971), p.)

Rittner, Christian/Paul, Norbert (ed.), Ethik der Lebendorganspende, Mainz 2005 (Author, in Rittner/Paul (ed.) (2005), p.)

Robert Koch-Institut, Gesundheitsberichterstattung des Bundes, Heft 17, Organtransplantation und Organspende, 2003 (Robert Koch-Institut (2003), p.)

Roche, Lexikon Medizinrecht, 5. edition, München 2003 (Roche (2003), p.)

Roff, Sue Rabbitt, Self-interest, self-abnegation and self-esteem: towards a new moral economy of non-directed kidney donation, Vol. 33 Journal of Medical Ethics 437–441 (2007) (Roff, Vol. 33 Journal of Medical Ethics 437, p. (2007))

Roodnat, Jeroen/Kal-van Gestel, Judith/Zuidema, Willij/van Noord, Marian/van de Wetering, Jodie/IJzermans, Jan/Weimar, Willem, Successful expansion of the Living Donor Pool by Alternative Living Donation Programs, Vol. 9 American Journal of Transplantation 2150–2156 (2009) (Roodnat et al., Vol. 9 American Journal of Transplantation 2150, p. (2009))

Roodnat, Jeroen/Zuidema, Willij/van de Wetering, Jodie/de Klerk, Marry/Erdman, Ruud/Massey, Emma/Hilhorst, Medard/IJzermans, Jan/Weimar, Willem, Altruistic donor triggered domino-paired kidney donation for unsuccessful couples from the kidney-exchange program, Vol. 10 American Journal of Transplantation 821–827 (2010) (Roodnat et al., Vol. 10 American Journal of Transplantation 821, p. (2010))

Roodnat, Joke/van de Wetering, Jacqueline/Zuidema, Willij/van Noord, Marian/Kal-van Gestel, Judith/Ijzermans, Jan/Weimar, Willem, Ethically Diverse Populations and Their Participation in Living Kidney Donation Programs, Vol. 89 Transplantation 1263–1269 (2010) (Roodnat et al., Vol. 89 Transplantation 1263, p. (2010))

Ross, Lainie Friedmann/Rubin, David/Siegler, Mark/Josephson, Michelle/Thistlethwaite, Richard/Woodle, Steve, Ethics of a Paired-Kidney-Exchange Program, Vol. 336 The New England Journal of Medicine 1752–1755 (1997) (Ross et al., Vol. 336 The New England Journal of Medicine 1752, p. (1997))

Ross, Lainie/Friedmann/Woodle, Steve, Ethical Issues in Increasing Living Kidney Donations by Expanding Kidney Paired Exchange Programs, Vol. 69 Transplantation 1539–1543 (2000) (Ross/Woodle, Vol. 69 Transplantation 1539, p. (2000))

Ross, Lainie Friedmann, Solid organ donation between strangers, Vol. 30 Journal of law, medicine and ethics 440–446 (2002) (Ross, Vol. 30 Journal of Law, Medicine and Ethics, 440, p. (2002))

Ross, Lainie Friedmann/Glannon, Walter/Josephson, Michelle/Thistlethwaite, Richard, Should all living donors be treated equally, Vol 74 Transplantation 418–421 (2002) (Ross et al., Vol 74 Transplantation 418, p. (2002))

Ross, Lainie Friedmann/Zenios, Stefanos, Practical and Ethical Challenges to Paired Exchange Programs, Vol. 4 American Journal of Transplantation 1553–1554 (2004) (Ross/Zenios, Vol. 4 American Journal of Transplantation 1553, p. (2004))

Ross, Lainie Friedmann, The ethical limits in expanding living donor transplantation, Vol. 16 Kennedy Institute of Ethics Journal 151–172 (2006) (Ross, Vol. 16 Kennedy Institute of Ethics Journal 151, p. (2006))

Ross, Lainie Friedmann/Thistlethwaite, Richard, Minors as Living Solid-Organ Donors, Vol. 122 American Academy of Pediatrics 454–461 (2008) (Ross/Thistlethwaite, Vol. 122 American Academy of Pediatrics 454, p. (2008))

Rotondo, Roberto, Veröffentlichung zur Organspende und Transplantation, Cross-over Lebendspenden (Überkreuzspenden) doch zulässig?, 2004, retrieved 24 March, 2011 from http://www.transplantation-information.de/organspende_organspender/lebendorganspende/cross_over_lebendspende_ueberkreuzspende_2003.htm (Rotondo (2004))

Roxin, Claus/Schroth, Ulrich (ed.), Handbuch des Medizinstrafrechts, 4. edition, Stuttgart 2010 (Author, in Roxin/Schroth (ed.) (2010), p.)

Russel, Gordon/Mentzel, Robert, Sympathy and Altruism in Response to Disasters, Vol. 130 The Journal of Social Psychology 309–316 (1990) (Russel/Mentzel, Vol. 130 The Journal of Social Psychology 309, p. (1990))

Sacco, Rodolfo, Einführung in die Rechtsvergleichung, Aus dem italienischen übersetzt von Dr. Jacob Joussen, Baden-Baden 2001 (Sacco (2001), at)

Sachs, Michael (ed.), Grundgesetz Kommentar, 6. edition, München 2011 (Author, in Sachs (ed.) (2011), Art. at)

Sándor, Judit, International Encyclopedia of Laws, Hungary, The Hague 2003 (Sándor (2003), p.)

Sándor, Judit/Drakopoulou, Aikaterini/Bárd, Petra, The Legal Regulation of Biobanks, National Report Greece, Celab Paper Series, Hungary 2009 (Sándor/Drakopoulou/Bárd (2009), p.)

Sándor, Judit/Bárd, Petra, The Legal Regulation of Biobanks, National Report Hungary, Celab Paper Series, Hungary 2009 (Sándor/Bárd (2009), Hungary, p. 9)

Schabel, Bettina, Das Selbstbestimmungsrecht des Patienten in Österreich, Wien 2007 (Schabel (2007), p.)

Schachter, Oscar, The Twilight Existence of Nonbinding International Agreements, Vol. 71 The American Journal of International Law 296–304 (1977) (Schachter, Vol. 71 The American Journal of International Law 296, p. (1977))

Schmidt am Busch, Birgit, Die Gesundheitssicherung im Mehrebenensystem, Tübingen 2007 (Schmidt am Busch (2007), p.)

Schönke, Adolf/Schröder, Horst (ed.), Strafgesetzbuch Kommentar, 28. edition, München 2001 (Author, in Schönke/Schröder (ed.) (2001), Sec. at)

Schneewind, Klaus/Ney, B./Hammerschmidt, H./Oerter, R./Pabst, Ob./Schultz-Gambard, E., Veränderungserwartungen und faktische Veränderungen der Lebensgestaltung bei Nierentransplantation: Ein Vergleich zwischen verwandten und nichtverwandten Spender-Empfänger-Paaren, Vol. 12 Transplantationsmedizin 164–173 (2000) (Schneewind et al., Vol. 12 Transplantationsmedizin 164, p. (2000))

Schneider, Ingrid, Lebendspende: Kommerzialisierung des Unbezahlbaren?, 2004, retrieved 21 June, 2011 from http://www.gen-ethisches-netzwerk.de/gid/163/mensch/schneider/lebendspende-kommerzialisierung-des-unbezahlbaren (Schneider (2004))

Schnitzer, Adolf, Vergleichende Rechtslehre, Vol. I, 2. edition, Basel 1961 (Schnitzer (1961), p.)

Schreiber, Hans-Ludwig/Wolfslast, Gabriele, Ein Entwurf für ein Transplantationsgesetz, Vol. 10 Medizinrecht 189–195 (1992) (Schreiber/Wolfslast, Vol. 10 Medizinrecht 189, p. (1992))

Schreiber, Markus, Die gesetzliche Regelung der Lebendspende von Organen in der Bundesrepublik Deutschland, Frankfurt on the Main 2004 (Schreiber (2004), p.)

Schroder, Nina/McDonald, Laurie/Etringer, Geri/Snyders, Michele, Consideration of psychosocial factors in the evaluation of living donors, Vol. 18 Progress in Transplantation 41–49 (2008) (Schroder et al., Vol. 18 Progress in Transplantation 41, p. (2008))

Schroth, Ulrich, Stellungnahme zu dem Artikel von Bernhard Seidenath: „Lebendspende von Organen – Zur Auslegung des § 8 I 2 TPG“, MedR 1998, 253, Vol. 17 Medizinrecht 67–68 (1999) (Schroth, Vol. 17 Medizinrecht 67, p. (1999))

Schroth, Ulrich, Anmerkung zum Urteil des BVerfG, Vol. 9 Juristenzeitung 469–472 (2004) (Schroth, Vol. 9 Juristenzeitung 469, p. (2004))

Schroth, Ulrich/König, Peter/Gutmann, Thomas/Oduncu, Fuat (ed.), Transplantationsgesetz, München 2005 (Author, in Schroth et al. (ed.) (2005), Sec. at)

Schroth, Ulrich/Schneewind, Klaus/Gutmann, Thomas/Fateh-Moghadam, Bijan (ed.), Patientenautonomie am Beispiel der Lebendorganspende, Göttingen 2006 (Author, in Schroth et al. (ed.) (2006), p.)

Schutzeichel, Corinna Iris, Geschenk oder Ware? Das begehrte Gut Organ. Nierentransplantation in einem hochregulierten Markt, Ethik in der Praxis, Münster 2002 (Schutzeichel (2002), p.)

Schwarze, Jürgen (ed.), EU-Kommentar, 2. edition, Baden-Baden 2009 (Author, in Schwarze (ed.) (2009), Art. at)

Schwarze, Jürgen/Hatje, Armin (ed), Europarecht, Der Reformvertrag von Lissabon, Baden-Baden (2009) (Author, in Schwarze/Hatje (ed.) (2009), p.)

Scottish Executive Health Department, Human Tissue (Scotland) Act 2006: A guide to its implications for NHS Scotland, Scotland 2006 (Scottish Executive Health Department (2006), p.)

Secretary General of the Council of Europe, Convention for the Protection of Human Rights and Dignity of the Human Being with regard to the Application of Biology and Medicine: Convention on Human Rights and Biomedicine, Explanatory Report, 1996 (Secretary General of the Council of Europe (1996))

Segev, Dorry/Montgomery, Robert, Regional and Racial Disparities in the Use of Live Non-Directed Kidney Donors, Vol. 8 American Journal of Transplantation 1051–1055 (2008) (Segev/Montgomery, Vol. 8 American Journal of Transplantation 1051, p. (2008))

Seidenath, Bernhard, Lebendspende von Organen – Zur Auslegung des § 8 Abs. 1 S.2 TPG, Vol. 16 Medizinrecht 253–256 (1998) (Seidenaht, Vol. 16 Medizinrecht 253, p. (1998))

Senden, Linda, Soft law in the European Community Law, Oxford 2004 (Senden (2004), p.)

Shaffer, Gregory/Pollack, Mark, Hard vs. Soft Law: Alternatives, Complements and Antagonists in International Governance, Vol. 94 Minnesota Law Review 706, 712 (2010) (Shaffer/ Pollack, Vol. 94 Minnesota Law Review 706, p. (2010))

Sheper-Hughes, Nancy, The Tyranny of the Gift: Sacrificial Violence in Living Donor Transplants, Vol. 7 American Journal of Transplantation 507–511 (2007) (Sheper-Hughes, Vol. 7 American Journal of Transplantation 507, p. (2007))

Siegmund-Schultze, Nicola, Organtransplantation, Reinbek 1999 (Siegmund-Schultze (1999), p.)

Singer, Marcus, On Duties to Oneself, Vol. 69 Ethics 202–205 (1959) (Singer, Vol. 69 Ethics 202 ff. (1959))

Solomon, Olga/Pavlou, Anna, Joint Action on Organs, Assisting Member States in reaching the full potential of deceased and living organ donation, 2011, retrieved 25 April, 2012 from http://ec.europa.eu/eahc/documents/health/Workshop_JA_2011/04_Pavlou_organ_transplantation.pdf (Solomon/Pavlou (2011))

Snyder, Francis, The Effectiveness of European Community Law: Institutions, Processes, Tools and Techniques, Vol. 56 Modern Law Review 19–54 (1993) (Snyder, Vol. 56 Modern Law Review 19, p. (1993))

Sommerer, Claudia/Morath, Christian/Andrassy, Joachim/Zeier, Martin, The long-term consequences of living-related or unrelated kidney donation, Vol. 19 Nephrology Dialysis Transplantation iv45-iv47 (2004) (Sommerer et al., Vol. 19 Nephrology Dialysis Transplantation iv45, p. (2004)

Spickhoff, Andreas, Aktuelle Rechtsfragen des medizinischen Behandlungsverhältnisses, München 2004 (Spickhoff (2004), p.)

Spickhoff, Andreas, Die ärztliche Aufklärung vor der altruistisch motivierten Einwilligung in medizinische Eingriffe, Vol. 59 Neue Juristische Wochenschrift 2075–2076 (2006) (Spickhoff, Vol. 59 Neue Juristische Wochenschrift 2075, p. (2006))

Spickhoff, Andreas, Ärztliche Aufklärungsfehler, haftungsrechtliche Zurechnung und hypothetisch Einwilligung, Entscheidung des House of Lords vom 14. Oktober 2004, mit Anmerkung von Andreas Spickhoff, Regensburg, Vol. 14 Zeitschrift für Europäisches Privatrecht 400–414 (2006) (Spickhoff, Vol. 14 Zeitschrift für Europäisches Privatrecht 400, p. (2006))

Spickhoff, Andreas, Informed Consent – Recent Developments, Vol. 54 World Medical Journal 9–11 (2008) (Spickhoff, Vol. 54 World Medical Journal 9, p. (2008))

Spickhoff, Andreas, Die Entwicklung des Arztrechts 2010/2011, Vol. 64 Neue Juristische Wochenschrift 1651–1658 (2011) (Spickhoff, Vol. 64 Neue Juristische Wochenschrift 1651, p. (2011))

Spickhoff, Andreas (ed.), Medizinrecht, München 2011 (Author, in Spickhoff (ed.) (2011), p.)

Spickhoff, Andreas, Patientenrechte und Gesetzgebung, Rechtspolitische Anmerkungen zum geplanten Patientenrechtegesetz, Vol. 45 Zeitschrift für Rechtspolitik 65–70 (2012) (Spickhoff, Vol. 45 Zeitschrift für Rechtspolitik 65, p. (2012))

Spiegel, Licht ins Dunkel, 1993, retrieved 07 April, 2011 from http://www.spiegel.de/spiegel/print/d-9289521.html (Spiegel (1993))

Spielberg, Petra, EU-Richtlinie zur Qualität und Sicherheit von Organtransplantationen: Mehr Solidarität bei Organspenden, Vol. 107 (22) Deutsches Ärzteblatt A-1102/B-973/C961 (2010) (Spielberg, Vol. 107 (22) Deutsches Ärzteblatt A-1102/B-973/C961 (2010))

Spital, Aaron, Ethical issues in living organ donation, Vol. 38 American Journal of Kidney Diseases 189–195 (2001) (Spital, Vol. 38 American Journal of Kidney Diseases 189, p. (2001))

Spital, Aaron, Public attitudes toward kidney donation by friends and altruistic strangers in the United States, Vol. 71 Transplantation 1061–1064 (2001) (Spital, Vol. 71 Transplantation 1061, p. (2001))

Spintal, Aaron, Response to "Do Genetic Relationships Create Moral Obligations in Organ Transplantation?" by Walter Glannon and Lainie Friedman Ross (CQ Vol 11, No 2), Vol. 12 Cambridge Quarterly of Healthcare Ethics 116–118 (2003) (Spital, Vol. 12 Cambridge Quarterly of Healthcare Ethics 116, p. (2003))

Spital, Aaron, Should people who donate a kidney to a stranger be permitted to choose their recipients? Views of the United States public, Vol. 76 Transplantation 1252–1256 (2003) (Spital, Vol. 76 Transplantation 1252, p. (2003))

Stanford University, Kidney Transplantation: Past, Present, Future, retrieved 24 March, 2011 from http://www.stanford.edu/dept/HPS/transplant/html/hla.html (Stanford University)

Stefan, Oana Andreea, European Competition Soft Law in European Courts: A Matter of Hard Principles, Vol. 14 European Law Journal 753–772 (2008) (Stefan, Vol. 14 European Law Journal 753, p. (2008))

Steinberg, David, The Allocation of Organs Donated by Altruistic Strangers Vol. 145 Annals of Internal Medicine 197–202 (2006) (Steinberg, Vol. 145 Annals of Internal Medicine 197, p. (2006))

Steinbrook, Robert, Public Solicitation of Organ Donors, Vol. 353 New England Journal of Medicine 441–444 (2005) (Steinbrook, Vol. 353 New England Journal of Medicine 441, p. (2005))

Sterner, Kimberly/Zelikovsky, Nataliya/Green, Cynthia/Kaplan, Bernard, Psychosocial evaluation of candidates for living related kidney donation, Vol. 21 Pediatric Nephrology 1357–1363 (2006) (Sterner et al., Vol. 21 Pediatric Nephrology 1357, p. (2006))

Streinz, Rudolf, Europarecht, 9. edition, Heidelberg 2012 (Streinz (2012), p.)

Streinz, Rudolf, EUV/AEUV, Vertrag über die Europäische Union und Vertrag über die Arbeitsweise der Europäischen Union, 2. edition, München 2012 (Author, in Streinz (ed.) (2012), Art. at)

Swiss Federal Office of Public Health, Glossar Domino-Transplantation, 2011, retrieved 24 March, 2011 from http://www.bag.admin.ch/glossar/index.html?action=id&id=80&lang=de (Swiss Federal Office of Public Health (2011))

Taupitz, Jochen (ed.), Regulations of Civil Law to Safeguard the Autonomy of Patients at the End of Their Life, Berlin 2000 (Author, in Taupitz (ed.) (2000), p.)

(The) Ethics Committee of the Transplantation Society, The Consensus Statement of the Amsterdam Forum on the Care of the Live Kidney Donor, Vol. 78 Transplantation 491–492 (2004) (The Ethics Committee of the Transplantation Society, Vol. 78 Transplantation 491, p. (2004))

Terasaki, Paul (ed.), Clinical Transplants, Los Angeles 1995 (Author, in Terasaki (ed.) (1995), p.)

Terasaki, Paul/Cecka, Michael/Gjertson, David/Takemoto, Steven, High survival rates of kidney transplants from spousal and living unrelated donors, Vol. 333 The New England Journal of Medicine 333–336 (1995) (Terasaki et al., Vol. 333 The New England Journal of Medicine, 333, p. (1995))

Terhechte, Jörg Philipp, Der Vertrag von Lissabon: Grundlegende Verfassungsurkunde der europäischen Rechtsgemeinschaft oder technischer Änderungsvertrag, Vol. 43 Zeitschrift Europarecht 143–189 (2008) (Terhechte, Vol. 43 Zeitschrift Europarecht 143, p. (2008))

Terhechte, Jörg Philipp (ed.), Verwaltungsrecht der Europäischen Union, Baden-Baden 2011 (Author, in Terhechte (ed.) (2011), p.)

Tettinger, Peter/Stern, Klaus (ed.), Europäische Grundrechte-Charta, München 2006 (Author, in Tettinger/Stern (ed.) (2006), Art. at)

Teubner, Andreas, Aufgaben und Umfang der Tätigkeit der Lebendspendekommission nach § 8 Abs. 3 TPG, Jena 2006 (Teubner (2006), p.)

Thiel, Gil/Nolte, Christa/Tsinalis, Dimitrios, Gender Imbalance in Living Kidney Donation in Switzerland, Vol. 37 Transplantation Proceedings 592–592 (2005) (Thiel/Nolte/Tsinalis, Vol. 37 Transplantation Proceedings 592, p. (2005))

Thiele, Alexander, Europäisches Prozessrecht, München 2007 (Thiele (2007), p.)

Thiruchelvam, Paul/Willicombe, Michelle/Hakim, Nadey/Taube, David/Papalois, Vassilios, Renal Transplantation, Vol. 343 British Medical Journal (2011) (Thiruchelvam, Vol. 343 British Medical Journal (2011))

Toellner, Richard (ed.), Organtransplantation – Beiträge zu ethischen und juristischen Fragen, Stuttgart 1991 (Author, in Toellner (ed.) (1991), p.)

Transplant Living, 2012, retrieved 09 March, 2012 from http://www.transplantliving.org (Transplant Living (2012))

Transplant Procurement Management, International Registry, 2011, retrieved 22 March, 2011 from http://www.tpm.org/ (Transplant Procurement Management (2011))

Trotter, James/Wisniewski, Karen/Terrault, Norah/Everhart, James/Kinkhabwala, Milan/ Weinrieb, Robert/Fair, Jeffrey/Fisher, Robert/Koffron, Alan/Saab, Sammy/Merion, Robert/ and the A2ALL Study Group, Outcomes of Donor Evaluation in Adult-to-Adult Living Donor Living Transplantation, Vol. 46 Hepatology 1476–1484 (2007) (Trotter et al., Vol. 46 Hepatology 1476, p. (2007))

Truog, Robert, The Ethics of Organ Donation by Living Donors, Vol. 353 New England Journal of Medicine 444–446 (2005) (Truog, Vol. 353 New England Journal of Medicine 444, p. (2005))

Ugowski, Patrick, Rechtsfragen der Lebendspende von Organen, Münster 1998 (Ugowski (1998), p.)

United Network for Organ Sharing, 1991 Ethics Committee Ethics of organ transplantation from living donors, Vol. 24 Transplant Proceedings 2236–2237 (1992) (United Network for Organ Sharing, Vol. 24 Transplantation Proceedings 2236, p. (1992))

United Network for Organ Sharing, 2011, retrieved 11 November, 2011 from www.unos.org (United Network for Organ Sharing (2011))

United Network for Organ Sharing, OPTN, 2012, retrieved 09 March, 2012 from http://www.unos.org/donation/index.php?topic=optn (United Network for Organ Sharing (2012))

University Hospital St.Ekaterina, 2004, retrieved 24 March, 2011 from http://www.uhsek.com/index.php?lang=eng_&s=3&id=news1 (University Hospital St.Ekaterina (2004))

University of Alaska, Unit one lesson plans, 2003, retrieved 11 July, 2011 from http://www.ktoo.org/gavel/guide/classGuide1-plan.htm (University of Alaska (2003))

U.S. Department of Health & Human Services, Living Donation, 2009, retrieved 09 March, 2012 from http://optn.transplant.hrsa.gov/resources/professionalResources.asp?index=7 (U.S. Department of Health & Human Services (2009))

US Department of Health & Human Services, data, view data reports > build advanced, 2011, retrieved 09 August, 2011 from http://optn.transplant.hrsa.gov/latestData/rptData.asp (US Department of Health & Human Services (2011))

U.S. National Library of Medicine, Medical Subject Heading Histocompatibility, 1999, retrieved 24 March, 2011 from http://www.nlm.nih.gov/cgi/mesh/2010/MB_cgi?mode=&term=Histocompatibility&field=entry (U.S. National Library of Medicine (1999))

Valapour, Maryam, The Live Organ Donor's Consent: Is It Informed and Voluntary?, Vol. 22 Transplantation Reviews 196–199 (2008) (Valapour, Vol. 22 Transplantation Reviews 196, p. (2008))

Veatch, Robert, Egalitarian and Maxim in Theories of Justice: Directed Donation of Organs for Transplant, Vol. 23 Journal of Medicine and Philosophy 456–476 (1998) (Veatch, Vol. 23 Journal of Medicine and Philosophy 456, p. (1998))

Veatch, Robert, Transplantation Ethics, Washington D. C. 2000 (Veatch (2000), p.)

Virzì, Antonino/Signorelli, Maria/Veroux, Pierfrancesco/Giammarresi, Giusi/Maugeri, Sonya/ Nicoletti, Alessanra/Veroux, Massimiliano, Depression and Quality of Life in Living Related Renal Transplantation, Vol. 39 Transplantation Proceedings 1791–1793 (2007) (Virzì et al., Vol. 39 Transplantation Proceedings 1791, p. (2007))

Wagner, Elke/Fateh-Moghadam, Bijan, Freiwilligkeit als Verfahren, Zum Verhältnis von Lebendorganspende, medizinischer Praxis und Recht, Vol. 56 Soziale Welt 73–98 (2005) (Wagner/Fateh-Moghadam, Vol. 56 Soziale Welt 73, p. (2005))

Walter, Marc/Dammann, Gerhard/Küchenhoff, Joachim/Frommer, Jörg/Schoeneich, Frank/ Danzer, Gerhard/Klapp, Burghard, Psychosocial situation of living donors: moods, complaints, and self-image before and after liver transplantation, Vol. 11 Medical Science Monitor 503–509 (2005) (Walter et al., Vol. 11 Medical Science Monitor 503, p. (2005))

Waterman, Amy/Stanley, Sara/Covelli, Tonie/Hazel, Erik/Hong, Barry/Brennan, Daniel, Living donation decision making: recipients concerns and educational needs. Vol. 16 Progress in Transplantation 17–23 (2006) (Waterman et al., Vol. 16 Progress in Transplantation 17, p. (2006))

Weatherill, Stephen, Cases & Materials on EU Law, 9. edition Oxford 2010 (Weatherill (2010), p.)

Weber, Helmut, Spezielle Moraltheologie, Graz 1999 (Weber (1999), p.)

Weigend, Ewa/Zielinska, Eleonora, Das neue polnische Transplantationsgesetz, Vol. 14 Medizinrecht 445–451 (1996) (Weigend/Zielinska, Vol. 14 Medizinrecht 445, p. (1996))

Weigend, Thomas/Küpper, Georg (ed.), Festschrift für Hans Joachim Hirsch zum 70. Geburtstag am 11. April 1999, Berlin 1999 (Author, in Weigend/Küpper (ed.) (1999), p.)

Weil, Prosper, Towards relative normativity in international law, Vol. 77 American Journal International Law 413–442 (1983) (Weil, Vol. 77 American Journal International Law 413, p. (1983))

Weimar, Willen/Zuidema, Willij/de Klerk, Marry/Haase-Kromwijk, Bernadette/IJzermans, Jan, Altruistic kidney donation, Vol. 368 The Lancet 987 (2006) (Weimar et al., Vol. 368 The Lancet 987 (2006))

Weimar, Willem/ Bos, Michael/van Busschbach, Jan (ed.), Organ Transplantation: Ethical, Legal and Psychological Aspects, Towards a Common European Policy, Lengerich 2008 (Author, in Weimar/Bos/Busschbach (ed.) (2008), p.)

Weimar, Willem/Bos, Michael/van Busschbach, Jan (ed.), Organ Transplantation: Ethical, Legal and Psychosocial Aspects, Expanding the European Platform, Lengerich 2011 (Author, in Weimar/Bos/Busschbach (2011), p.)

Wenzel, Frank, Handbuch des Fachanwalts Medizinrecht, Cologne 2007 (Wenzel (2007), p.)

Winterton, House of Commons Hansard Debates, col. 987, 15 January 2004, retrieved 26 March, 2011 from http://www.publications.parliament.uk/pa/cm200304/cmhansrd/vo040115/debtext/40115-14.htm#40115-14_spnew3 (Winterton (2004))

(de) Witte, Bruno/Hanf, Dominic/Vos, Ellen (ed.), The Many Faces of Differentiation in EU Law, 2001 (Author, in de Witte/Hanf/Vos (ed.) (2001), p.)

Wittreck, Fabian (ed.), 60 Jahre Grundgesetz, Verfassung mit Zukunft?, Baden-Baden 2010 (Author, in Wittreck (ed.) (2010), p.)

Wolfe, Robert/Ashby, Valarie/Milford, Edgar/Ojo, Akinlolu/Ettenger, Robert/Agodoa, Lawrence/ Held, Philip/Port, Friedrich, Comparison of Mortality in All Patients on Dialysis, Patients on Dialysis Awaiting Transplantation, and Recipients of a First Cadaveric Transplant, Vol. 341 The New England Journal of Medicine 1725–1730 (1999) (Wolfe et al., 341, New England Journal of Medicine 1725, p. (1999))

World Health Organization, Human Organ Transplantation. A report on developments under the auspices of the WHO (1987–1999), Genf 1991 (WHO (1991), p.)

World Health Organization, About WHO, 2011, retrieved 12 April, 2011 from http://www.who.int/about/en/ (World Health Organization (2011))

Wright, Linda/Faith, Karen/Richardson, Robert/Grant, David, Ethical guidelines for the evaluation of living organ donors, Vol. 47 Canadian Journal of Surgery 408–413 (2004) (Wright et al., Vol 47 Canadian Journal of Surgery 408, p. (2004))

Zalta, Edward (ed.), Informed Consent, The Stanford Encyclopedia of Philosophy, 2011 (Author, in Zalta (ed.) (2011))

Zweigert, Konrad, Die Rechtsvergleichung im Dienste der Europäischen Rechtsvereinheitlichung, Vol. 16 Rabels Zeitschrift für ausländisches und internationales Privatrecht 387–397 (1951) (Zweigert, Vol. 16 Rabels Zeitschrift für ausländisches und internationales Privatrecht 387, p. (1951))

Zweigert, Konrad/Kötz, Hein, Einführung in die Rechtsvergleichung, 3. edition, Tübingen 1996 (Zweigert/Kötz (1996), p.)

II. Cited experts

Anulyte, Vita, Lithuania, Specialist for transplantation, National Transplantation Bureau under the Ministry of Health of the Republic of Lithuania

Bos, Michael, The Netherlands, Secretary to the Health Council of the Netherlands in the Hague

Bruckmueller, Karin, Austria, Assistant at the Department for Criminal Law and Criminology, University of Vienna

Bruzzone, Paolo, Italy, Assistant Professor of Surgery

Burnapp, Lisa, United Kingdom, Living Organ Donation Guy's and St. Thomas' NHS Foundation Trust

Coppen, Remco, The Netherlands, Secretary of the NIVEL quality council

Fateh-Moghadam, Bijan, Germany, University of Muenster, Department of Law

Füszl, Sylvia, Austria, Federal Ministry of Health

Gutmann, Thomas, Germany, University of Muenster, Department of Law

Kjeldsen, Tove, Denmark, Ministry of the Interior and Health

Kozyga, Kornelia, Austria, Gesundheit Österreich GmbH

Lennerling, Annette, Sweden, Clinical Transplant Coordinator, Sahlgrenska University Hospital

Lewandowska, Dorota, Poland, Poltransplant

Madden, Deidre, Ireland, Senior Lecturer Faculty of Law

Marino, Ignazio, Italy, Professor of Surgery

Möller, Charlotte, Sweden, National Board of Health and Welfare, Department for Organ and Tissue Donation

Monnier, Marcel, Switzerland, Federal Ministry of Health

Nikolova, Cveta, Bulgaria, PhD student of Prof. Dr. Gutmann

Pascalev, Assya, Bulgaria, Executive Director, Bulgarian Center for Bioethics

Pereira, Andre, Portugal, Assistant Professor, Faculty of Law, University of Coimbra

Price, David (†), United Kingdom, De Montfort University, Department of Law

Rigg, Keith, United Kingdom, Consultant Surgeon, City Hospital Campus, Nottingham University Hospital NHS Trust

Rudge, Chris, United Kingdom, UK Transplant

Sándor, Judit, Hungary, Central European University, Professor of law and political science

Sénémaud, Béatrice, France, Agence de la biomédecine

Wilmotte, Regine, Belgium, Federal public service Health, Food Chain Safety and Environment

Zardoya Martínez, Maria, Spain, Transplant Services Foundation, Hospital Clínic de Barcelona

MIX
Papier aus verantwortungsvollen Quellen
Paper from responsible sources
FSC® C105338

If you have any concerns about our products,
you can contact us on
ProductSafety@springernature.com

In case Publisher is established outside the EU,
the EU authorized representative is:
Springer Nature Customer Service Center GmbH
Europaplatz 3, 69115 Heidelberg, Germany

Printed by Libri Plureos GmbH
in Hamburg, Germany